STRUCTURAL ADHESIVES:

Developments in Resins and Primers

STRUCTURAL ADHESIVES:

Developments in Resins and Primers

Edited by

A. J. KINLOCH
Department of Mechanical Engineering,
Imperial College of Science and Technology,
London, UK

ELSEVIER APPLIED SCIENCE PUBLISHERS
LONDON and NEW YORK

ELSEVIER APPLIED SCIENCE PUBLISHERS LTD
Crown House, Linton Road, Barking, Essex IG11 8JU, England

Sole Distributor in the USA and Canada
ELSEVIER SCIENCE PUBLISHING CO., INC.
52 Vanderbilt Avenue, New York, NY 10017, USA

WITH 44 TABLES AND 135 ILLUSTRATIONS

British Library Cataloguing in Publication Data

Structural adhesives: developments in resins
and primers.
1. Building materials 2. Adhesives
I. Kinloch, A. J.
691′.99 TA455.A34

Library of Congress Cataloging in Publication Data

Structural adhesives.

Bibliography: p.
Includes index.
1. Adhesives. I. Kinloch, A. J.
TP968.S84 1986 668′.37 86-6340

ISBN 1-85166-002-X

The selection and presentation of material and the opinions expressed in this publication are the sole responsibility of the authors concerned

Special regulations for readers in the USA

Typeset and printed at The Universities Press (Belfast) Ltd

Preface

The use of adhesives in general industry continues to develop at a rapid pace and this is particularly the case for structural adhesives. Such adhesives are based upon resin compositions that polymerise to give high-modulus, high-strength adhesives so that a load-bearing joint is formed. As might be expected, developments in the chemistry and properties of structural adhesives have led and supported the spread of these materials from the high-technology aerospace industries into all types of general engineering applications. It appeared therefore to be an appropriate time to review some of the developments in our understanding of the chemistry, structure and properties of resins which are used in the technology of structural adhesives bonding.

The resins which form the mainstay of the industry are the epoxies and acrylics and several chapters consider various aspects of these materials. In Chapter 1 the effect of cure conditions on the final properties of the resin are discussed and this chapter extends the most useful concepts of time–temperature–transformation diagrams to rubber-toughened epoxy resins. In the next two chapters authors from two of the world's leading adhesives companies review recent developments in acrylic- and epoxy-based adhesives. In Chapter 4 the comparatively new bismaleimide resins are discussed. These materials possess relatively high thermal resistance but are readily processable and, since they are new materials, their use as matrix materials in fibre composites as well as in structural adhesives is considered. Structural adhesive formulations usually contain additives such as rubbers and rigid-particulate fillers and Chapters 5 and 6 review the relationships between the microstructure and properties of such materials. In the next chapter the contentious issue of whether there is a definite microstructure in simple epoxy resins is addressed, and whether any microstructure would affect the properties of the material. Finally, in adhesives technology the problem of attaining good resistance to moisture for the bonded assembly always represents a challenge and the use of silane resin coupling agents to achieve this goal is increasing. Thus, the book concludes with a review chapter on this important area in adhesives technology.

A. J. KINLOCH

Contents

List of Contributors

J. Comyn

School of Chemistry, Leicester Polytechnic, PO Box 143, Leicester LE1 9BH, UK

E. W. Garnish

Ciba–Geigy, Bonded Structures Division, Duxford, Cambridge CB2 4QD, UK

J. K. Gillham

Polymer Materials Program, Department of Chemical Engineering, Princeton University, Princeton, New Jersey 08544, USA

A. J. Kinloch

Department of Mechanical Engineering, Imperial College of Science and Technology, Exhibition Road, London SW7 2BX, UK

F. R. Martin

Loctite (Ireland) Ltd, Whitestown Industrial Estate, Tallaght, Co. Dublin, Republic of Ireland

H. S. Stenzenberger

Technochemie GmbH, Verfahrenstechnik, Postfach 40, D-6901 Dossenheim, Federal Republic of Germany

G. C. STEVENS

Central Electricity Generating Board, Central Electricity Research Laboratories, Kelvin Avenue, Leatherhead, Surrey KT22 7SE, UK

R. J. YOUNG

Department of Materials, Queen Mary College, Mile End Road, London E1 4NS, UK (Present address: Department of Polymer Science and Technology, University of Manchester Institute of Science and Technology, PO Box 88, Sackville Street, Manchester M60 1QD, UK)

1

Cure and Properties of Thermosetting Polymers

J. K. GILLHAM

Polymer Materials Program,
Department of Chemical Engineering,
Princeton University,
New Jersey, USA

1. THE TIME–TEMPERATURE–TRANSFORMATION CURE DIAGRAM AND PROPERTIES OF THERMOSETTING SYSTEMS

A time–temperature–transformation (TTT) cure diagram may be used to provide an intellectual framework for understanding and comparing the cure and physical properties of thermosetting systems.[1–10] The main features of an isothermal TTT cure diagram such as that shown in Fig. 1 can be obtained by measuring the times to events which occur during isothermal cure vs temperature (T_{cure}). These events include the onset of phase separation, gelation, vitrification, full cure, and devitrification. Gelation corresponds to the incipient formation of an infinite molecular network which gives rise to viscoelastic behavior in the macroscopic fluid. Vitrification occurs when the glass transition temperature rises to the isothermal temperature of cure. Devitrification occurs when the glass transition temperature decreases through the isothermal temperature (as in degradation). The diagram displays the distinct states encountered on cure due to chemical reactions. These states include liquid, sol/gel rubber, elastomer, ungelled (sol) glass, gelled glass, and char. The gelled glass region in the TTT cure diagram is divided into two parts by the full cure line:[8] in the absence of degradation (Fig. 1, devitrification and char), the top and lower parts could be designated fully cured (gel) glass and undercured (sol/gel) glass region, respectively. The terms A-, B- and C-stage

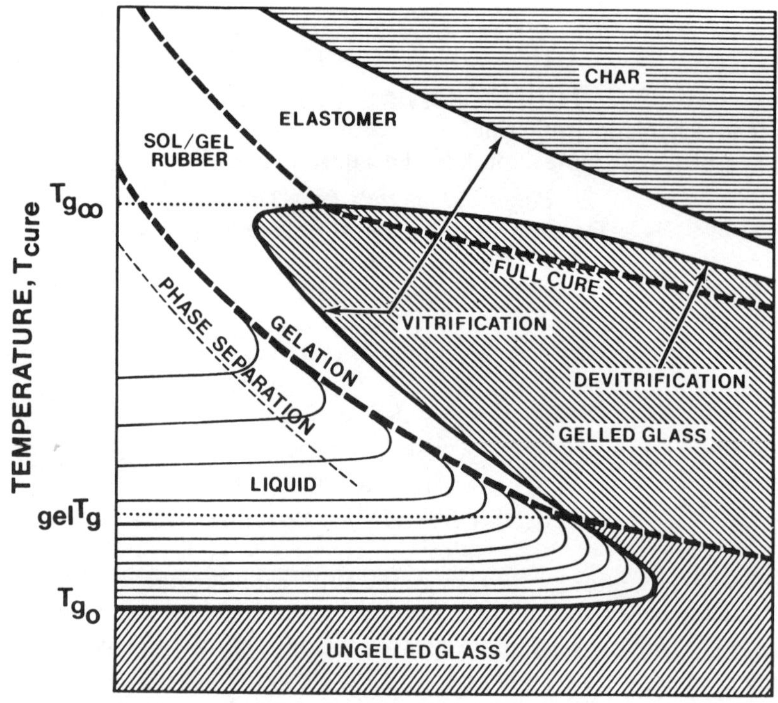

FIG. 1. Time–temperature–transformation (TTT) cure diagram for a reactive thermosetting polymer system showing the different states encountered during isothermal reaction. Such a diagram is useful for designing plastic systems to replace metals. T_{g0}, $_{gel}T_g$ and $T_{g\infty}$ are the glass transition temperature of the reactants, the temperature at which the times to gelation and vitrification are the same, and the glass transition temperature of the fully reacted system, respectively. The states are liquid, sol/gel rubber, elastomer, gelled glass, ungelled (or sol) glass, and char. The full cure line (i.e., $T_g = T_{g\infty}$) divides the gelled glass region into sol/gel glass and gel (fully cured) glass regions. Successive isoviscous contours shown in the liquid region differ by a factor of ten. The diagram also shows the locus of the onset of phase separation in which a second phase (e.g. rubber) separates during cure: the presence of a dispersed phase modifies the properties of the cured material.

resins, which are used in the technological literature, correspond to sol glass, sol/gel glass and gel glass, respectively. The diagram also displays the critical temperatures $T_{g\infty}$, ${}_{gel}T_g$, and T_{g0} which are, respectively, the glass transition temperature of the fully cured system, the temperature at which gelation and vitrification occur simultaneously, and the glass transition temperature of the reactants.

The schematic isothermal TTT cure diagram of Fig. 1 includes a series of isoviscous contours in the liquid region, successive contours differing by a factor of ten.[3] Visual extrapolation of the isoviscous contours suggests that the vitrification process below ${}_{gel}T_g$ is an isoviscous one.

Much of the behavior of thermosetting materials can be understood immediately in terms of the TTT cure diagram through the influence of the gelation, vitrification, and devitrification events on properties: gelation retards macroscopic flow, and retards growth of a dispersed phase (as in rubber-modified systems); vitrification retards chemical conversion; devitrification due to thermal degradation marks the limit in time for the material to support a substantial load. The remainder of this section amplifies these statements.

The ungelled glassy state is the basis of commercial molding materials since, on heating, the ungelled (sol) material can flow before gelling. Formulations can be processed as solids (e.g. molding compositions) when T_{g0} is above ambient temperature; they can be processed as liquids (e.g. casting fluids) when T_{g0} is below ambient temperature.

Parameter ${}_{gel}T_g$ is a critical temperature in determining the upper temperature for storing reactive materials to avoid gelation. The glass transition temperature of the material at the composition corresponding to gelation is ${}_{gel}T_g$ since at the point of gelation the glass transition temperature has risen to ${}_{gel}T_g$.

The morphology developed in a two-phase system (e.g. those in which rubber-rich domains precipitate as a dispersed phase) depends on the temperature of cure. The reaction temperature determines the competition between thermodynamic and kinetic (transport) factors. For optimum mechanical properties, a two-phase system will be cured at one temperature to control the morphology, and subsequently cured, usually at higher temperatures, to full cure to complete the reactions of the matrix.

Shrinkage stresses due to cure begin to develop with adhesion to a rigid substrate at gelation above ${}_{gel}T_g$ and at vitrification below ${}_{gel}T_g$.

The tensile stresses in the resin and the corresponding compressive stresses on a substrate are important in composite behavior.

Prolonged isothermal cure at T_{cure} below $T_{g\infty}$ would lead to $T_g = T_{cure}$ if the reactions were quenched by the process of vitrification. In practice $T_g > T_{cure}$ for three principal reasons: (1) T_g can increase during the heating scan used to measure it; (2) although vitrification is defined to occur when $T_g = T_{cure}$, T_g as usually measured does not correspond to the glassy state but rather to approximately halfway between the rubbery and glassy states, and so isothermal reactions will proceed beyond the assigned time of vitrification; and (3) reactions do proceed in the glassy state to extents depending on the influence of the glassy state on the reaction mechanism.

Correlations between macroscopic behavior and molecular structure of the reactants result most clearly for fully cured materials. Full cure is attained most readily by reacting above $T_{g\infty}$; it can be achieved more slowly by curing below $T_{g\infty}$ to the full cure line of the TTT cure diagram.

For cure below ${}_{gel}T_g$ the cure reactions lead eventually to gelation. At high temperatures other chemical reactions can lead to thermal degradation. Thermal degradation can result in devitrification as the glass transition temperature decreases through the isothermal temperature due to decrease in crosslinking, or due to the formation of low molecular weight plasticizing material. Degradation can also result in vitrification (e.g. char formation) as the glass transition temperature rises to the isothermal temperature due to increase in crosslinking or volatilization of low molecular weight plasticizing materials.[4] For high-$T_{g\infty}$ systems there is competition between cure and thermal degradation.

The limiting viscosity in the fluid state is controlled by gelation above ${}_{gel}T_g$, and by vitrification below ${}_{gel}T_g$. At gelation the weight-average molecular weight and zero-shear-rate viscosity become infinite. Viscosity in the vicinity of vitrification is described by the Williams–Landel–Ferry (WLF) equation.[11]

Recording the time to reach a specified viscosity (Fig. 1) is often used as a practical method for measuring gelation times: above temperature ${}_{gel}T_g$ the apparent activation energies obtained from the temperature dependence of the time to reach a specified viscosity approach the true activation energy for the chemical reactions leading to gelation with increase of the specified viscosity.[1]

The times to gelation and to vitrification each can be computed from the reaction kinetics, the conversion at gelation (which is constant according to Flory's theory of gelation),[12] and the conversion at vitrification (which increases with T_{cure}), respectively. Since vitrification occurs when the glass transition temperature rises to the temperature of cure, computation of the time to vitrify requires knowledge of the relationship between T_g and conversion (see Fig. 2, which shows T_g increases with conversion at an increasing rate). In the absence of diffusion control, the general kinetic equation describing the reaction is $dX/dt = A \exp(-E_A/\mathbf{R}T)f(X)$ where X is the extent of reaction and the other symbols have their usual meanings. The times to gelation and (in the absence of diffusion control) to vitrification at different temperatures can be computed using this equation, knowledge of X_{gel} (for gelation) and a relationship between X and T_g (for vitrification) (Fig. 2), and the reaction kinetics.[3,7,13] The influence of diffusion control on the reaction rate can be deduced in principle from the experimentally measured vitrification curve.

The 'S'-shaped vitrification curve obtained experimentally (in the

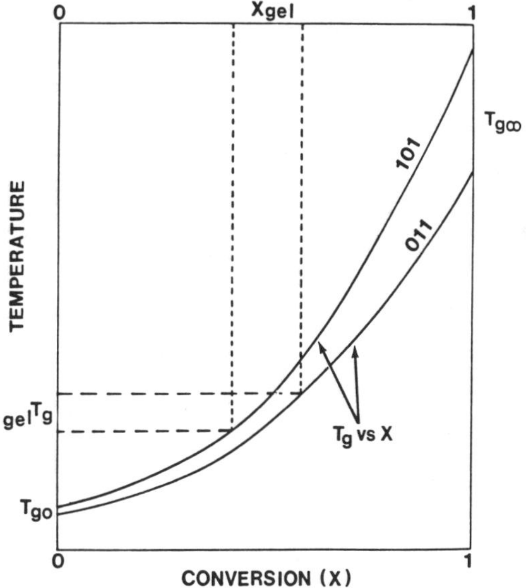

FIG. 2. Schematic diagram of T_g vs conversion at vitrification for reactants differing in functionality (101 > 011). The conversions at gelation are also included. The diagram is useful for demonstrating the effect of increasing functionality on gelation, vitrification, and the temperatures ${}_{gel}T_g$, T_{g0}, and $T_{g\infty}$. (Reproduced with permission from *Journal of Applied Polymer Science*.[4])

absence of thermal degradation) has been matched computationally for one epoxy system from temperature T_{g0} to temperature $T_{g\infty}$.[3]

The vitrification curve is generally 'S'-shaped for both network- and linear-forming step-growth reactions and for linear chain-growth reactions.[7] At temperatures immediately above T_{g0}, the time to vitrification passes through a maximum in consequence of the opposing influences of the temperature dependence of the viscosity and the reaction rate constant. Immediately below $T_{g\infty}$, the time to vitrification passes through a minimum in consequence of the opposing influences of the temperature dependence of the reaction rate constant and the decreasing concentration of reactive sites at vitrification as $T_{g\infty}$ is approached. Knowledge of the minimum time and the corresponding temperature is economically useful in molding technology where molds can only be opened after solidification.

The conversion at vitrification can be computed in principle by relating the glass transition temperature to contributions from the molecular weight and from the crosslinking density, both of which vary with conversion.[7,13] For linear polymerization the computation is simplified by the absence of crosslinking.[14]

The fractional extent of reaction at vitrification and the time to vitrify, like gelation, decrease with increasing functionality of the reactants.[4] The effect of increasing functionality on gelation, vitrification, and the temperatures ${}_{gel}T_g$, T_{g0}, and $T_{g\infty}$ can be understood from consideration of X vs T_g and X_{gel} relationships such as those of Fig. 2.[4]

Increasing cure time at any temperature leads to increasing conversion, T_g, and crosslinking density. Prior to vitrification the modulus and density at the curing temperature also increase. However, on cooling intermittently from the curing temperature to a temperature well below T_g (e.g. room temperature for high-T_g materials), the modulus and density are found to decrease whereas absorption of water is found to increase with increasing extent of cure.[9] A common basis for these interrelated phenomena is the increasing free volume at room temperature (RT) with increasing extent of cure.[10,15]

2. TORSIONAL BRAID ANALYSIS/TORSION PENDULUM: A TECHNIQUE FOR CHARACTERIZING THERMOSETTING SYSTEMS

A freely oscillating torsion pendulum can be used in two ways to characterize polymeric systems, namely in the conventional torsion pendulum (TP) mode and in the torsional braid analysis (TBA) mode.[1,2,16–22] Section 3 includes application of the TP/TBA technique to an investigation of the cure and properties of rubber-modified epoxy systems.[5,6]

Figure 3 shows a schematic diagram of the pendulum, which consists of a specimen held in place with clamps attached to two rods. The upper supporting rod is held vertically in alignment with a 2 in (5 cm) Teflon sleeve and is rigidly attached to a gear. The lower extender rod hangs freely and is magnetically coupled to a polaroid disc at its lower end.

The pendulum is enclosed in an air-tight cylindrical chamber (0·5 in (1·3 cm) diameter), whose atmosphere can be closely controlled and monitored: inert, water-doped, and reactive gases have been used. There are no electronic devices within the specimen chamber. Dry helium, rather than nitrogen, is usually employed as an inert atmosphere because of its higher thermal conductivity at cryogenic temperatures. An on-line electronic hygrometer can be used to continuously monitor the water vapor content of atmospheres from <20 to 20 000 ppm H_2O. A cylindrical copper block, round which cooling coils for liquid nitrogen and band heaters are wound, surrounds the pendulum. Excellent temperature control is achieved by virtue of the large thermal mass of the copper block, with a temperature spread of <1°C over a 2 in (5 cm) specimen. A temperature programmer/controller system permits experiments to be performed in isothermal (±0·1°C above 30°C) and dynamic modes from −190 to 400°C; in the dynamic mode the temperature may be increased or decreased linearly at rates of 0·05–5°C/min. Cooling is achieved by controlling the flow of liquid nitrogen from a pressurized container. Routine temperature scans are made at rates of change of temperature of ±1·5°C/min. Measurements have been made as low as 4K and as high as 700°C in modified apparatus.

A key factor in the instrumentation is the non-drag optical transducer which produces an electrical response that varies linearly with angular displacement. A polarizing disc is used as part of the inertial member of the pendulum, and a stationary second polarizer is positioned in front of a linearly-responding photo-cell. An analog electrical signal is obtained from a light beam passing through the pair of polarizers.

The clamps on the pendulum allow a variety of specimens to be used: film, fiber, or coating on glass braid or foil substrate. Depending on the specimen, the system can be used as a conventional torsion

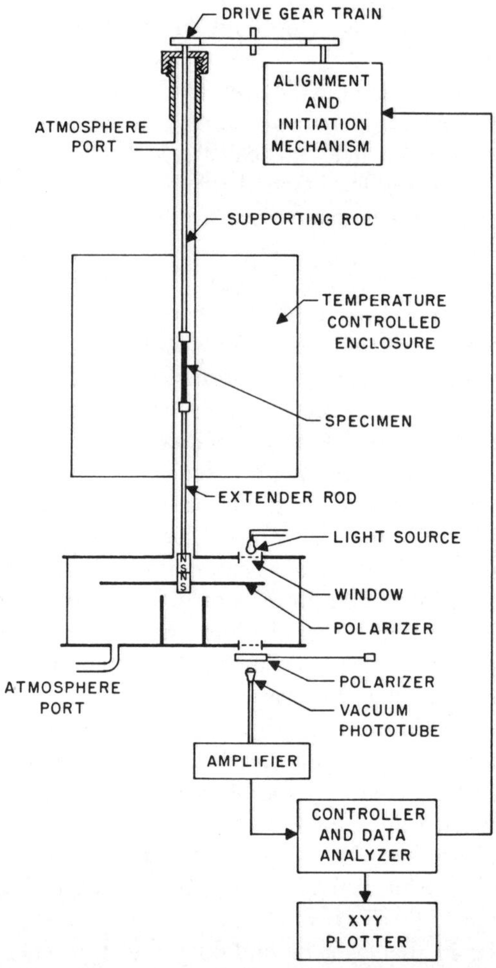

FIG. 3. Schematic diagram of automated torsion pendulum and TBA instrument. An analog electrical signal results from using a light beam passing through a pair of polarizers, one of which oscillates with the specimen. The pendulum is aligned for linear response of the transducer and oscillations are initiated using a gear train controlled by a computer. The computer also processes the damped waves to provide the elastic modulus and mechanical damping data, which are plotted against temperature or time.

pendulum or in the supported TBA mode. The substrate generally used in a TBA experiment is a loose, heat-cleaned glass braid containing about 3600 filaments. The specimen is prepared by simply dipping the braid into the neat liquid or into a solution of the material of interest dissolved in a solvent.

Quantitative values of the shear and loss moduli of polymer films are obtained from the natural frequency (~1 Hz) and damping of the induced oscillations of the torsion pendulum. Some advantages of the torsion pendulum method for the dynamic mechanical thermal analysis of polymers are its fundamental simplicity, its utilization of the resonance frequency of the pendulum (which means that the measurement is always made at the maximum sensitivity, since the specimen is a primary component of the pendulum), its use of a non-drag optical transducer (a pair of polarizers) which permits specimens to be small (<10 mg), and its free movement (because the pendulum is fixed only at one end, no adjustments for thermal expansion or contraction are required). In the torsional braid analysis (TBA) mode a glass braid (or other substrate) is impregnated with a polymer solution or polymer 'melt', and measurements are made on the composite specimen. Although the relative rigidity of the specimen is obtained instead of its shear modulus, the transition temperatures can readily be identified. The advantages of the TBA technique include ease of specimen fabrication, vertical self-alignment (by gravity), and the capability of monitoring the physical properties of a specimen from the liquid to the solid state (as in the cure by chemical reaction of a thermosetting resin or on cooling through the glass transition region of a thermoplastic system).

The pendulum is intermittently set into oscillation to generate a series of damped waves as the material properties of the specimen change with temperature and/or time. The free oscillations are initiated by step-displacement of the gear attached to the upper rod of the pendulum, and the damped oscillations are converted to electrical analog signals by the optical transducer.

The shear modulus, G', is given by

$$G' = KI(2\pi f)^2\left[1 + \left(\frac{\Delta}{2\pi}\right)^2\right] \qquad (1)$$

where f is the frequency (Hz) of the oscillation, I is the moment of inertia, Δ is the logarithmic decrement $\{\Delta = \ln[A_i/A_{i+1}]$, A_i is the amplitude of the ith oscillation}, and K is a geometric constant. Equation (1) is usually approximated by

$$G' \approx KI(2\pi f)^2 \tag{2}$$

when $\Delta/2\pi < 0{\cdot}1$. The loss modulus, G'', is

$$G'' = \frac{KI\Delta}{\pi}(2\pi f)^2 \tag{3}$$

The loss tangent, $\tan\delta$, is a characteristic measure of the ratio of the energy dissipated per cycle to the maximum potential energy stored during a cycle:

$$\frac{G''}{G'} = \tan\delta \approx \frac{\Delta}{\pi} \approx \frac{\alpha}{\pi f} \tag{4}$$

where α is the damping coefficient. During a dynamic mechanical thermal analysis experiment, the moduli can be monitored by observing the changes in the frequency and logarithmic decrement of the oscillations. In a torsion pendulum experiment, where the geometric constants are known, the absolute moduli are obtained: for example, for a rectangular film,

$$G' = \frac{(2\pi f)^2 IL}{N}\left[1 + \left(\frac{\Delta}{2\pi}\right)^2\right] - \frac{mgb^2}{12N} \tag{5}$$

in which the form factor, N, is

$$N = \frac{a^3 b}{3}\left(1 - 0{\cdot}63\,\frac{a}{b}\right) \tag{6}$$

where a, b and L are the thickness, width ($a < b/3$), and length respectively of the film, g is the gravitational constant, and m is the mass. I can be obtained in a calibration experiment using a wire of known modulus and dimensions. Although the geometric constants are not known in a TBA experiment, the frequency can be used to calculate a relative modulus, since $G' \propto f^2$. The plots in Section 3 show the relative rigidity as $1/P^2$, where P is the natural period of the oscillations.

The natural frequency of the oscillations range from 0·1 Hz (for a liquid) to 10 Hz (for a solid), and the logarithmic decrement ranges from 0·001 to 3·0.

The data acquisition system consists of a desk-top computer (with printer, plotter, and disc drives), multiplexed switch, and a digital

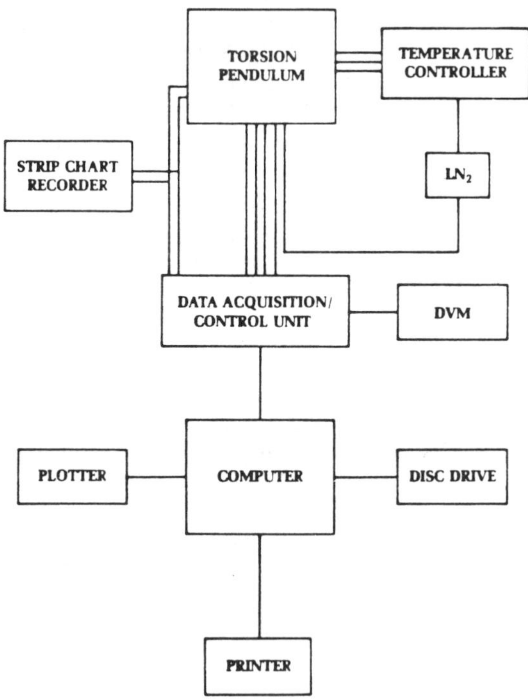

FIG. 4. Automated torsion pendulum and TBA instrument: system schematic for interfacing with a digital desk-top computer. The relay-activated motors, which align the specimen and initiate the waves, are under computer control, as is the temperature controller. The wave and thermocouple analog signals reach the computer digitized via a digital voltmeter (DVM). A scanner supervises the I/O (input/output) activity. Upon receiving the digitized raw data the computer calculates the frequency and damping parameters and plots the dynamic mechanical properties of the specimen as a function of temperature and time.

voltmeter, as shown in Fig. 4. The computer monitors the temperature and torsion pendulum oscillation signals, controls the initiation of the oscillations, and derives the frequency and logarithmic decrement of the oscillations from the raw data. The reduced data are stored and are available for plotting and/or printing or further massaging.

The computer controls the pendulum in a repetitive sequence, as shown in Fig. 5: after monitoring the wave until it has damped out, the computer aligns the pendulum for a linear response by rotating one of the polarizers and then initiates a new oscillation by cocking the pendulum against a spring (via a gear train), again monitoring the

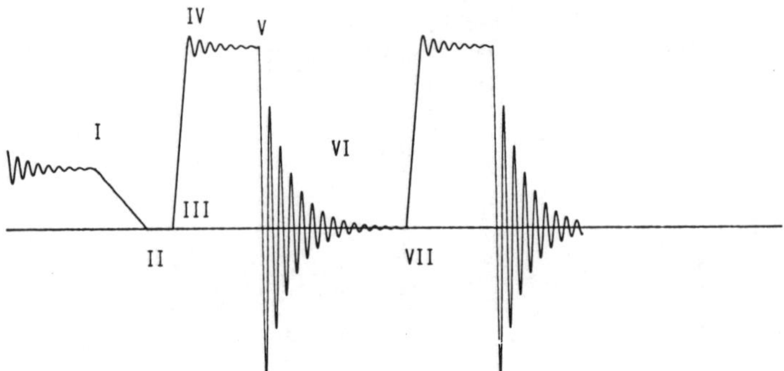

FIG. 5. Automated torsion pendulum and TBA instrument. Control sequence: (I) previous wave decays, drift detected and correction begins; (II) reference level of polarizer pair reached; (III) wave-initiating sequence begins; (IV) decay of transients; (V) free oscillations begin; (VI) data collected; (VII) control sequence repeated.

wave until the oscillations have decayed, and then releasing it so that the oscillations are centered about the center of the linear region. The wave is digitized and stored until the oscillations have decayed. The frequency and logarithmic decrement are extracted from the digitized wave by, for example, a non-linear least-squares fit to the solution of the differential equation of motion:

$$\theta = \theta_0(\exp(-\alpha t))\cos(2\pi f t + \phi) + Bt + C \tag{7}$$

where θ is the angular deformation of the pendulum as a function of time t, θ_0 is the initial deformation, ϕ is the phase angle, B is the drift coefficient, and C is the offset.

The Torsion Pendulum/Torsional Braid Analysis system is manufactured by Plastics Analysis Instruments, Inc., PO Box 408, Princeton, NJ 08540, USA.

3. RUBBER-MODIFIED EPOXIES: CURE, TRANSITIONS, MORPHOLOGY AND MECHANICAL PROPERTIES

3.1. Introduction

This section provides an example of polymer characterization which is based on the concept of the TTT cure diagram and the experimental TBA/TP approach.

Unmodified cured epoxy resins are brittle materials. The crack resistance can sometimes be improved by the addition of reactive liquid rubber to uncured neat systems. *In situ* phase separation occurs during cure; the cured rubber-modified resins consist of finely dispersed rubber-rich domains (~0·1–5 μm) bonded to the epoxy matrix. Improvement of fracture energy is dependent on the particle size, volume fraction, and size distribution of the dispersed phase, and on the chemical structures and the glass transition temperatures of the matrix and the dispersed phase. Shear deformation, void formation, crazing, and rubber tear have been proposed as toughening mechanisms, as discussed in detail in Chapter 5.

Cure of an initially homogeneous solution of an epoxy resin/curing agent/rubber formulation generally involves the sequential processes of phase separation, gelation, and vitrification. Phase separation of the rubber occurs during cure as a result of the general increase in molecular weight, which lowers the compatibility of the components of the epoxy system. Phase separation is eventually quenched by the high viscosity accompanying gelation (above ${}_{\mathrm{gel}}T_g$) or vitrification (below ${}_{\mathrm{gel}}T_g$). The development of morphology is dependent on the rates of nucleation and growth of the dispersed phase, compatibility of the rubber, and the rate of cure reactions. Because these factors are temperature dependent, cure conditions may have a profound effect on the morphology and, consequently, on the mechanical behavior of the cured resin. The temperature dependence could be compounded if the cure of the rubber-modified epoxy/curing agent system or of the corresponding neat system involves competing chemical reactions, since the properties of the matrix would then be cure-path dependent. Despite the potential implications, the effect of cure conditions on the properties of rubber-modified epoxies has received scant study.

The main objective of this section is to consider the cure process and its relationship to the development of morphology, transitions and mechanical properties for two rubber-modified epoxy systems using an unmodified neat system as a control sample. In an attempt to develop fully cured but distinct cure dependent morphologies, the systems were cured isothermally at different temperatures (T_{cure}) to well beyond gelation and vitrification. They then were postcured by heating above the maximum glass transition temperature (${}_{\mathrm{E}}T_{g\infty}$) of the system to complete the reactions of the matrix. An aromatic tetrafunctional diamine-cured diglycidyl ether of bisphenol A (DGEBA)-type epoxy resin was selected as the neat system because of its high ${}_{\mathrm{E}}T_{g\infty}$ (167°C). The two rubbers were a prereacted carboxyl-terminated rubber (DTK system) and an amino-terminated rubber (ATBN system); both were made from the same copolymer (CTBN) of butadiene and acrylonitrile (AN) containing 17% AN. The chemistry of cure would be expected to be the same for the neat and prereacted carboxyl-terminated modified systems, whereas competing cure reactions should be introduced with the amino-terminated rubber. A full report has been published.[5,6]

A methodology for understanding and comparing cure behavior and properties of the cured state for rubber-modified thermosets has been developed on the basis of the TTT cure diagram and the TBA technique. The first part of the methodology involves monitoring cure to give gelation and vitrification times at different cure temperatures, assessing the kinetics of phase separation by complementary turbidity measurements, and summarizing the data in a time–temperature–transformation (TTT) isothermal cure diagram. The second part summarizes transition temperatures after extended isothermal cure and also after postcure, in plots of transition temperatures vs T_{cure}. The third part involves assessing the relative amounts of dissolved and phase-separated rubber from the depression of the glass transition temperature of the matrix of the fully cured system (${}_{E}T_{g\infty}$). Independent estimates can also be obtained from comparison of the moduli of rubber-modified and neat epoxy specimens. The amount of the phase-separated rubber is then compared with that obtained from morphological examination with transmission electron microscopy (TEM). An apparent inconsistency in the amount of dissolved vs phase-separated rubber for the amino-terminated modified system led to considerations of the influence of time–temperature cure paths on the chemistry of cure. The fourth part of the methodology involves measuring mechanical properties vs temperature of fully cured specimens prepared from different time–temperature paths of cure. In an attempt to separate from the analysis improvements due to plasticization of the matrix by dissolved rubber, the data for the fracture energy are normalized relative to the ${}_{E}T_{g\infty}$ of the matrix. Qualitative information on the fracture process at different temperatures can be obtained from scanning electron micrographs (SEMs) of fractured surfaces.[6] Discussion of the ATBN system is emphasized here.

3.2. Materials

The chemical formulae of the difunctional epoxy resin (DER 331), the tetrafunctional aromatic amine curing agent (TMAB), and the two

NEAT SYSTEM

EPOXY: Diglycidyl Ether of Bisphenol A (DER 331)

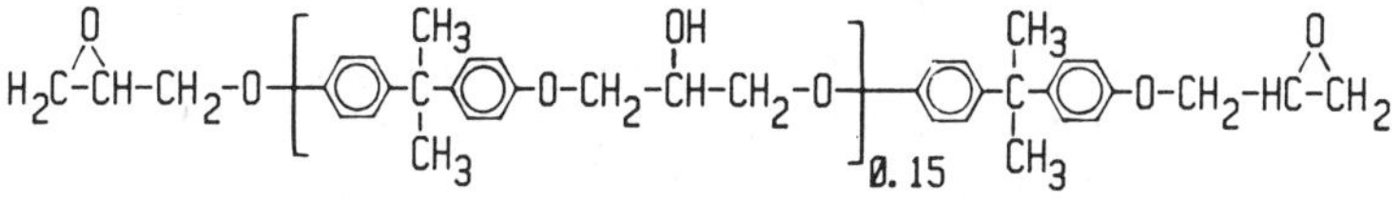

$$H_2C\overset{O}{-}CH-CH_2-O\left[-C_6H_4-C(CH_3)_2-C_6H_4-O-CH_2-\overset{OH}{\overset{|}{C}H}-CH_2-O-\right]_{0.15}-C_6H_4-C(CH_3)_2-C_6H_4-O-CH_2-HC\overset{O}{-}CH_2$$

AMINE: Trimethylene Glycol Di-p-aminobenzoate (TMAB)

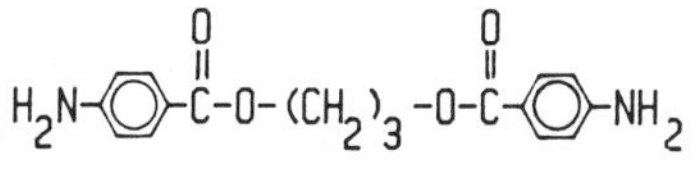

$$H_2N-C_6H_4-\overset{O}{\overset{||}{C}}-O-(CH_2)_3-O-\overset{O}{\overset{||}{C}}-C_6H_4-NH_2$$

REACTIVE LIQUID RUBBER

i) Prereacted Carboxyl-Terminated Copolymer of Butadiene and Acrylonitrile (AN content of rubber = 17 %) (K-293)

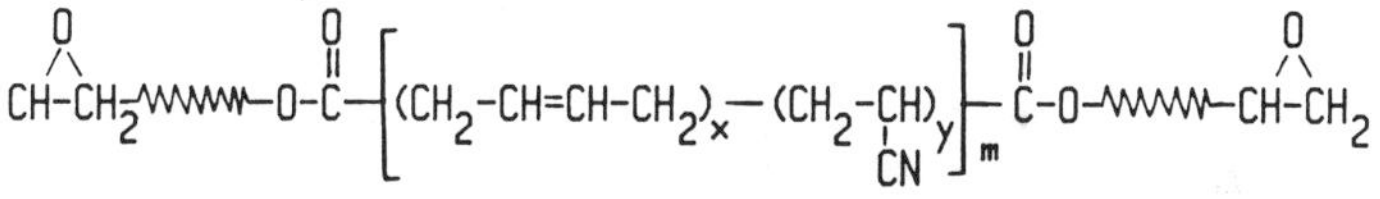

$$\overset{O}{CH-CH_2}\sim\sim\sim-O-\overset{O}{\overset{||}{C}}-\left[(CH_2-CH{=}CH-CH_2)_x-(CH_2-\underset{CN}{\underset{|}{C}H})_y\right]_m-\overset{O}{\overset{||}{C}}-O-\sim\sim\sim-\overset{O}{CH-CH_2}$$

$$+ \quad \text{excess} \quad \overset{O}{CH-CH_2}\sim\sim\sim-\overset{O}{CH-CH_2}$$

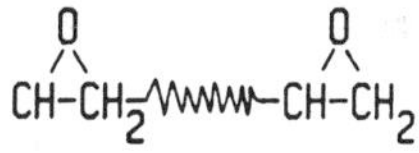

ii) Amino-Terminated Copolymer of Butadiene and Acrylonitrile (AN content = 17 %) (ATBNx16)

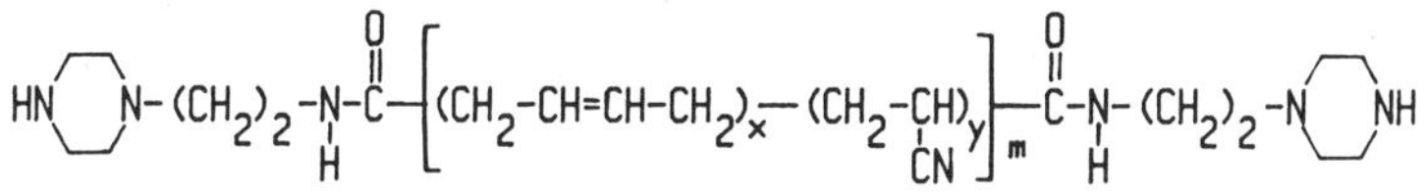

$$HN\langle\rangle N-(CH_2)_2-\underset{H}{\underset{|}{N}}-\overset{O}{\overset{||}{C}}-\left[(CH_2-CH{=}CH-CH_2)_x-(CH_2-\underset{CN}{\underset{|}{C}H})_y\right]_m-\overset{O}{\overset{||}{C}}-\underset{H}{\underset{|}{N}}-(CH_2)_2-N\langle\rangle NH$$

$$+ \quad 3\,\% \quad NH\langle\rangle N-(CH_2)_2-NH_2$$

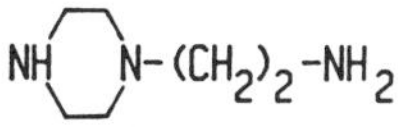

FIG. 6. Chemical formulae of reactants.

end-reactive rubbers (DTK-293 and ATBN × 16) are included in Fig. 6. Stoichiometric formulations (1 epoxy to 1 amine hydrogen), and for the rubber-modified formulations 15 parts of rubber per 100 parts of epoxy resin, were used. The preparation of the DTK rubber had involved reaction of CTBN with excess diglycidyl ether of bisphenol A. The preparation of the ATBN rubber had involved the reaction of CTBN with *N*-(2-aminoethyl)piperazine (AEP), a residual amount (3%) of which remained in the ATBN rubber. In the graphs of physical testing parameters vs temperature, the temperature of isothermal precure is included in parentheses after the designation of the resin system, as in DTK-293 (200°C).

3.3. Torsional Braid Analysis (TBA) and the TTT Cure Diagram

The times to gelation and vitrification were measured by monitoring the dynamic mechanical behavior (~1 Hz) of a composite specimen of reactive formulation on a glass braid in a torsion pendulum apparatus. Representative TBA isothermal cure spectra for the ATBN system are shown in Fig. 7. Gelation and vitrification times were obtained from

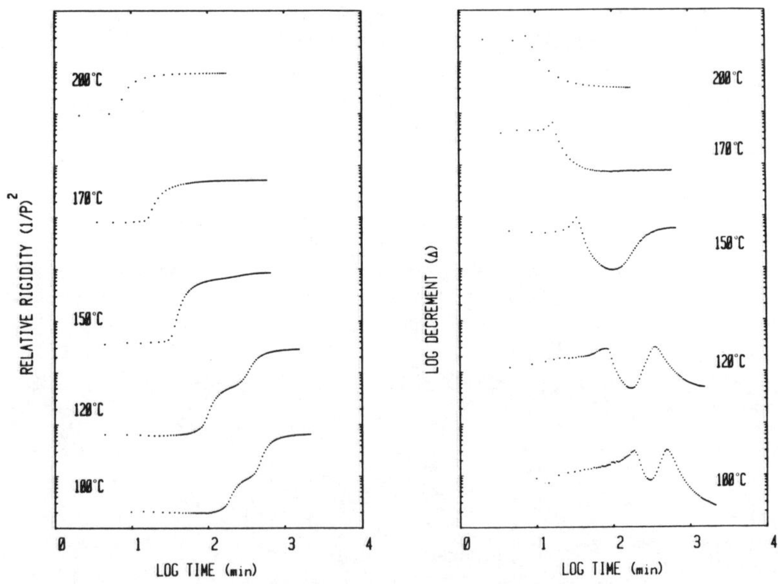

FIG. 7. Representative TBA isothermal cure spectra for the rubber-modified DTA × 16 system. (Reproduced with permission from Ref. 5 by courtesy of the American Chemical Society.)

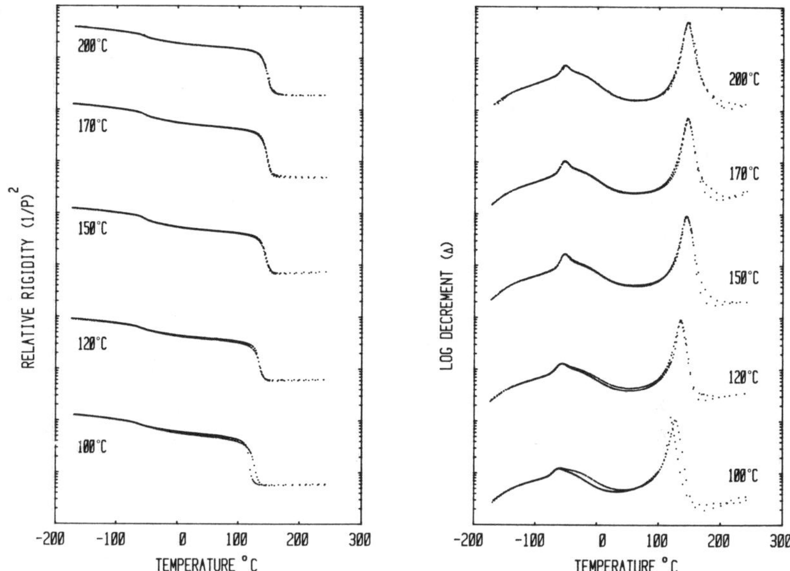

FIG. 8. Representative TBA thermomechanical spectra after isothermal cure for the rubber-modified DTA × 16 system. (Reproduced with permission from Ref. 5 by courtesy of the American Chemical Society.)

the peaks in the logarithmic decrement damping plots. After isothermal cure, temperature scans (T_{cure} to −180°C to 240°C to −180°C) were used to obtain the transition temperatures of the material after cure at T_{cure}, and after postcure to 240°C (which was considered to provide full cure). Representative TBA thermomechanical spectra after isothermal cure for the ATBN system are shown in Fig. 8. The glass transition temperatures of the matrix (${}_{E}T_g$ and ${}_{E}T_{g\infty}$) and of the rubber-rich dispersed phase (${}_{R}T_g$) were obtained from peaks in the logarithmic decrement plots.

The TTT cure diagram for the ATBN system (Fig. 9) includes the gelation and vitrification times (from TBA), the development (onset, 90%, and end) of phase separation (from cloud-point measurements), and the range of values of the maximum values of glass transition temperature of the matrix, ${}_{E}T_{g\infty}$ (from TBA), which result from curing at different temperatures (from Fig. 11). Figure 10 includes the TTT cure diagram of the ATBN and that of the corresponding neat system. It is apparent that the presence of ATBN rubber influences the

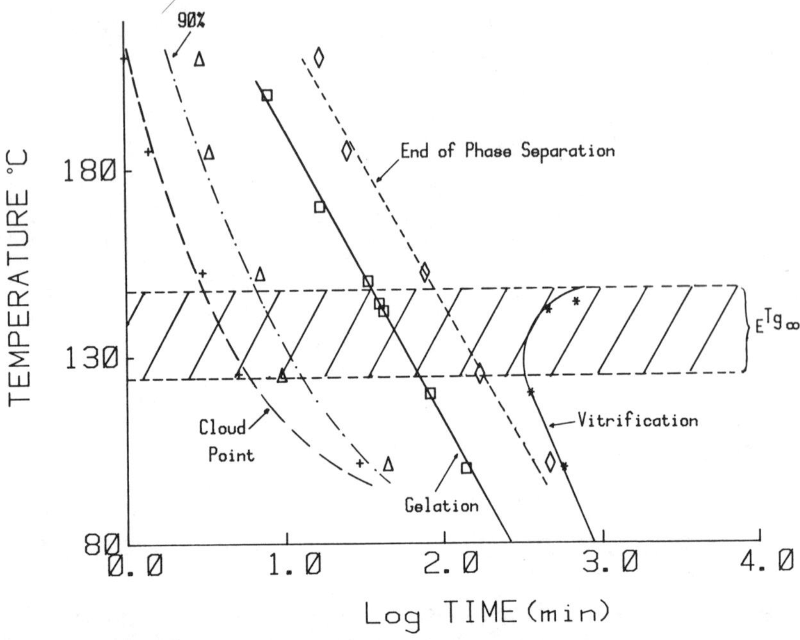

FIG. 9. TTT isothermal cure diagram including phase separation for the rubber-modified DTA × 16 system. Key: □, gelation; ∗, vitrification; +, cloud point; △, 90% decrease of light intensity; ◇, end of phase separation. The range of values of $T_{g\infty}$ which result from curing at different temperatures is shown by the hatched region. (Reproduced with permission from Ref. 5 by courtesy of the American Chemical Society.)

gelation and vitrification times, and depresses $_E T_{g\infty}$ to extents depending on the cure conditions. Figure 11 shows the relationship to T_{cure} of the glass transition temperature of the matrix after prolonged cure ($_E T_g$) and after postcure ($_E T_{g\infty}$) for both the ATBN and neat systems. T_{cure} determines $_E T_g$ (when vitrification occurs during cure) and also determines the depression of $_E T_{g\infty}$ from the value for the neat system.

3.4. Phase-Separated Versus Dissolved Rubber

Values of $_E T_{g\infty}$ range from 125°C ($T_{cure} = 100$°C) to 148°C ($T_{cure} =$ 200°C) for the fully cured ATBN system. In comparison, the value of $_E T_{g\infty}$ for the neat system was 167°C and was independent of the temperature of cure. To cause a reduction of $_E T_{g\infty}$ equivalent to 42°C for the specimen cured at 100°C, the epoxy matrix must contain 10%

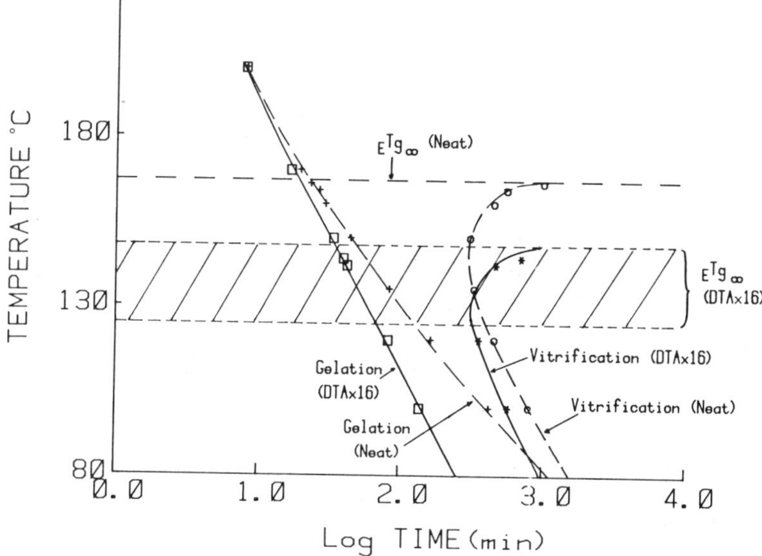

FIG. 10. TTT isothermal cure diagram for the rubber-modified DTA × 16 system vs the neat system. (Reproduced with permission from Ref. 5 by courtesy of the American Chemical Society.)

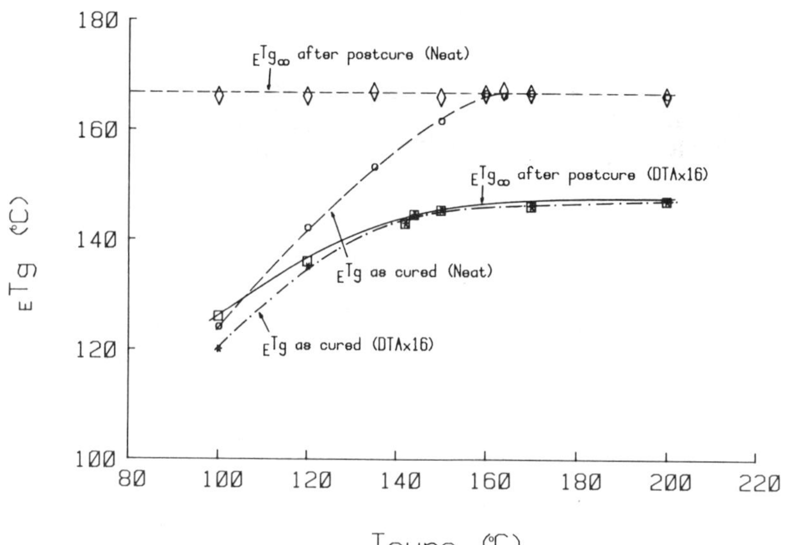

FIG. 11. $_{\mathrm{E}}T_{\mathrm{g}}$ and $_{\mathrm{E}}T_{\mathrm{g}\infty}$ vs T_{cure} for the rubber-modified DTA × 16 system and the neat system. (Reproduced with permission from Ref. 5 by courtesy of the American Chemical Society.)

rubber by weight as estimated by the Fox equation.[5] However, the formulation contained only 9·7% rubber by weight and yet the volume fraction of phase-separated rubber (V_f) obtained from TEM micrographs appeared to be extraordinarily high (Fig. 12). Thus the large decrease in ${}_{E}T_{g\infty}$ could not have been caused by the dissolved rubber alone. Comparison of TEMs of a specimen cured at 100°C and a specimen cured at 200°C show that the 100°C-cured specimen had a much higher V_f (34%) than the 200°C-cured specimen (15%), in spite of the ${}_{E}T_{g\infty}$ for the 100°C specimen being much lower. The rubber-rich domains must have contained significant amounts of epoxy, which was supported by the translucency of the specimens. In contrast, for the DTK system the volume fraction of phase-separated rubber obtained from the depression of ${}_{E}T_{g\infty}$ was more consistent with that obtained from TEMs (11–13%); specimens of it were opaque.

Aliphatic amines promote homopolymerization of epoxy resins. In fact, DGEBA-type epoxy resin can be cured at room temperature with commercial ATBN rubber if more than 50 phr of rubber is used. To study the level of reactivity of the ATBN rubber in the absence of TMAB, 15 phr of the commercially available ATBN rubber was mixed

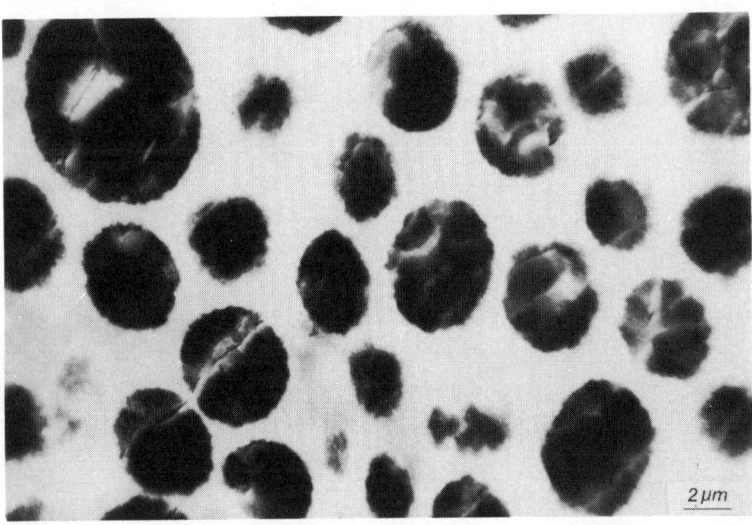

FIG. 12. Transmission electron micrograph (TEM) of rubber-modified DTA × 16 specimen cured at 100°C. (Reproduced with permission from Ref. 5 by courtesy of the American Chemical Society.)

with 100 parts of the diepoxide resin and heated at 200°C for 2 h. The T_g of the mixture increased by 62°C (from −12°C to 50°C) as determined from TBA thermomechanical spectra obtained before and after cure. A sample of ATBN rubber with negligible AEP content was also examined: the T_g of the mixture of 100 parts diepoxide with 15 phr residue-free ATBN rubber increased by 23°C (from −12 to 11°C) after cure at 200°C for 2 h. The chemistry of cure for the ATBN system is therefore more complex than for the neat system. The latter is considered to involve a stoichiometric reaction between epoxy and NH groups. In contrast, homopolymerization of the diepoxide, and reactions between the NH groups of the ATBN and diepoxide, AEP and diepoxide, and TMAB and diepoxide, can all occur competitively for the ATBN rubber-modified epoxy system. The differences in cure chemistry between the ATBN and neat systems were reflected in the reaction kinetics by the times of gelation (Fig. 10). Epoxy networks formed by stoichiometric reaction of epoxy resin and tetrafunctional aromatic diamine are highly crosslinked: for example, the neat diepoxide–TMAB system gives a value for ${}_E T_{g\infty}$ of 167°C. In contrast, cure of epoxy resins through homopolymerization usually produces more loosely crosslinked networks with low values of ${}_E T_{g\infty}$; for example, ${}_E T_{g\infty} = 100$°C for a (5 phr) piperidine-cured DGEBA-type epoxy system.[5] Therefore, the large decrease of ${}_E T_{g\infty}$ for the ATBN system may reflect the extent of homopolymerization rather than the amount of dissolved rubber. To isolate the influence of homopolymerization on ${}_E T_{g\infty}$ in the absence of rubber, the values of ${}_E T_{g\infty}$ versus T_{cure} for a formulation of diepoxide–TMAB–0·76 phr of AEP were obtained. In the formulation, the NH contribution was made equivalent to the NH contribution from the commercial ATBN rubber (including its residual AEP). The values of ${}_E T_{g\infty}$ are included in Fig. 13. As was found for the ATBN rubber-modified system, values of ${}_E T_{g\infty}$ were also dependent on T_{cure} and ranged from 137°C when cured at 100°C to 161°C when cured at 200°C. At low T_{cure} the reactivity of TMAB is low; therefore, significant amounts of epoxy could have homopolymerized to produce a low value of ${}_E T_{g\infty}$. At high T_{cure}, TMAB is more reactive; therefore, the 1:1 reaction of epoxy with amine hydrogen would be expected to occur more competitively. A consequence of homopolymerization of epoxy in the ATBN system is exclusion of amine hydrogen of TMAB from reaction; this will have some deleterious effects.

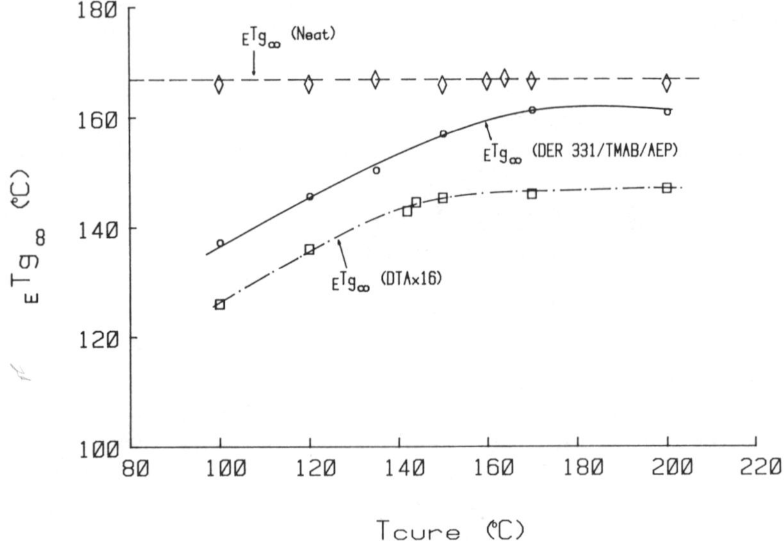

FIG. 13. ${}_{E}T_{g\infty}$ vs T_{cure} for the DER 331/TMAB/AEP system, the rubber-modified DTA × 16 system, and the neat system. (Reproduced with permission from Ref. 5 by courtesy of the American Chemical Society.)

3.5. Shear Modulus Versus Temperature

Values of the shear modulus vs test temperature for the fully cured ATBN system after isothermal cure at 100 and 200°C, and corresponding data for the neat system, are shown in Fig. 14. The data were obtained from machined strips using the TBA instrument as a conventional freely oscillating torsion pendulum. The values for all systems are expected to be similar below the glass transition temperature of the dispersed phase (${}_{R}T_{g}$), but at higher temperatures should reflect the extent of phase separation and values of ${}_{E}T_{g\infty}$. In general, between the temperatures ${}_{R}T_{g}$ and ${}_{E}T_{g\infty}$, the moduli of the rubber-modified specimens were lower than for the neat specimens. The moduli for the neat specimens precured at 100 and 200°C were about the same throughout the test temperature range because of the identical values of ${}_{E}T_{g\infty}$. In contrast, different values of the moduli were obtained at test temperatures above 80°C for the ATBN specimens precured at 100 and 200°C because of the different values of ${}_{E}T_{g\infty}$ for them.

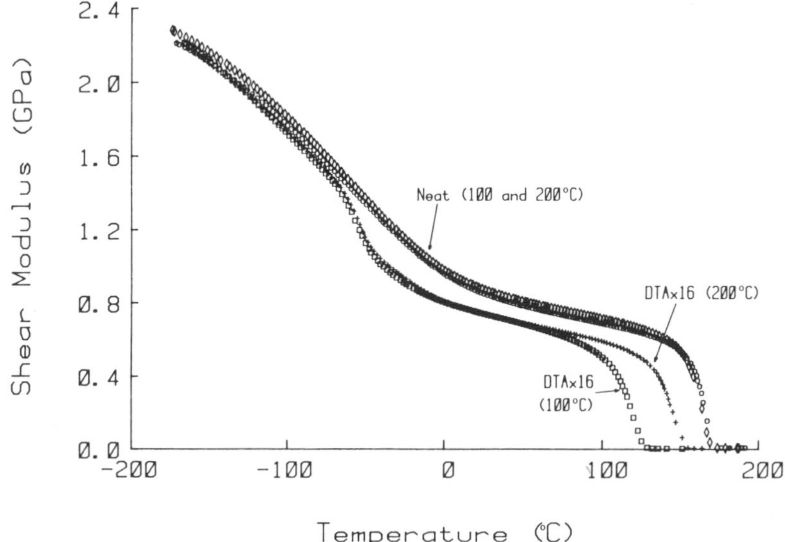

FIG. 14. Shear modulus vs test temperature for the DTA × 16 and neat systems. (Reproduced with permission from Ref. 6 by courtesy of the American Chemical Society.)

3.6. Fracture Behavior

The fracture energy, G_{Ic}, vs test temperature for the systems is shown in Fig. 15. For the neat system, the properties of the specimens fully cured after precure at 100 and 200°C should be relatively independent of their cure histories. The fracture energies for the two DTK specimens (precured at 100 and 200°C) were also relatively insensitive to their cure histories because of their similar values of ${}_{E}T_{g\infty}$ and volume fractions of dispersed phase. In contrast, the fracture energies for the ATBN specimens (precured at 100 and 200°C) were dependent on their cure histories since the cure chemistry and volume fraction of dispersed phase of this system are dependent on the time–temperature path of cure. The fracture energy of the ATBN specimen precured at 100°C, which has a higher volume fraction of dispersed phase and lower ${}_{E}T_{g\infty}$, was higher than that of the 200°C-cured specimen at all temperatures.

A small improvement in G_{Ic} above that for the neat specimens was observed for the DTK specimens throughout the test temperature range. On the other hand, a larger improvement in fracture energy

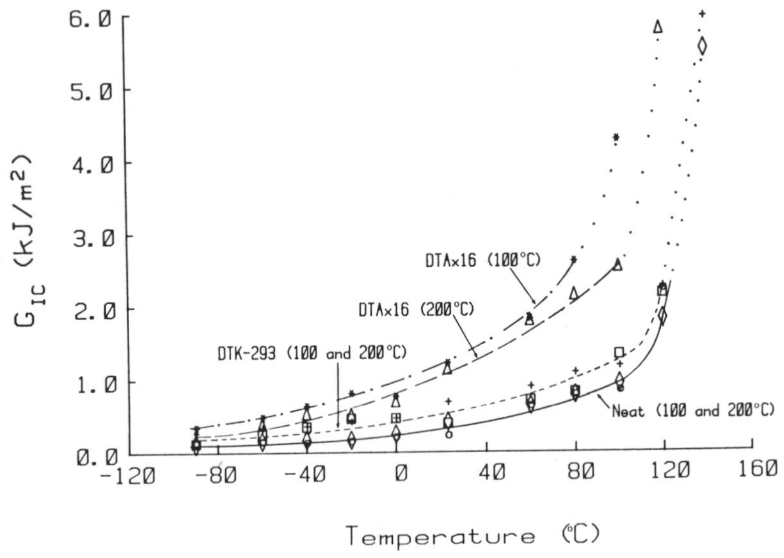

FIG. 15. Fracture energy (G_{Ic}) vs test temperature for the DTK-293, DTA × 16, and neat systems. (Reproduced with permission from Ref. 6 by courtesy of the American Chemical Society.)

was observed for the ATBN specimens, especially at high temperatures. The fracture energies of the rubber-modified specimens decreased in the order ATBN (100°C precure), ATBN (200°C precure), and DTK (100 and 200°C precure). This is the order of increasing values of ${}_{E}T_{g\infty}$ and decreasing volume fraction of the dispersed phase. The ratios of fracture energies of the rubber-modified specimens to those of the neat specimens at RT varied from 1·7 for DTK (precured at 100 and 200°C), to 3·5 for ATBN precured at 200°C, and to 3·7 for the ATBN precured at 100°C. Fracture energy has been related to the volume fraction of the dispersed phase by other researchers.

For all of the systems, the fracture energy increased at an accelerating rate with test temperature because of increasing ductility of the matrix. At low temperatures, the crack is relatively sharp and the fracture energy is low because the yield stress is high. Thus, the extent of plastic deformation and associated crack-tip blunting is relatively limited. As the temperature increases, the yield stress decreases so more crack-tip yielding is possible; as a result, the crack becomes blunter. This blunting results in higher failure loads and

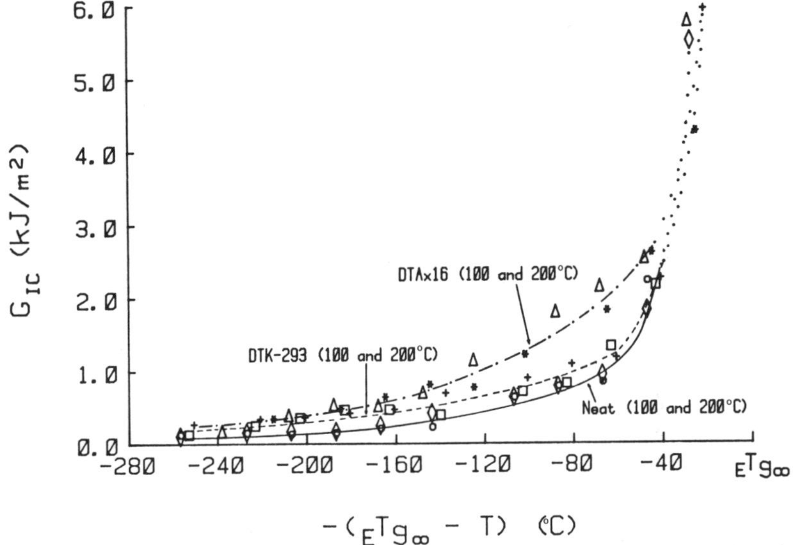

FIG. 16. Fracture energy vs test temperature normalized to ${}_ET_{g\infty}$ for the DTK-293, DTA × 16, and neat systems. (Reproduced with permission from Ref. 6 by courtesy of the American Chemical Society.)

higher fracture energies at higher temperatures. The increase of ductility of the matrix with temperature is apparent in SEM fractographs of the ATBN system.[6]

The effect of decreasing ${}_ET_{g\infty}$ is equivalent to the effect of increasing test temperature; therefore, the fracture energy of the ATBN precured at 100°C specimen should be the highest of all the specimens because of its low value of ${}_ET_{g\infty}$. However, a second factor is the volume fraction of dispersed phase, and this specimen also has the highest V_f. The relatively poor improvement in fracture energy for the DTK specimens results from the similar values of ductility of the neat (${}_ET_{g\infty} = 167$°C) and the rubber-modified matrices (${}_ET_{g\infty} = 161$ and 163°C) and/or from the relatively low volume fraction of dispersed phase. To investigate further the effect of ${}_ET_{g\infty}$ on the fracture energy, the fracture energies were compared after normalizing the test temperature (T) relative to the ${}_ET_{g\infty}$ of each system. The normalized results of G_{Ic} vs $[-({}_ET_{g\infty} - T)]$ (Fig. 16) show that some of the improvement in G_{Ic} for the ATBN specimens is due to the increased ductility of the matrix which arises from their lower ${}_ET_{g\infty}$ values.

However, as would be expected, the V_f of the dispersed phase also plays a role in improving the fracture energy.

It should be noted that the yield behavior of the matrix is controlled by the details of molecular structure of the matrix. The value of ${}_{E}T_{g\infty}$ is just a convenient 'first-order' parameter to reflect the ductility of the matrix.

For high-T_g matrix materials, the improvement of fracture energy at room temperature by rubber modification, with the inclusions produced in the present work, is not substantial. For example, at 100°C below ${}_{E}T_{g\infty}$ the maximum improvement in fracture energy over the neat system was about threefold.

4. CONCLUSIONS

To summarize Section 3, the mechanical properties of two fully cured rubber-modified epoxy systems were investigated vs test temperature as a function of different cure conditions using the neat system as control. Those of the ATBN-modified system were found to be more sensitive to cure history than those of the prereacted CTBN-modified system. This was attributed to the higher sensitivity of the volume fraction of the dispersed phase and of the maximum glass transition temperature, ${}_{E}T_{g\infty}$, of the matrix to cure conditions for the ATBN-modified system which resulted in a high volume fraction of dispersed rubber-rich phase as well as high depression of the maximum glass transition temperature. For example, ${}_{E}T_{g\infty}$ values followed the order: neat (167°C) > prereacted carboxyl-terminated rubber (161–163°C) > amino-terminated rubber (125–148°C). (Additional cure reactions were invoked for the amino-terminated rubber.) The fracture energy increased with temperature (−90 to 140°C) for all of the systems as a result of increasing ductility of the matrix. In general, the fracture energy followed the order: amino-terminated rubber > prereacted carboxyl-terminated rubber > neat system. The fracture energies of the different systems followed the same order after normalizing the test temperature to the ${}_{E}T_{g\infty}$ of each system.

ACKNOWLEDGEMENT

The research represented by this chapter has been sponsored in part by the Office of Naval Research and the US Army Research Office.

REFERENCES

1. GILLHAM, J. K., in *Developments in Polymer Characterisation—3,* Ed. J. V. Dawkins, Applied Science Publishers, London, 1982, Ch. 5, p. 159.
2. ENNS, J. B. and GILLHAM, J. K., in *Polymer Characterization: Spectroscopic, Chromatographic, and Physical Instrumental Methods,* Ed. C. D. Craver, Amer. Chem. Soc. Adv. Chem. Series, **203** (1983) 27.
3. ENNS, J. B. and GILLHAM, J. K., *J. Appl. Polym. Sci.,* **28** (1983) 2567.
4. CHAN, L. C., NAÉ, N. H. and GILLHAM, J. K., *Amer. Chem. Soc. Div. Organic Coatings Plastics Chem. Prepr.,* **48** (1983) 566–70; *J. Appl. Polym. Sci.,* **29** (1984) 3307.
5. CHAN, L. C., GILLHAM, J. K., KINLOCH, A. J. and SHAW, S. J., in *Rubber-Modified Thermoset Resins,* Ed. C. K. Riew and J. K. Gillham, Amer. Chem. Soc. Adv. Chem. Series, **208** (1984) 235.
6. CHAN, L. C., GILLHAM, J. K., KINLOCH, A. J. and SHAW, S. J., in *Rubber-Modified Thermoset Resins,* Ed. C. K. Riew and J. K. Gillham, Amer. Chem. Soc. Adv. Chem. Series, **208** (1984) 261.
7. ARONHIME, M. T. and GILLHAM, J. K., *J. Coatings Technol.,* **56** (718), (1984) 35.
8. PENG, X. and GILLHAM, J. K., *J. Appl. Polym. Sci.,* **29** (1985) 4685.
9. ENNS, J. B. and GILLHAM, J. K., *J. Appl. Polym. Sci.,* **28** (1983) 2831.
10. ARONHIME, M. T., *Formation and Properties of Thermosetting Systems,* Ph.D. Thesis, Princeton University, Princeton, NJ, USA, 1985.
11. FERRY, J. D., *Viscoelastic Properties of Polymers,* 3rd edn, Wiley, New York, 1980.
12. FLORY, P. J., *Principles of Polymer Chemistry,* Cornell University Press, Ithaca, New York, 1953.
13. ADABBO, H. E. and WILLIAMS, R. J. J., *J. Appl. Polym. Sci.,* **27** (1982) 1327.
14. ARONHIME, M. T. and GILLHAM, J. K., *J. Appl. Polym. Sci.,* **29** (1984) 2017.
15. SHIMAZAKI, A., *J. Polym. Sci., Part C,* **23** (1968) 555.
16. LEWIS, A. F. and GILLHAM, J. K., *J. Appl. Polym. Sci.,* **6** (1962) 422.
17. GILLHAM, J. K., *Crit. Rev. Macromol. Sci.,* **1** (1972) 83.
18. GILLHAM, J. K., *Amer. Inst. Chem. Eng. J.,* **20**(6) (1974) 1066.
19. BELL, C. L. M., GILLHAM, J. K. and BENCI, J. A., *Amer. Chem. Soc. Polym. Prepr.,* **15**(1) (1974) 542; see also *Soc. Plast. Eng. Techn. Papers,* **20** (1974) 598.
20. GILLHAM, J. K., STADNICKI, S. J. and HAZONY, Y., *J. Appl. Polym. Sci.,* **21**(2) (1977) 401.
21. ENNS, J. B. and GILLHAM, J. K., in *Computer Applications in Applied Polymer Science,* Ed. T. Provder, Amer. Chem. Soc. Symposium Series **197** (1982) Ch. 20, p. 329.
22. ENNS, J. B. and GILLHAM, J. K., *N. Amer. Thermal Analysis Soc. Proc., Philadelphia, Pa, 23–26 September, 1984,* p. 474.

2

Acrylic-Based Adhesives

F. R. MARTIN

Loctite (Ireland) Ltd, Dublin, Republic of Ireland

1. INTRODUCTION

Acrylic adhesives today are a large class of specifically designed products made to meet the needs of industry in the assembly of a wide variety of components. Some of these adhesives have found use as well in the consumer market, where ease of use, strength, speed of bonding and a bit of technical 'magic' have led to large sales in the past decade.

These adhesives are solvent-free 'reactive' engineering adhesives and include cyanoacrylate, anaerobic and modified acrylic adhesives. In 1978 Holappa[1] estimated sales in North America, Europe and Japan for cyanoacrylates, anaerobics and modified acrylics at 685, 1150 and 141 metric tonnes respectively. Compared with the volume of products sold by the rest of the adhesives industry these quantities are modest indeed. Even the other engineering adhesives—epoxies and polyurethanes—were estimated by Holappa[1] each to enjoy sales of thousands of tonnes. More traditional adhesives, such as urea–formaldehyde, phenolics, plastisols and so on, have sales which dwarf all of these.

However, the selling prices of acrylic adhesives are 10 to 100 times those of older bonding technologies and therefore these materials represent a commercially very significant and growing part of the adhesive industry.

The reasons for the commercial success of the acrylic technologies and for the high value put on these products by end-users lie in the specific benefits they offer:

(a) *Structural bonding*: If a bonded assembly depends on the integrity of the joint to keep that assembly together and functioning properly, then the adhesive must give strength and durability which inspires confidence in the end-user. Production and quality assurance engineers, and sales and marketing people, are very concerned about the quality of their (bonded) products and will demand high-performance bonding systems.
(b) *Production*: Cyanoacrylates, anaerobics and modified acrylics all lend themselves to being used in high-speed, highly automated assembly operations (see Fig. 1.) Precise application of the adhesive and rapid cure at normal ambient temperatures give the opportunity for large savings in production costs, often by eliminating cumbersome assembly with nuts and bolts, rivets, or other mechanical fasteners.
(c) *Ease of handling*: The elimination of the need for heat to cure the adhesive, the lack of solvents in the adhesives, simple storage at room temperature and the option of automatic application make possible safe, low-energy assembly operations. This has become particularly important in recent years with increased awareness of the need for control of environmental factors in industry.

Whilst cyanoacrylates, anaerobics and modified acrylics have inherent properties which make them particularly suitable for industrial use, constant research activity has led to a number of significant developments during the past several years. Before these are discussed, however, the basic chemistry of the materials will be briefly reviewed.

2. CHEMISTRY OF ACRYLICS

The acrylic adhesives are liquid compositions made up of a number of ingredients, including monomers which polymerize *in situ* in the bond line to form the cured material which forms the adhesive joint. The physical characteristics of the cured adhesive determine the strength and durability of the joint, but the industrially important properties of the speed and ease of bond formation depend on the chemistry of the polymerization process.

FIG. 1. Automated adhesive system.

2.1. Cyanoacrylates

Cyanoacrylate monomers can be made to polymerize by a free-radical mechanism, but cyanoacrylate adhesives exploit the benefits of an anionic curing mechanism.

The cyanoacrylate monomers have the structure:

$$CH_2{=}\underset{\displaystyle COOR}{\overset{\displaystyle CN}{\overset{|}{\underset{|}{C}}}}$$

where R is usually an alkyl group. These monomers are made by the Knoevenagel condensation polymerization of formaldehyde and the relevant cyanoacetate, followed by thermal depolymerization and distillation of the monomer.[2]

The strongly electronegative nitrile and carboxyl groups attached to the α-carbon make the cyanoacrylate monomer very susceptible to anionic polymerization, which can be initiated by relatively weak bases.

The processes of polymerization initiation and propagation proceed by the reactions:[2]

$$CH_2{=}C(CN)(COOR) \xrightarrow{A^-} CH_2^{\delta+}{\cdots}C(C{\equiv}N^{\delta-})(C{\cdots}O^{\delta-}(OR)) \longrightarrow A{-}CH_2{-}C^-(CN)(COOR)$$

$$A{-}CH_2{-}C^-(CN)(COOR) + CH_2{=}C(CN)(COOR) \longrightarrow A{-}CH_2{-}C(CN)(COOR){-}CH_2{-}C^-(CN)(COOR)$$

and so on.

Such a polymerization occurs very rapidly and, in practice, can be initiated by the presence of moisture on the substrate to be bonded. This reactivity obviously also makes manufacture and storage of the liquid adhesive difficult, and much of the research activity on cyanoacrylates has been concerned with making rapid-curing one-component adhesives that can still be packaged for sale with a reasonable shelf-life.

2.2. Methacrylates and Acrylates

Polymers made from the esters of methacrylic and acrylic acids have been widely used for many years. As a result of this, many methacrylate and acrylate monomers are available commercially for use in making liquid monomeric adhesives.[3] In addition, the technology for making other monomers as required is well established,[4] so it is now possible to have monomers specifically designed to meet particular adhesive requirements.

The basic monomers have the form:

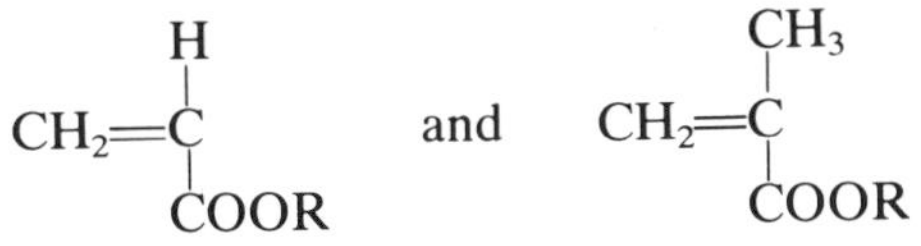

$$CH_2{=}\underset{\substack{|\\ COOR}}{\overset{\substack{H\\ |}}{C}} \qquad \text{and} \qquad CH_2{=}\underset{\substack{|\\ COOR}}{\overset{\substack{CH_3\\ |}}{C}}$$

where R is often an alkyl group such as methyl, ethyl, 2-ethylhexyl, but can include a wide range of other functional groups.

Polymerization is started by a free-radical initiator, often a peroxide. The monomers then polymerize rapidly, adding onto the growing chain to form the final polymer:

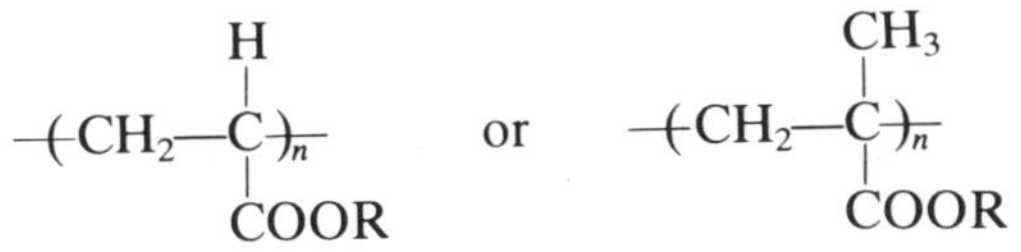

$$-\!\!\left(CH_2{-}\underset{\substack{|\\ COOR}}{\overset{\substack{H\\ |}}{C}}\right)_{\!n}\!\!- \qquad \text{or} \qquad -\!\!\left(CH_2{-}\underset{\substack{|\\ COOR}}{\overset{\substack{CH_3\\ |}}{C}}\right)_{\!n}\!\!-$$

There are many ways of initiating the polymerization process and these form the basis for commercially important adhesive product variations which will be discussed later.

The bond properties are determined by the physical properties of the cured adhesive, which in turn are determined by the nature of the monomers used.

Major property changes are determined by the glass transition temperature, T_g, of the final polymer composition,[3] and also by crosslinking. Crosslinking can be effected by polyfunctional (difunctional or higher) methacrylate or acrylate monomers. These can be used at low concentrations to give improved resistance to creep or flow of flexible adhesives, or at high concentrations to give a hard, temperature- and solvent-resistant, three-dimensional structure typical of 'thermoset' polymers.

Highly crosslinked structures are typical of cured anaerobics, while modified acrylics have more flexible structures formed from mixtures of monofunctional monomers and polyfunctional monomers and polymers (resins).

3. ADHESIVES—DEVELOPMENTS

The acrylic adhesives derive their bonding properties from their ability to wet the substrates to be bonded, then to polymerize rapidly in the bond line to form a strong material joining the two substrates. The adhesives—cyanoacrylates, anaerobics and modified acrylics—differ greatly in how they are formulated to achieve this and in the molecular structures they form on polymerizing.

3.1. Cyanoacrylates

3.1.1. General Properties

Cyanoacrylate adhesives consist almost completely of cyanoacrylate monomers, with small amounts of thickening agents to control viscosity [usually high molecular weight polymers such as poly(methyl methacrylate)] and very small amounts of free-radical and ionic stabilizers. These formulations have the ability to wet a wide range of materials—including metals, rubbers and plastics—and polymerization or bond formation can be initiated without the use of heat or activator primers, or the mixing-in of another component. (Existing levels of moisture on the substrate are sufficient to start the process.)

The material formed in the bond line is a high molecular weight uncrosslinked polymer (usually a homopolymer) which has bond strengths, and resistance to temperature and solvents, determined by the nature of the starting monomer.

Table 1 shows typical bond strengths and humidity resistance for

TABLE 1
BOND PROPERTIES OF CYANOACRYLATE ADHESIVES[a]

Substrate	*Tensile shear strength [ASTM D1002] (MN/m²)*		
	Methyl cyanoacrylate	*Ethyl cyanoacrylate*	*Butyl cyanoacrylate*
Grit-blasted mild steel	22 (20–40)[b]	17 (10–30)	15 (60–80)
Polycarbonate	15 (40–50)	15 (10–15)	9 (40–50)
Butyl rubber	7 (4–6)	7 (3–4)	7 (20–30)

[a] N.B. Data are from the commercial literature[5] and are believed to be typical. Some variation in bond properties with different grades of substrate material will be observed.
[b] Fixture time (s) in parentheses.

TABLE 2

MOISTURE RESISTANCE OF CYANOACRYLATE ADHESIVES[a]

Substrate	*Tensile shear strength (% retained after 8 weeks at 40°C/95% RH)*		
	Methyl cyanoacrylate	*Ethyl cyanoacrylate*	*Butyl cyanoacrylate*
Grit-blasted mild steel	60	60	60
Polycarbonate	95	95	95

[a] N.B. Data are from the commercial literature[5] and are believed to be typical. Some variation with different grades of substrate material will be observed.

adhesives made from methyl, ethyl, and butyl cyanoacrylate. It can be seen from this table that bond strengths of these adhesives are very high, and that speed of bond formation is rapid. Given these properties, the range of substrates that can be bonded, and their one-component nature, it is not surprising that cyanoacrylate adhesives are often considered to be ideal.

However, Table 2 points to significant limitations. Moisture resistance on metal substrates is somewhat limited (although for rubber and plastic substrates it can be very good). Recent studies[6] suggest that this may be due to ingress of moisture along the metal oxide layer on the substrate, which in turn could cause hydrolytic degradation of the cyanoacrylate polymer.[7] The much better moisture resistance of, for example, polycarbonate bonds would be due to the absence of any easy route of ingress for moisture into the bond line.

Temperature resistance is also limited. Because the cured polymer is not crosslinked it softens and loses bond strength as the glass transition temperature is approached. In practice, the adhesives are normally suitable for use up to 80°C, although short excursions to higher temperatures can be tolerated.

Modifying cyanoacrylates to vary or improve on these basic properties is inherently difficult. The high reactivity of the liquid adhesive makes it very susceptible to even parts per million (ppm) contamination by acids (which prevent polymerization) or by bases (which destabilize the adhesive and cause premature polymerization during storage).

However, in recent years a number of commercially important developments have taken place which significantly extend the capability of these materials.

3.1.2. *Thermal Resistance*

Adhesives based on allyl cyanoacrylates have been known for some time.[8] These are capable of giving substantial resistance to temperatures, apparently through the formation of a secondary network by curing the allylic function. These have the disadvantage, however, of requiring long times at the elevated temperature to form the network.

The addition of dimethacrylates or diacrylate esters is claimed[9] to give significant strength improvement at temperatures as high as 150°C, by promoting crosslinking.

Another approach has been to add maleic anhydride to the adhesive just before use.[10] This raised the useful temperature of ethyl cyanoacrylate to about 120°C. A further recent improvement was to add phthalic anhydride to form a stable one-part formulation.[11]

Improvement in thermal shock resistance (−20°C to +60°C) has been found by Shiraishi *et al.*[12] upon adding up to 20% of certain esters (for example, tetraoxyethylene methyl ester acrylate). Small amounts of polyhydroxybenzoic acids were also beneficial.

3.1.3. *Toughness*

Cyanoacrylates are polymeric glasses and under shock loading or in bond geometries which have large stress concentrations (e.g. cleavage) resistance is relatively low. Adding plasticizers is a well known method of increasing flexibility of polymeric materials, but this flexibility is usually achieved at the expense of strength.

Gleave[13] describes the addition of flexible tougheners such as ABS and MBS (methacrylate–butadiene–styrene terpolymer) to cyanoacrylates to increase resistance to peel forces by factors of 4–5. Bond strengths are not impaired by this method, which is similar to the technology of rubber toughening widely used in the plastics industry. O'Connor[14] describes similar compositions which include acrylic rubbers as additives. It has been found with these adhesives that significant improvements in moisture resistance over normal cyanoacrylates is obtained.

Yamada and Kimura[15] have found that polyfunctional carboxylic acids, anhydrides or esters, when added at quite low levels in cyanoacrylates, can give improvements in impact strength by a factor of up to three. For example, 1000 ppm of 1-hexene-2,4,6-tricarboxylic acid added to ethyl 2-cyanoacrylate increased impact strength on steel/steel by a factor of 3 (ASTM D950-54).

Another approach is described by Teramoto *et al.*,[16] who found substantial improvements in peel strength by the addition of 1,1-disubstituted diene (e.g. 1-cyano-1-carbethoxybuta-1,3-diene) at levels of about 30%. Good speed of cure and shear strength were maintained.

3.1.4. Odour

Cyanoacrylates normally have a sharp odour, which in an industrial environment can be dealt with by suitable ventilation.

A more difficult problem can be 'blooming', a fine white precipitate near a cyanoacrylate bond caused by cyanoacrylate vapour being polymerized by moisture in the air. While this effect helps to minimize odour the fine white precipitate, even though it is very localized, can be unacceptable for bonding decorative articles such as cosmetic packages.

Adhesives made from cyanoacrylate monomers with low vapour pressures have been known for some time. Monomers such as β-ethoxyethyl 2-cyanoacrylate and β-methoxyethyl 2-cyanoacrylate have been made into commercially available adhesives which offer virtually no odour and a much reduced tendency to 'bloom'. The adhesives tend to be slower to polymerize than other cyanoacrylates, but offer real commercial benefits. Kimura and Sugiura[17] claim an optimum balance of speed/shelf stability of most of this class of cyanoacrylate adhesives when the monomer has a water content of 300 to 2000 ppm.

3.1.5. Adhesive Strength

Significant improvements in bond strengths to metals were found by Schoenberg and Ray-Chaudhuri[18] by the addition of about 5% of acetic acid. Strength enhancement of about 50% was obtained without losing cure speed.

3.1.6. Porous Substrates

Cyanoacrylate adhesives generally do not bond well to substrates such as wood, leather, paper, ceramics or fabrics. This is partly because such substrates can be acidic, which inhibits polymerization, and also because the porous nature of such materials requires the surface-initiated polymerization to extend across relatively large gaps. Both problems can be overcome by applying 'activator' solutions containing small amounts of basic initiating species such as amines. Recently, however, developments have taken place which allow one-component adhesive formulations to be made with greatly enhanced bonding to porous or acidic substrates.

Crown ethers,[19] polyethylene glycols,[20] mixtures of aromatic and aliphatic polyols and polyethers[21] and polyorganosiloxanes[22] have all been disclosed as giving enhanced bonding properties to cyanoacrylates on these substrates. More recently new additives[23] have been developed which give the performance shown in Table 3, which compares cyanoacrylates using these additives with normal cyanoacrylates. The benefit seen is clear and substantial.

3.1.7. Viscosity

Cyanoacrylates normally are thickened by adding polymers of high molecular weight. Because of the highly reactive nature of cyanoacrylates, adding fillers to make pastes or gels has been difficult. In a recent development,[24] however, silica was added to cyanoacrylates to give a clear gel. This offers the considerable benefit of easy handling, without contamination of parts by excess adhesive.

3.1.8. Medical Uses

Because of the extraordinary ability of cyanoacrylates to bond skin and other tissue an enormous amount of work[25] has gone into finding

TABLE 3
BONDING TO POROUS SUBSTRATES

Substrate	*Time to fixture*[a] (*s*)	
	Ethyl cyanoacrylate[b]	*Ethyl cyanoacrylate + additive*
Wood (spruce)	>180	20–30
Leather	>180	<10
Paper	No bond	<5

[a] Data are believed to be typical. Differences in time for bond formation will occur with variations in substrate.
[b] Normal commercial adhesive.

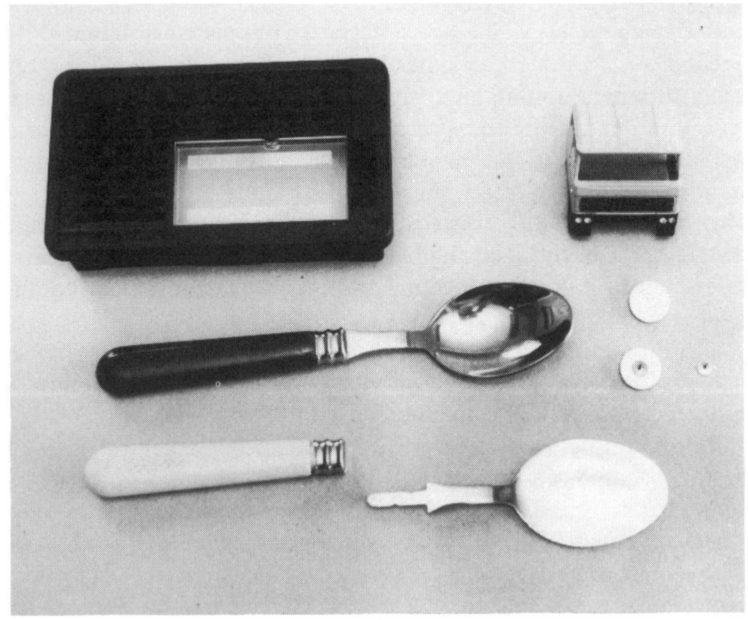

FIG. 2. Bonding applications using cyanoacrylate adhesives.

potential medical uses for these materials. At present the commercial potential in this area is unclear and will remain so until a full evaluation of using cyanoacrylates for medical purposes is performed.

For cyanoacrylate adhesives, therefore, the past several years have been a period of high activity. Technologically important developments have occurred to give commercially available adhesives with improved thermal resistance, toughness, moisture resistance, adhesion to metal and to substrates such as wood, leather and paper, as well as substantial improvements in handling characteristics such as odour and rheology.

These one-component adhesives still enjoy the benefits of speed, strength and ease of use peculiar to cyanoacrylates, but the range of potential uses for these products has been greatly increased. Figure 2 shows a few of the many current applications.

3.2. Anaerobics

Currently available liquid acrylate and methacrylate adhesives can be seen as logical developments of the 'anaerobic' machinery adhesives introduced about 25 years ago.[26] These materials were based on low molecular weight dimethacrylates which exploited the stabilizing effect of oxygen, which inhibits methacrylate polymerization.[3] In an air-free metal joint, and with the presence in the adhesive formulation of metal-sensitive redox initiation systems, the dimethacrylate composition cures to a strong, rigid 'thermoset' structure.

This basic technology was first exploited in bonding nuts and bolts and studs to prevent loosening by vibration. It thus replaced expensive and less effective locking systems and was an extraordinary commercial success.

While threadlocking undoubtedly contributes to the integrity of a bolted system, it is not structural in the sense that the assembly does not depend primarily on the adhesive joint.

Adhesive retaining of cylindrical parts can be structural, however, and has become an increasingly important element in machinery assembly. A typical example is shown in Fig. 3. Such systems offer the mechanical benefits of full load distribution combined with sealing

FIG. 3. Retaining cylindrical parts by means of anaerobic adhesives.

against corrosion, as well as the considerable economic advantages gained by not having to maintain very close tolerances on the machined parts. Schwaiger and Schuch[27] have recently described engineering studies which demonstrate the benefits of bonding in retaining slip-fitted, press-fitted and shrink-fitted parts. The latter two showed strength increases of respectively 110–180% and 70–160% when an anaerobic retaining adhesive was used.

Adhesives are now commercially available which have been designed to be integrated into automatic assembly operations and which can fixture parts in a matter of seconds.

The commercial success of anaerobic technology has resulted in large-scale research activity, with over 200 separate patents (not including duplicate foreign filings) issued in this field to date. While nearly 50 companies hold at least one patent in this area, most are held by only two or three companies. As almost all of this work relates to formulations designed primarily for threadlocking and thread or flange sealing, it will not be discussed here. There are, however, important extensions to this technology which have led to the development of a class of fast-curing resin based acrylic adhesives for rapid assembly operations.

3.3. Modified Acrylic Adhesives

Since the introduction of methacrylate-based adhesives there have been two main strands of development. In one, the low molecular weight dimethacrylates used in anaerobics have been replaced with increasingly sophisticated polymeric polyfunctional methacrylate or acrylate resins to give improved flexibility and toughness. In the other strand, high molecular weight rubbery polymers have been added to monofunctional methacrylate monomers to give the toughness required.

3.3.1. Modified Acrylic Adhesives—Resin-Based

3.3.1.1. Introduction. Resin-based adhesives described by Gorman and Toback[28] included urethane methacrylates which could be either low molecular weight to give hard, rigid adhesives or higher molecular weight low-T_g resins to give flexible, tougher adhesives.[3] Properties of the cured adhesives could be varied by using different amounts of 'hard' and 'flexible' resins in the formulation.

Other resin types have also been used. Owston and Howard[29] described polyisocyanate resins capped with hydroxyalkyl methacrylates, allyl alcohol or vinyl alcohols.[3] Kusayama *et al.*[30] describe materials based on epoxy resins capped with acrylic or methacrylic acid.

Polymerization of the liquid adhesives requires the generation of free radicals in the joint upon assembly of the parts. Free-radical generation can be achieved by heat, UV light irradiation, or reaction of free-radical generators with catalysts. There are many possible ways of doing this. For example, Toback[31] described curing systems which involve a reducing agent with a catalyst in a primer solution which is applied to one substrate. The resin-based adhesive containing a hydroperoxide is applied either to the same surface or to the mating surface. Depending on the nature of the redox curing system used, and because of the high functionality of the resin-based adhesive, handling strength of the bonded joint can be achieved in remarkably short times (as little as 10 s).

Because the resins used have relatively low molecular weights compared with polymeric materials, the viscosity of the liquid adhesives tends to be low (1000 to 20 000 mPa s, typically). These adhesives also include methacrylates such as hydroxyalkyl methacrylates to carry the resins, and small amounts of acrylic or methacrylic acids for adhesion.

The resulting adhesives are easy to apply in precisely controlled quantities by automatic application equipment, are non-flammable and easy to handle and provide very rapid bond formation to give high-strength bonds to metals, glass, ferrites and similar materials where relatively close-fitting, clean joints can be attained.

Recent developments in this technology have been aimed at improving the toughness of the adhesives, various ways of achieving polymerization or 'cure' in the bond line, and adhesion.

3.3.1.2. Toughness. Baccei[32,33] describes reactive resins which combine in one molecule the 'hard' and 'flexible' moieties described earlier. On polymerization this molecular structure gives a very tough and impact resistant adhesive, while retaining the speed of cure and ease of use of previous resin-based adhesives.

3.3.1.3. Polymerization. Melody *et al.*[34] describe a novel cure system which provides very rapid curing while maintaining the ability to form bonds through reasonably large gaps. A problem with surface-activated adhesive systems is that polymerization throughout the bond line requires diffusion of the activating species through the adhesive. With very rapidly polymerizing systems this diffusion is limited and therefore bonding through large gaps is difficult. Variations of this rapid cure system have a resin-based formulation to carry the catalyst system instead of a primer solution. This (non-volatile) component is applied to the first adherand, either as a drop or as a continuous bead. The second component (also non-volatile) of the adhesive is placed beside or on top of the first. The joint is then assembled; the action of the two adherands coming together forces the two adhesive components to mix. This method improves gap filling and gives the industrially important benefit of reducing the adhesive/ activator application step to one rapid automatic operation.

Another possible variation is to make the adhesive UV-curable. When the adhesive is applied a slight excess or 'fillet' is deliberately left outside the bond line. This can be polymerized in seconds with UV light, giving a very rapid initial fixturing, and enabling the assembled part to proceed to the next stage in the production line.

Table 4 outlines typical performance parameters for such adhesives. Alternative approaches to achieving rapid-curing adhesives have been described. Ukita *et al.*[36] cite a reactive polybutadiene resin-based formulation which uses a hydroperoxide/amine in one part of the system and a cobalt salt of an organic acid as a catalyst in the second part. Bond formation occurs in a few minutes.

Skoultchi[37] describes a two-part curing system with good speed and shelf-life which consists of copper-saccharin (or saccharin with a copper salt) in one part and an α-hydroxysulphone or amine-sulphone in the second part.

Improving curing through thicker bond lines is described by Bachmann.[38] All methacrylate and acrylate monomer polymerization can be inhibited by the presence of oxygen, as described earlier. In this development improvements have been obtained in the bond line thickness that can be polymerized by means of surface activation.

3.3.1.4. Adhesion. Adhesion to various substrates of modified acrylic adhesives can be achieved by selection of monomers (e.g. hydrophilic monomers for high-surface-energy substrates, or monomers such as methyl methacrylate for plastics) and by the addition of small amounts of adhesion promoters such as acrylic or methacrylic acid or silanes (see Chapter 8).

TABLE 4

PERFORMANCE OF RESIN-BASED MODIFIED ACRYLIC ADHESIVES[a]

Property	*High-strength adhesive*	*Very fast adhesive*
Speed		
Time to fixture (minimum bond gap)	2–5 min	30–60 s
Strength		
Tensile shear of GBM steel (ASTM D1002-64)	22 N/mm^2	20 N/mm^2
Toughness		
T-peel of GB aluminium (ASTM D1876-69T)	5 N/mm	2·5 N/mm
Maximum bond line thickness	0·7 mm	0·5–1 mm
Recommended substrates	Primarily metals, ferrites	Metals, ferrites
Viscosity (25°C)	20 000 mPa s	15 000 mPa s
Flashpoint	>90°C	>100°C
Method of cure	Surface activator	Two-component (bead-on-bead)

[a] Data are taken from the technical literature[35] and are believed to be typical.

Zalucha *et al.*[39,40] have described the use of phosphoric acid and organic derivatives of phosphoric, phosphonic or phosphinic acid (e.g. 2-methacryloxyethyl phosphate). Substantial improvements in adhesion to solvent-wiped steel and aluminium on that obtained with just methacrylic acid were obtained.

The benefits offered by resin-based acrylic adhesives make them ideally suited for the rapid automatic assembling of small parts such as ferrites, which are used in many electrical or electronic systems, loudspeakers and small motors, or for glass bonding—where UV light curing is very fast and effective. Examples of ferrite-bonding applications are shown in Fig. 4.

Many of the industrial applications of these adhesives are those where epoxy adhesives were traditionally used. Table 5 outlines the relative merits of cyanoacrylates, resin-based modified acrylics and epoxies. It can be seen that cyanoacrylates are nearly ideal for this type of assembly operation, but where high-temperature, environment

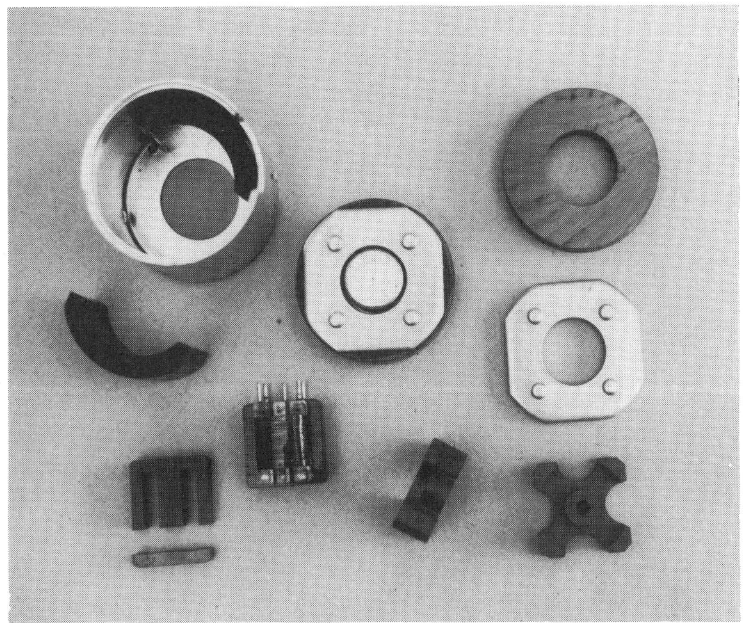

FIG. 4. Typical bonding applications for fast resin-based acrylic adhesives.

TABLE 5
COMPARATIVE BENEFITS OF INDUSTRIAL ADHESIVES[a]

Property	*Epoxies*		*Cyanoacrylates*	*Resin-based modified acrylics*
	Two-part mixed	*One-part heat-cured*		
Bond strength				
Metals/ferrites	+	+ +	+	+ +
Plastics	0	−	+ +	0
Toughness	−	+ +	0	+ +
Temperature/ humidity	+	+ +	0	+
Speed of bonding	0	− −	+ +	+
Ease of use				
Automatic application	−	−	+ +	+ +

[a] + +, Great benefit; +, significant benefit; 0, no benefit; −, significant disadvantage; − −, great disadvantage.

or impact resistance are required then tough modified acrylics are used. Epoxy adhesives offer good bond performance, but are difficult to adapt to really high-speed, automatic, production systems.

However, the bonding of larger components, components made of a wider variety of materials such as wood, plastic and unprepared metals, or flexible substrates such as sheet metal, requires in each case a different type of modified acrylic adhesive.

3.3.2. Modified Acrylic Adhesives—Rubber-Toughened

3.3.2.1. Introduction. The principle of toughening hard plastics such as polystyrene by the incorporation of high molecular rubbers into the monomers before polymerization has long been well established. The extension of this technology to epoxy and acrylic adhesives was a logical way to give toughness, while retaining strength, as is described in detail in Chapter 5.

Owston[41] describes adhesive formulations in which polybutadiene rubbers, styrene–butadiene or acrylonitrile–butadiene copolymers are used in monomethacrylates such as methyl methacrylate with methacrylic acid to provide adhesion. Cure is effected by incorporating reducing agents in the adhesive, with a free-radical source such as benzoyl peroxide in the activator portion. When such a composition polymerizes in the bond line the rubbery molecules separate out to form discrete rubber domains or particles. These provide the necessary resistance to impact or cleavage while the shear strength and resistance to solvents and temperature are provided essentially by the polymerized monomethacrylate mixture. A similar adhesive system based on polyurethanes was disclosed by Wolinski.[42]

Such adhesive systems and similar adhesives are now widely used.[43,44] With methyl methacrylate used as a monomer, high bond strengths and toughness with a wide variety of substrates is achieved, although the odour and flammability of the monomer does require special handling procedures. Although cure speed can be improved by mixing in accelerators before use, most of these adhesives will fixture a joint in 5–10 min.

Briggs and Muschiatti[45] described adhesive compositions based on chlorosulphonated polyethylene as a toughening agent. With suitable activators these adhesives can be cured more rapidly than the previous types (1–2 min) and still provide high strength and toughness. Figure 5 shows the rubber particle structure obtained with these adhesives. This

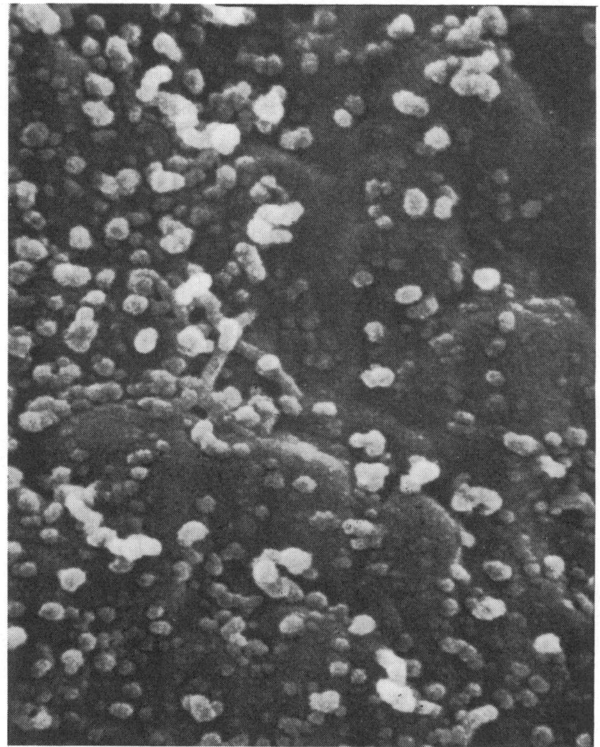

FIG. 5. Rubber particles in a rubber-toughened acrylic adhesive.[46]

technology was hailed as the 'Second Generation' Acrylic Adhesives[47] and has since become widely used both industrially and in the consumer market. Table 6 shows typical performance parameters of these adhesives.

Speed, ease of use and the availability of a low volatility, non-flammable formulation have led to the application of this type of adhesive in general purpose bonding, where bond line thickness or substrate type preclude the use of resin-based adhesives (Fig. 6.)

As with other acrylic adhesives commercial success, primarily in industrial bonding, has led in recent years to many significant new developments.

3.3.2.2. Curing in the bond line. The 'Second Generation' adhesives (described above) include as substrate activators formulations

TABLE 6
PERFORMANCE OF RUBBER-TOUGHENED ACRYLIC ADHESIVES[a]

Property	Typical second-generation acrylic	
	Non-volatile	*Volatile*
Speed		
Time to fixture (min)	1–2	1–2
Strength		
Tensile shear of GBM steel (N/mm^2) (ASTM D1002-64)	15–20	20–25
Toughness	3	3–4
T-peel of GB aluminium (N/mm) (ASTM D1876-69T)		
Maximum bond line thickness (mm)	0·5–1	0·5–1
Recommended substrates	Metals, wood, concrete, glass, ferrites, some plastics	Metals, wood, concrete, glass, ferrites, plastics
Method of curing	Surface activator	Surface activator
Maximum recommended temperature (°C)	100–120	100–120

[a] Data are from the commercial literature.[48,49]

of Schiff's bases similar to those described by Toback.[31] They are easier to work with than the peroxides and amines used in the earlier adhesive systems.

Wolinski has approached the problem with these early systems in two ways. In one,[50] he describes the use of micro-encapsulation to provide a stable one-component adhesive, with curing being initiated by rupturing the microcapsules. In the second,[51] he describes reacting a relatively volatile amine accelerator with an epoxy to form a tertiary amine accelerator with low vapour pressure.

Another, general, approach is to incorporate the activator in a separate portion of the monomer/toughener formulation. Depending on the composition, curing in the two-component system can be

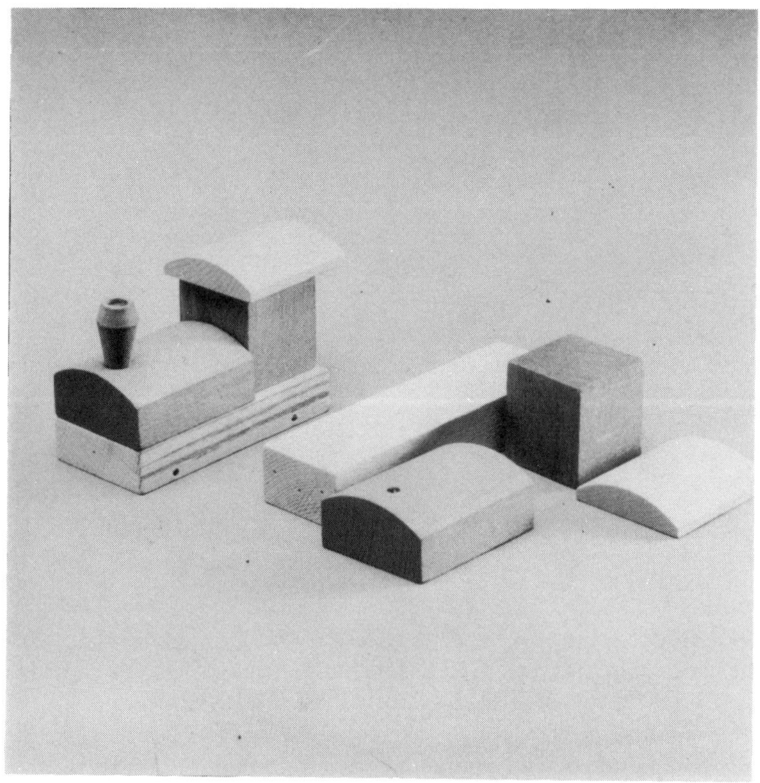

Fig. 6. Bonding wood toys with rubber-toughened acrylic adhesives.

initiated by premixing the parts before the joint is made or by applying a bead of the first part on top of the second, with mixing in the bond line and initiation beginning when the components to be bonded are joined.

3.3.2.3. Adhesion. Adhesion promoters based on phosphoric acid and derivatives of phosphoric, phosphonic and phosphinic acids have been mentioned in Section 3.3.1.4.[39,40] The same technology can be used with rubber-toughened acrylic adhesives, to provide alternatives to the commonly used acrylic and methacrylic acid adhesion promoters.

3.3.2.4. Tougheners. Recently Charnock[52,53] has disclosed the use

FIG. 7. Spot-welded and bonded sheet metal joints.

of polyisoprene and polyacrylate elastomers as tougheners in acrylic adhesives. This has led to the development of adhesives capable of bonding as received (oiled) sheet steel.[54,55] Applications of such adhesives include bonding steel panel stiffeners and brackets to sheet steel panels in office equipment, computer cabinets and various automotive areas (trucks, buses, agricultural machinery).

The most obvious benefit offered by bonding over the traditional spot-welding technique is illustrated in Fig. 7. Spot-welding leaves unsightly marks on the surface of the welded parts, which must be removed by hand and refinished. Bonding eliminates this obviously expensive process. To achieve such bonds, however, the adhesive must bond through waxy protective and lubricating oils on the steel surface and then later survive paint stoving temperatures up to 200°C. These new acrylic adhesives meet these requirements and also offer the usual benefit of room-temperature curing.

Table 7 summarizes some key properties of these adhesives. Bond line thicknesses up to 2 mm can be accommodated by using a 'bead-on-bead' adhesive application, but above that limit mixing is recommended. In studies of flexural stiffness[55,56] it has been demonstrated that bonded structures can actually be made stiffer than welded structures, by distributing loads more evenly.

TABLE 7

PERFORMANCE OF STEEL-BONDING ACRYLIC ADHESIVE[a]

Property	Nominal bond line thickness	
	0 mm	*2 mm*
Speed of cure		
Time to fixture		
Room temperature	5 min	30 min
Spot-heat[b]	5 s	3 min
Strength		
Tensile shear (N/mm^2) (ASTM D1002-64)	11	9
T-peel (N/mm) (ASTM D1876-69T)	7	15
Environmental resistance		
Strength (%) retained after 1000 h at		
40°C/95% RH	95	95
150°C/air	80	80

[a] Data after Charnock.[55]
[b] Application of a hot clamp (>200°C) to sheet metal.

TABLE 8

TOUGH ACRYLIC VERSUS TOUGH EPOXY ADHESIVE

Property	*Toughened acrylic*[c]	*Toughened epoxy*[c]
Speed		
Time to fixture (min)		
Room temperature	5–30	No cure
Heating (200°C)	1–3	5–10
Strength[a]		
Tensile shear (N/mm^2)		
GBM steel	10	30
Oiled steel[b]	10	15
T-peel (N/mm)		
GBM steel	7–15	4
Oiled steel[b]	7–15	0
Maximum temperature (paint-bake) (°C)	200	230

[a] All strengths after a 'paint-bake' heat cycle of 30 min at 185°C.
[b] Astrolan 17/40 oil.
[c] Data after Charnock.[55]

Table 8 compares tough acrylics with tough epoxies for bonding applications such as sheet metal bonding. For some substrates, or where very high paint-bake temperatures are required, or where a heat cure is actually preferred then epoxy adhesives offer a good strong bonding solution. For this type of application, the two adhesive types should be seen as complementing one another, with the acrylic adhesives offering the greater number of benefits on as-received sheet steel.

4. CONCLUSION

The acrylic adhesives (cyanoacrylates, anaerobics and modified acrylics) are commercially very successful, because of the particular benefits they offer. Strong bonds can be formed rapidly under conditions which are well suited to industrial assembly operations.

In recent years new developments have extended the capabilities of these systems.

(a) *Cyanoacrylates*: Significant improvements have been made in thermal resistance, toughness and moisture resistance, adhesion to metal and to materials like wood, leather and paper. In addition, low odour and highly thixotropic adhesives are now available.

(b) *Anaerobics*: Research activity has remained high in this commercially important area. Recently, retention of cylindrical machined parts in engineering assemblies by bonding has become increasingly important.

(c) *Modified acrylics*: Easily applied, fast-bonding resin-based acrylic adhesives are rapidly displacing older adhesive technologies in the high-speed automatic assembly of small components. Recent developments give even faster bonding, with improved adhesion and toughness.

Rubber-toughened acrylics have gained widespread acceptance for use in bonding many different materials. Improvements in speed, odour and adhesion have led to increased use in industrial operations. Recent developments in bonding as-received steel make possible practical methods to replace spot-welding techniques in the sheet steel fabrication industry.

All of these developments in acrylic adhesives can be expected to contribute to the continued growth of these materials in the future.

REFERENCES

1. HOLAPPA, H. S., in *Designing with Today's Engineering Adhesives,* Papers presented at the Spring Seminar, 11–14 March, 1979, Cherry Hill, NJ, Adhesive and Sealant Council, Arlington, Virginia, 1979.
2. ROONEY, J. M., *Polym. J.,* **13** (1981) 975–8.
3. MARTIN, F. R., 'Acrylic adhesives', in *Developments in Adhesives—1,* Ed. W. C. Wake, Applied Science Publishers, London, 1977, pp. 157–179.
4. BRINDLEY, W. H., CALDER, G. V. and WETZEL, L. A., in *Applied Polymer Science,* Eds J. K. Craver and R. W. Tass, American Chemical Society, Washington DC, 1975.
5. ANON., *Cyanoacrylate Adhesives, Technical Data Sheet 60* Locite Corp., Conn., USA, 1980.
6. DRAIN, K. F., GUTHERIE, J., LEUNG, C. L., MARTIN, F. R. and OTTERBURN, M. S., *J. Adhesion,* **17** (1984) 71–82.
7. LEONARD, F. R., KULKARNI, R. K., BRANDES, G., NELSON, J. and CAMERON, J. J., *J. Appl. Polym. Sci.,* **10** (1966) 259–72.
8. COOVER, H. W. and MCINTIRE, J. M., 'Cyanoacrylate Adhesives', in *Handbook of Adhesives,* 2nd edn, Ed. I. Skeist, Van Nostrand–Reinhold, New York, 1977.
9. SETSUDA, K. and SUGIYAMA, I., US Patent 3 692 752, 1972.
10. O'SULLIVAN, D. J. and MELODY, D. P. (Loctite Corp.), US Patent 3 832 334, 1974.
11. HARRIS, S. J. (Loctite Corp.), US Patent 4 450 265, 1984.
12. SHIRAISHI, Y., NAKAZAWA, K., NAKATA, C. and OHASHI, K. (Sumitomo Chem. Co.), US Patent 4 307 216, 1981.
13. GLEAVE, E. R. (Loctite Corp.), US Patent 4 102 945, 1978.
14. O'CONNOR, J. T. (Loctite Corp.), US Patent 4 440 910, 1984.
15. YAMADA, A. and KIMURA, K. (Toa Gosei Chem. Co.), US Patent 4 196 271, 1980.
16. TERAMOTO, T., IJUIN, N. and KOTANI, T. (Japan Synthetic Rubber Co.), US Patent 4 313 865, 1982.
17. KIMURA, K. and SUGIURA, K. (Toa Gosei Chem. Co.), US Patent 4 321 180, 1982.
18. SCHOENBERG, J. E. and RAY-CHAUDHURI, D. K. (National Starch and Chemical), US Patent 4 125 494, 1978.
19. ANON. (Toa Gosei Chem. Co.), De-Os 2 816 836, 1978.
20. AKIRA, M. and KAORU, K. (Toa Gosei Chem. Co.), US Patent 4 170 585, 1979.
21. REICH, K. and TOMASCHEK, H. (Terason), US Patent 4 386 193, 1983.
22. ANON. (Three Bond), Japan Kokai Tokkyo Koho 82-70171, 1982.
23. HARRIS, S. J., MCKERVEY, M. A., MELODY, D. P., WOODS, J. and ROONEY, J. (Loctite Corp.), US Patent 4 556 700, 1985.
24. LITKE, A. (Loctite Corp.), US Patent 4 477 607, 1984.
25. LEE, H., Ed., *Cyanoacrylate Resins—The Instant Adhesives,* Pasadena Technology Press, Pasadena, 1981.

26. Krieble, V. K. (The American Sealants Co.), US Patent 2 895 950, 1959.
27. Schwaiger, M. and Schuch, F., 'Experimental investigation of adhesives to augment cylindrical joints,' presented at *IAVD Congress on Vehicle Design,* 4–6 March, 1985, Geneva.
28. Gorman, J. W. and Toback, A. S. (Loctite Corp.), US Patent 3 425 988, 1969.
29. Owston, W. J. and Howard, D. D. (Lord Corporation), US Patent 3 873 640, 1975.
30. Kusayama, S., Ohashi, K. and Takada, K. (Sumitomo Chemical Corporation), US Patent 3 870 675, 1975.
31. Toback, A. (Loctite Corp.), US Patent 3 616 040, 1971.
32. Baccei, L. J. (Loctite Corp.), US Patent 4 309 526, 1982.
33. Baccei, L. J. (Loctite Corp.), US Patent 4 295 909, 1981.
34. Melody, D. P., Doherty, D. A., O'Grady, J. O. and Rich, R. (Loctite Corp.), US Patent 4 180 640, 1979.
35. Anon., *Adhesives,* Technical Data Sheet, Loctite Corp., Conn., USA, pp. 324, 327, 1985.
36. Ukita, K., Nakano, T. and Kishi, I. (Denki Kagaku Kogyo Kabushiki), US Patent 4 331 795, 1982.
37. Skoultchi, M. M. (National Starch and Chemical Corp.), US Patent 4 081 308, 1978.
38. Bachmann, A., US Patent 4 348 504, 1982.
39. Zalucha, D. J., Sexsmith, F. H., Hornaman, E. C. and Dawdy, T. H. (Lord Corp.), US Patent 4 223 115, 1980.
40. Zalucha, D. J., Sexsmith, F. H., Hornaman, E. C. and Dawdy, T. H. (Lord Corp.), US Patent 4 293 665, 1981.
41. Owston, W. J. (Lord Corp.), US Patent 3 832 274, 1974.
42. Wolinski, L. E. (Pratt and Lambert Inc.), US Patent 3 994 764, 1976.
43. Anon., *Adhesive Agomet 310,* Technical Data, De Gussa, West Germany.
44. Anon., *Hardloc Adhesives,* Technical Data, Denka–Kagaku, Japan, 1982.
45. Briggs, P. C. and Muschiatti, L. C. (E.I. du Pont de Nemours and Co.), US Patent 3 890 407, 1975.
46. Charnock, R. S. and Martin, F. R., Structure–property relationships of a rubber-modified acrylic adhesive, *International Adhesion Conference,* Durham, UK, 1980.
47. Anon., 'Second generation acrylic adhesives', in *Adhesives Age,* **19**(9) (1976) 21–4.
48. Anon., *Multibond,* Technical Data Sheet, Loctite, pp. 329, 330, 1979.
49. Anon., *Flexon,* Technical Data Sheet, Permabond, UK, p. 241.
50. Wolinski, L. E. and Berezak, P. D. (Pratt and Lambert, Inc.), US Patent 4 126 504, 1978; Wolinski, L. E. (Pratt and Lambert, Inc.), US Patent 4 080 238, 1978.
51. Wolinski, L. E. and Berezak, P. D. (Pratt and Lambert, Inc.), US Patent 4 155 950, 1979; US Patent 4 297 158, 1981; US Patent 4 212 921, 1980.

52. Charnock, R. S. (Loctite Corp.), US Patent 4 451 615, 1984.
53. Charnock, R. S. (Loctite Corp.), US Patent 4 442 267, 1984.
54. Charnock, R. S., 'Structural acrylic adhesives for the sheet steel fabrication industries', *PRI International Adhesion Conference,* Nottingham, UK, 1984.
55. Charnock, R. S., 'Toughened adhesives for sheet metal bonding', *IAVD Congress on Vehicle Design,* 4–6 March, 1985, Geneva.
56. Lees, W. A., 'Designing and producing toughened structural adhesives', *Adhesives Age,* (October 1984) 26–30.

3

Epoxy-Based Adhesives

E. W. Garnish

Ciba–Geigy, Bonded Structures Division, Cambridge, UK

1. INTRODUCTION

Epoxy technology is supported by a large body of knowledge of chemistry, formulation and application gathered over a period of 40 years of industrial use. However, the pursuit of new developments continues with vigour. A surveyor finds that each year there are many new compositions and significant increases in our knowledge of their chemistry, gleaned perhaps with new techniques, expansions in polymer physics and mechanics, and all are relevant to wider and more successful applications of these thermosetting resins.

Epoxy compositions are made up of the resins themselves plus the hardeners that produce the curing reactions. There are many other modifying adjuncts that can be present as well: fillers, polymers, rubbers, co-reacting resins. In the formulation of adhesives all these materials find full scope, imparting special advantages. It is the work reported in these areas of epoxy resin technology, in the past decade, that are surveyed in the present review.

2. RESINS

The first commercial epoxy resins, and still the most important, are those from the diglycidyl ether of bisphenol A (DGEBA resins). A formal representation of the structure of these resins is shown in Fig. 1.

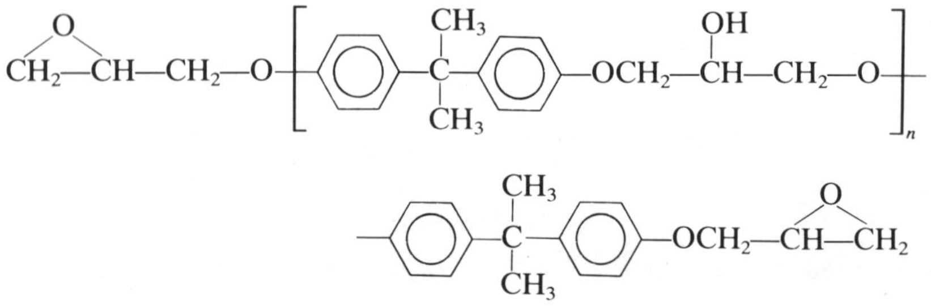

FIG. 1. General formula for DGEBA resins.

However, many other polyglycidyl materials have been synthesised and tested and a number are commercially marketed. Indeed, new epoxy materials of wide variety continue to be suggested for use as adhesives and interesting developments include the following.

(i) For example, glycidyl ethers from 3,3′,5,5′-tetra-alkyl-4,4′-dihydroxybiphenyl[1] (Fig. 2), bis(hydroxyphenyl)phenylethane[2] (Fig. 3) or polyhydroxynaphthylene[3] have been suggested.

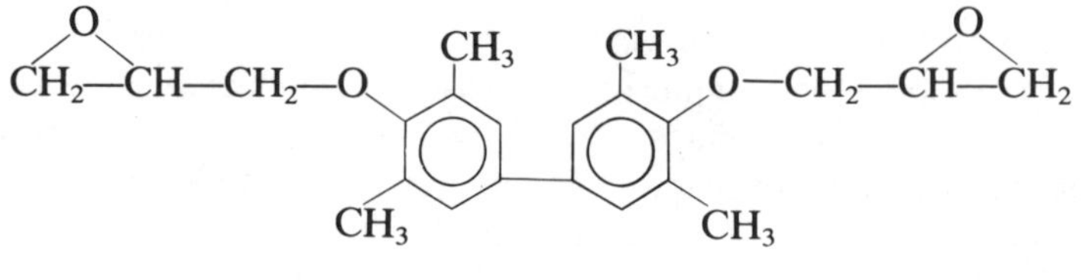

FIG. 2. 3,3′,5,5′-Tetramethyl-4,4′-dihydroxybiphenyl diglycidyl ether.

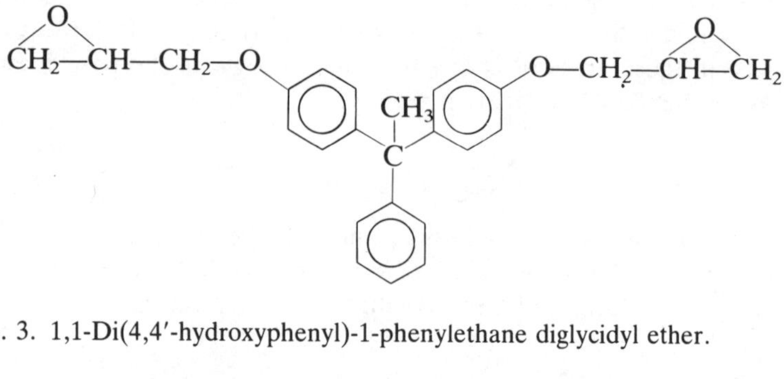

FIG. 3. 1,1-Di(4,4′-hydroxyphenyl)-1-phenylethane diglycidyl ether.

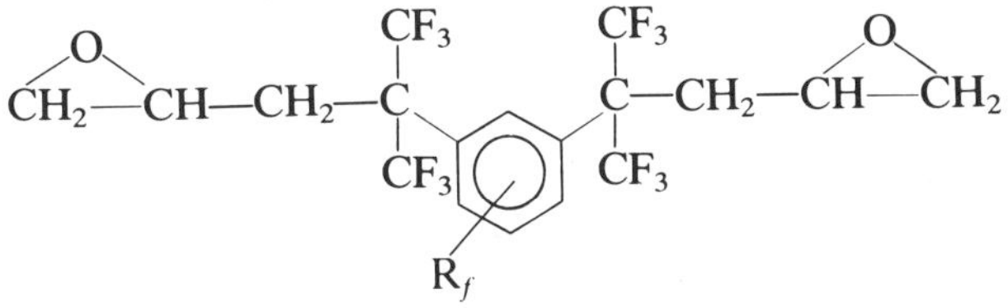

FIG. 4. Perfluoroalkyl derivative of di-1,3-(1,2-epoxy-4,4-di(trifluoromethyl)-butyl)benzene. R_f = fluoro(1–18C)alkylene.

(ii) Fluorinated epoxy molecules with up to 53% fluorine have been synthesised[4] with interesting low-energy water-repellent properties. However, compatability with hardeners can be limited; fluorinated anhydrides are satisfactory whilst conventional amines are not. This observation has led to amine-bearing siloxanes having been synthesised[5] and silanes with functional groups have been employed as hardeners, with perfluoroalkyl derivatives of di-1,3-(1,2-epoxy-4,4-di(trifluoromethyl)butyl)benzene (Fig. 4), for bonding fluorinated rubbers.[6] In this application, good resistance to high temperature was shown: the peel strength declined only 20% after 24 h at 230°C.

(iii) Glycidyl-1,2,4-triazolidine-3,5-diones (Fig. 5) have been suggested as the basis of adhesive binders for non-woven textiles.[7] An adduct of bishydantoin compounds with a hydantoin trisepoxide[8] or bisepoxide[9] gave more hydrophobic properties than previously known types.

(iv) Adducts from diazacrown ethers and bisphenol A epoxy resin cured with primary diamines have been claimed to result in improved shear and peel strength, due to increased flexibility and polarity.[10]

(v) Propenyl-substituted glycidyl ethers (Fig. 6) can react conventionally with hardeners whilst the unsaturated group can

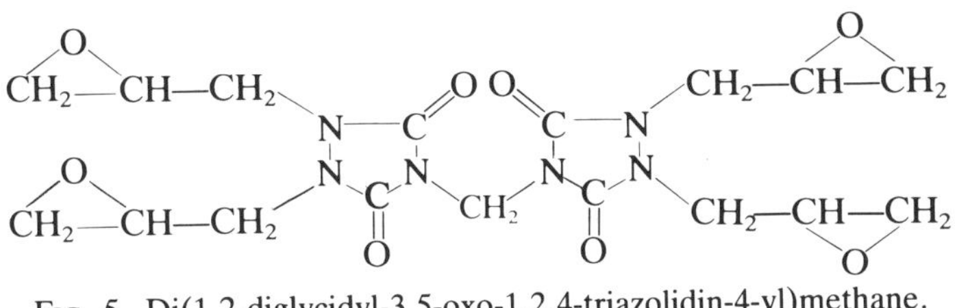

FIG. 5. Di(1,2-diglycidyl-3,5-oxo-1,2,4-triazolidin-4-yl)methane.

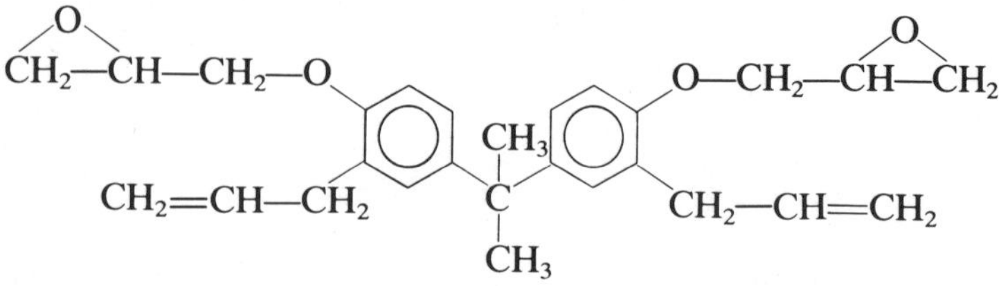

$$\begin{array}{c} \overset{O}{\overbrace{CH_2-CH}}-CH_2-O-C_6H_3(CH_2-CH=CH_2)-C(CH_3)_2-C_6H_3(CH_2-CH=CH_2)-O-CH_2-\overset{O}{\overbrace{CH-CH_2}} \end{array}$$

FIG. 6. Di-2,2-(3-allyl-4-glycidyl-2,2′-allylphenyl)propane.

be cured using radical polymerisation. The two reactions can be carried out simultaneously or successively.[11].

(vi) Glycidyl ethers with allyl groups can be co-reacted with conventional hardeners and bis-polyimides.[12] An analogous co-reaction can be achieved with a bis-polyimide, a polyglycidyl ether and a polyphenol with allyl substituents.[13]

(vii) Polyglycidylamines are already used as resins in composites and adhesives. More recently *N*-glycidylamides have been synthesised and found to be useful as adhesives for steel.[14]

3. HARDENERS

3.1. Amine-Based Hardeners

Aliphatic and aromatic amines are important classes of hardener for epoxide resins in adhesive and other applications. Aliphatic amines are usually regarded as curing agents for room-temperature or warm-cure regimes; aromatic amines for hot-cure. Compositions with the former have comparatively short useable lives, whilst the latter can give compositions with room-temperature lives of one month or more. The rates of cure reflect the different rate constants of the amine hydrogen addition reaction with epoxide in the two classes, whilst the relative ability of the two types to maintain long periods of fusibility before gelation depends on the ratio of rate constants for secondary-to-primary amine hydrogen with epoxide resins. With hexamethylenediamine this is 0·6–0·7, for 4,4-diaminodiphenylmethane (DDM) it is 0·35–0·45.[15] Also, steric effects can be used in aliphatic amines to alter the ratios. For example, with 2,5-dimethyl-2,5-hexanediamine the steric hindrance of the methyl groups causes the ratio to be reduced to about 0·017.[16] In the curing of amine–epoxides, side reactions can also take place. One of importance is epoxide etherification. It is slow, but

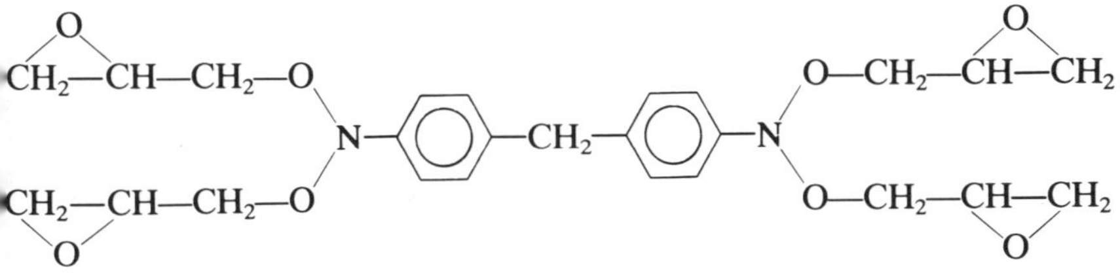

$$\begin{array}{c}CH_2(O)CH{-}CH_2{-}O\\ \quad N{-}C_6H_4{-}CH_2{-}C_6H_4{-}N \\ CH_2(O)CH{-}CH_2{-}O\end{array}\quad \begin{array}{c}O{-}CH_2{-}CH(O)CH_2\\ \\ O{-}CH_2{-}CH(O)CH_2\end{array}$$

FIG. 7. N,N,N′,N′-Tetraglycidoxydi(4-aminophenyl)methane.

model calculations show that gelation can take place even when the epoxy groups are three or more times in excess of amine hydrogen,[17] due to epoxide etherification. Also, for example, a non-specific hydroxyl–epoxide side reaction will take place between diethanolamine and an epoxy resin.[18] Aromatic amines with DGEBA resins react through specific stoichiometric addition of amine hydrogen to epoxide. However, in the polyfunctional system of 4,4′-diaminodiphenylsulphone (DDS) and tetraglycidyldiaminodiphenylmethane resin (TGDDM; Fig. 7) gelation occurs at about 30% conversion by primary amine reactions.[19] Thereafter epoxy–hydroxyl reactions tend to dominate,[20] contributing to an inhomogenous morphology[21] (see Chapter 7).

Some particular developments worthy of consideration are discussed below.

(i) The extent of internal stress in amine-cured epoxides has been studied[22,23] and with α-diamines the internal stress was reduced by increasing the chain length of the diamine.

(ii) Mechanical properties may show variations based on cure regime. For example, with triethanolamine and *o*-, *m*- or *p*-phenylenediamine, stepwise cure gave a 50% increase in impact strength over continuous cure.[24] With diethylenetriamine as a hardener for DGEBA resin, annealing below the glass transition temperature, T_g, caused an increase in yield stress and a decrease in ultimate elongation; quenching from above T_g lowered stress and increased elongation and the changes were reversible.[25] These effects arise because the cured network is not a continuum: there is free volume present. With aromatic amine hardeners it has been suggested that free volume has a distribution of hole size; the form of that distribution, as well as the fraction of free volume, influences the properties of the cured glassy composition.[26]

(iii) Postcuring is a common step in epoxy technology and this can increase the glass transition temperature, T_g, in some systems.[27] Such effects could be due to secondary reactions, i.e. irreversible, or to free volume effects, i.e. reversible. Further, slow changes due to either cause may occur during the service life of adhesives.

(iv) The technology of using aliphatic amine hardeners with epoxide resins is well established. When a novel useful type like the polyoxyalkylene polyamines is identified, the number of patents of composition required for protection, e.g. for adducts[28] or derivatives,[29] can extend to more than 20.

(v) The modifying effect of the amine structure on properties is shown by using aliphatic diamines $H_2N(CH_2)_nNH_2$ ($n =$ 2, 6, 10, 12) as hardeners for DGEBA resins in adhesives.[30] Fatigue strength, fracture energy and elongation at break increased with n, whilst glass transition temperature, shear modulus and tensile strength decreased with n. Polymers or elastomers that partially crosslink with the resin composition are important in adhesive technology. For example, nylons have been important and tests with diamine oligomers of nylon 6 and nylon 66 gave adhesives with only about 10% of the peel strength obtained with an epoxy–nylon adhesive.[31]

(vi) An amine group containing polysiloxane compound can act as an advantageous co-hardener with another amine by giving good peel strength (5 N/mm), and good water resistance, no change being observed after 90 days in water at 90°C.[32]

(vii) When the amine : epoxide ratio with aromatic amines was varied between 0·7 and 2·2, it gave comparatively small effects on tensile properties in castings, but the values of impact strength and fracture toughness varied directly with the ratio.[33]

(viii) There have been comparisons between aromatic amines as hardeners for epoxide adhesives. For example, alternative isomers to the accessible 4,4′-diaminodiphenyl methane and sulphone have been examined. The 3,3′-isomer of the former can have the advantage over the 4,4′-form,[34] either in shear strength or water resistance. Isomers of the sulphone were not found significantly to affect these properties. Also, no advantages were found in using the isomers of diaminobenzophenone.

(ix) Advantages have been claimed for employing polychlorinated benzene diamines as hardeners; for example they give good moisture resistance[35] and quinoxaline diamine compounds have resulted in materials showing good thermal stability, without extreme hardening conditions being necessary or brittleness being developed.[36]

(x) Special amine–epoxy resin compositions have been devised for particular advantages of adhesives. A synergistic combination of a cyanic acid ester with an aliphatic or aromatic amine gives rapid gelation of an epoxy adhesive.[37]

(xi) A polyamide hardener from dicarboxylic acids with 1,4-bis(aminoalkyl)piperazine end-groups gave, with an epoxy resin, rapid solidification, rather like a hot-melt adhesive, and resulted in high shear and peel strength after cure.[38]

(xii) A two-phase solid adhesive for hot-melt bonding with delayed cure has been formed by blending a liquid epoxy resin with an excess of aliphatic polyamine[39] or liquid aromatic amine[40] and then mixing in a finely powdered solid epoxy resin to complete the composition. The liquid constituents react to a solid adduct. 'Hot-melting' allows the reaction to go to completion.

(xiii) Finally, corrosion inhibition of metal substrates may be enhanced by reacting aqueous chromic acid with a polyaminoamide and using the grease-like product as a hardener.[41]

3.2. Latent Hardeners

Latent hardeners such as dicyandiamide (Fig. 8) were used in the first commercial epoxy adhesives and this compound is still used, often with an accelerator or co-hardener. Latent systems usually involve physical separation of the active ingredients, e.g. the high-melting

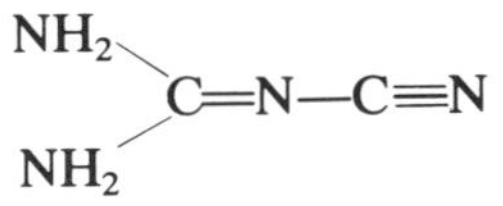

$$(NH_2)_2C{=}N{-}C{\equiv}N$$

FIG. 8. Dicyandiamide.

solid dicyandiamide may be dispersed in the resin. More recent developments are described below.

(i) A blend of epoxy resin and 2-methylimidazole has been given increased latency by adding methanol or ethanol, which suppresses the crosslinking reactions.[42]

(ii) Latent hardeners have been shown to cure at elevated temperature via complex reaction sequences. The curing reaction of *o*-tolylbiguanide has been studied[43] with epoxides. It was found that the hardener brings five active hydrogens to the reaction and model reactions of dicyandiamide and tertiary amine with glycidyl ether showed amine–epoxide addition, etherification with further epoxide of the hydroxyl formed, and polyether formation from the epoxide.[44]

(iii) A feature of accelerated latent compositions is that the reaction is strongly exothermic once initiated, and this often presents difficulties if thick adhesive bond lines are present, since charring or foaming may occur. With dicyandiamide, 1,1′-*o*-phenylenebis(3,3-dimethylurea) can be used as a co-hardener for epoxy resins. Together with high filler loadings, this results in a solder curable under an infrared lamp. The exotherm temperature may rise to 200°C after initiation at 70°C, but without accompanying degradation.[45]

(iv) An epoxide resin with an isocyanate prepolymer, Monuron [3-(*p*-chlorophenyl)-1,1-dimethylurea] and fillers gave a gasoline-resistant, high peel strength joint to steel substrates.[46]

(v) Cyanamide with co-ordination complexes of silicon[47] or titanium[48] with epoxide resins have been suggested as stable adhesive compositions that cure rapidly when heated. For example, they may cure in merely 3 min at 160°C.

(vi) A ternary co-hardener recipe may be used. Dicyandiamide with a urea derivative and imidazole have been suggested.[49] Dicyandiamide with a metal acetylacetonate, e.g. zinc plus butyltin dilaurate, is another example.[50] The cure time of the adhesive is often halved by these types of formulations.

(vii) The principle of separation is also used by blending a rather active amine, such as 2,5-bis(5-aminopentyl)pyrazine[51] (Fig. 9) or a diamine phenate[52] with a thermoplast. The blend is

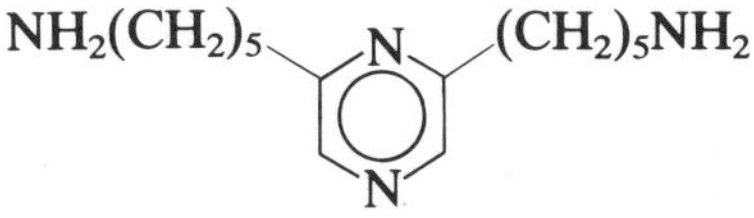

$NH_2(CH_2)_5$ — pyrazine ring — $(CH_2)_5NH_2$

FIG. 9. 2,6-Bis(5-aminopentyl)pyrazine.

powdered and then mixed with a solid powdered epoxide resin to form a heat-fusible and-curable adhesive. Some adducts, e.g. of *N*-methylpiperazine and epoxy resin,[53] can be powdered and blended with liquid resin and dicyandiamide to be stable at room temperature but to cure rapidly at 160°C.

(viii) A melt of a crystallising epoxy resin, e.g. diglycidyl terephthalate, with a latent catalyst may be used to impregnate a nylon carrier, then rapidly cooled, causing crystallisation, to give the latent composition.[54]

(ix) An aromatic iodonium salt and a co-catalyst releasing free radicals on heating can be used for thermal or combined photopolymerisation/thermal curing of epoxide resin for adhesive application[55]. A sulphonium compound, an iodonium compound and a copper salt can be used as co-catalysts for a mineral-filled epoxy composition, and this material has been employed as a substitute for solder in the manufacture of motor bodies.[56]

3.3. Miscellaneous Hardeners

(i) Polythiols are fast-reacting hardeners for epoxy resins and thioglycolates of polyols with basic catalysts have been examined for fast-curing adhesives for steel.[57] A polythiol–polyepoxide–polyene adduct as a hardener is said to have given good moisture resistance,[58] as is a propoxylated ether polythiol.[59] The problem of high odour in these types of hardener has been addressed by naming methylol derivatives of thiol–polyene adducts[60] or reaction products of a polyol with a mixture of mecaptoalkanoic and thiodialkanoic acids.[61] Also, tack can be developed in an epoxy adhesive by using the fast reaction of a polymercaptan with a glycol–maleic anhydride polyester, the latter incorporated in the resin, the former in a polyamine hardener.[62]

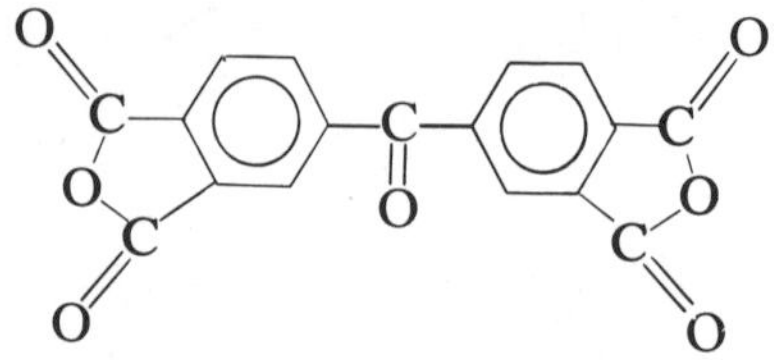

FIG. 10. Benzophenone-3,3′4,4′-tetracarboxylic dianhydride.

(ii) Anhydrides are usually slow hardeners and although they are generally used for casting or laminating, they can also be included in epoxy adhesive compositions. It has been reported that improved adhesives result from introducing metal ionic links using divalent metal salts of monohydroxyethyl phthalate as an additive.[63] The toughness of such systems may be increased by using carboxyl-terminated liquid rubber.[64–66]

(iii) The components of a particular composition, triglycidyl *p*-aminophenol with benzophenone-3,3′,4,4′-tetracarboxylic dianhydride (Fig. 10), were found to be mutually soluble and to cure at room temperature to give reasonable metal-to-metal strength at high temperature, namely 8·3 MPa at 150°C.[67]

(iv) A fast cure has been obtained (i.e. 10 min at 100°C) by using *p*-isopropenylphenyl glycidyl ether with maleic anhydride. This example combines epoxide-ring with unsaturated addition reactions.[68]

(v) A reaction product of a polyhydric phenol with a trivalent acid anhydride, e.g. resorcinol with trimellitic anhydride (Fig. 11), with a bisphenol A resin, gave an adhesive that cured in 2 min at 180°C.[69a] A thermoplast may also be included.[69b]

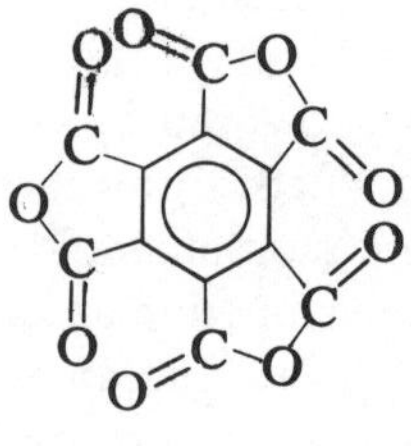

FIG. 11. Trimellitic anhydride.

(vi) Tetracarboxylic anhydrides with epoxide resins are used as high-temperature adhesives. They may be toughened by carboxyl-terminated butadiene acrylonitrile rubbers by pre-reacting the rubber with the resin.[70]

4. CO-REACTING COMPOSITIONS

Epoxy resins can co-react with other thermosetting systems to form useful 'alloys' as adhesives.

(i) For example, DGEBA resin may be pre-reacted with a poly-isocyanate and polyalkylene glycol to form a urethane-modified epoxy resin.[71] Urethane-modified epoxy resins in dicyandiamide-cured compositions gave high peel strengths.[72] Room-temperature curing systems may be made with epoxy resin polyol, polyisocyanate and catalyst.[73] With a micro-encapsulated Lewis acid catalyst and DGEBA resin, with a cationically polymerisable co-monomer added, a stable one-part but fast-reacting adhesive can be made.[74]

(ii) Alloys with an aliphatic bismaleimide prepolymer, an aromatic bismaleimide and an aromatic polyamine with cycloaliphatic[75] or aromatic[76] epoxy resins cure at room temperature. They may be used for critical field repairs of aircraft. Alloys of polyfunctional maleimide and of cyanic acid ester with epoxy resin gave good heat-resistant adhesives.[77]

(iii) An alloy for brake linings of a phenolic resol, acrylonitrile/acrylate copolymer together with a small amount of epoxy resin, is said to have given a better bond than resol polyvinyl–formal adhesives.[78]

(iv) An acrylate ester of a polyol in an epoxy–heterocyclic poly-amine adhesive can be used on wet surfaces for civil engineering application.[79]

(v) An epoxy adhesive film may be formed by incorporating a compound with an unsaturated double bond and crosslinking by irradiation with ultraviolet light, or by electrons with a suitable sensitiser present and an epoxy hardener.[80,81] The epoxy crosslinking is completed by a thermal cure of the adhesive assembly.

(vi) Co-curing can be achieved with an acrylate monomer and a diepoxide in the presence of a thermal initiator such as a BF_3 adduct in combination with a thermal initiator.[82]

(vii) An epoxy resin containing more than two hydroxyl groups can be reacted with a diol–polyisocyanate reaction product plus a latent hardener such as dicyandiamide to give a melt-processable adhesive making strong metal joints after cure.[83]

(viii) 'Phased' reactions of epoxy resins can also be arranged. For example, slow self-curing of an epoxy adhesive containing an aromatic iodonium catalyst and dye activator can be initiated on a substrate by visible light and continues when the bond assembly is completed.[84]

5. MORPHOLOGY

This topic is discussed in detail in Chapter 7, but a few comments are appropriate at this point. There have been extensive investigations to relate structure of cured compositions with properties.

Supermolecular structures have been reported in epoxy resins after cure. Nodular structures of *ca* 5 nm diameter that may aggregate on postcure to *ca* 40 nm have been described.[85] Two types of structure may be present in the cured compositions, high-crosslink-density particles in a low-crosslink-density continuum or low-crosslink-density particles in a high-crosslink-density continuum, and it has been suggested that the latter is more common.[86] The structures may form because of diffusional or steric restrictions or at an interface where the structures can be influenced by preferential absorption, changing their size.[87] Such nodular structures have been reported in DGEBA resins cured with aliphatic amine, tertiary amine or polyaminoamide hardeners[85] and with anhydrides,[88,89] and also for the TGDDM resin cured with alicyclic amine.[90] After cure, TGDDM compositions with 4,4′-diaminodiphenyl sulphone (DDS) may contain crystalline regions of unreacted hardener and these may then be eliminated during postcure to form microvoids.[91] Inclusions of unreacted DDS are also believed to occur in cured compositions based on resorcinol diglycidyl ether.[92] The presence of a two-phase morphology in epoxy resins cured with less than one equivalent of dicyandiamide has been suggested.[93] Finally, as discussed in Chapter 7, light-scattering techniques have demonstrated local order in unreacted DGEBA resin, of the order of 20 nm in a liquid resin and 70 nm in a solid resin.[94] Such ordered regions were attributed to an epoxy–hydroxyl group association.

6. MODIFYING CONSTITUENTS

Modifying constituents in epoxy compositions may be polymers, rubbers or diluents; they enhance adhesive properties by modification of the cured resin network.

(i) Phenoxy polymers, at up to 50 wt% of the epoxy resin, have been claimed as advantageous components in epoxy adhesives with latent hardeners used for metal-clad laminates.[95] Similar proportions of a polyvinyl acetal may also be used.[96] With a resol and an epoxy resin (4:1) an equal amount (by weight) of polyvinyl butyral may be used.[97] Polyamidimides, of reduced viscosity 0·1 to 1·0, may be blended with epoxy composites, 20:1 to 1:6, to give adhesives for electronic use.[98] Polyurethane addition at 5 to 10 wt% is said to improve the adhesion of an epoxy adhesive to PVC.[99] Sulphochlorinated polyolefin with an epoxy adhesive gives an adhesive for thermoplasts to metals.[100]

(ii) A one-part adhesive curing rapidly at 80–100°C, including a powdered epoxy resin, powdered bis(aminoalkyl)pyrazine and powdered ethylene–acrylate co-polymer, can be used for bonding polymers of low softening point.[101]

(iii) Carboxy-grafted polyolefins can be blended with epoxy adhesives to give a strong bond between metals and to polyolefins.[102]

(iv) Considering rubbers, the use of acrylonitrile–butadiene materials with functional groups to toughen epoxy compositions has become well established in the past few years. The liquid forms of low molecular weight, typically 3000 to 5000, with terminal groups such as carboxyl or amine groups have been used principally, as reviewed in Chapter 5. At first the liquid rubbers were simply added to the hardenable compositions. Later, advantages were found in pre-reacting the carboxyl-terminated rubbers with excess epoxy resin, sometimes with simultaneous chain extension with a bisphenol. Reaction of the carboxyl-terminated rubber during the preparation of a polyaminoamide hardener has been suggested[103] and a peel strength 12 times that of the conventional hardener was obtained. The significant increases in toughness are usually produced by small additions of rubber, typically 5–15%; the effect is exerted through a phase separation of a rubbery phase in the resin matrix.[104] In adhesives this phenomenon produces enhanced tensile lap-shear strengths and much improved impact and peel strengths.

Toughening by blending elastomers or thermoplasts to thermoset adhesives has long been recognised, e.g. in the nitrile– or vinyl–phenolic adhesives of the 1940s and 1950s, and the nitrile– or nylon–epoxies of the 1950s and 1960s.[105] These were blends with materials of high molecular weight, 10 000 to 100 000, at comparatively high additions, perhaps 15–30%. The adhesives have been used in solution or in film form. However, the new method of toughening with liquid rubbers, mentioned above, gives a range of liquid paste or film adhesive compositions. Their development is due to proprietary technology based on the chemistry and physics of the process, and aspects of this are illustrated in the open literature and patents. Basically, with the liquid rubbers small rubbery domains of a definite size and shape are formed *in situ* during cure[106] (see Chapters 1 and 5). The domains cease growing at gelation.[107] Therefore, after cure is complete the adhesive consists of an epoxy matrix with a glassy rubber transition temperature minimally affected (e.g. 50°C for a diaminopropylethoxy ether hardener) with the T_g of precipitated domains at about −40 to −50°C.[108–113] The formation of a disperse phase depends on a delicate balance between miscibility of the rubber, or its adduct with resin, in the initial resin/hardener mixture and appropriate fine precipitation during the crosslinking reactions.[114] Higher acrylonitrile content in the rubber gives less miscibility with resin, hence more rapid precipitation.[107] Adducts which increase the molecular weight give better phase separation.[109] The way an adduct is prepared may be important: it may sometimes be preferable to do it without catalyst.[115,116] With amine- and carboxyl-terminated rubbers chemically bound in two-component epoxy adhesives, the peel strength is enhanced, and not lost in thick bond lines. At the higher additions, for example 30 phr rubber, there is an enhancement in lap shear strength over unmodified resins in thick bond lines.[117] Toughening effects with these liquid acrylonitrile–butadiene rubbers have been most successful with bisphenol A or bisphenol F resins. Although some toughening has been shown with additions to tetrafunctional glycidylamine with 4,4′-diaminodiphenyl sulphone hardener, the gain in fracture energy and impact strength is much less marked.[118]

(v) Other methods of toughening rubber can also be demonstrated. Flexible epoxy resin adhesives have been made using a telechelic co-reaction of isoprene and acrylonitrile in the presence of bisphenol A diglycidyl ether.[119] Acrylic elastomers from *n*-butyl acrylate and acrylic acid can be adducted with DEGBA resin and cured with DDM, and show a two-phase composition.[120] A functionality of about

12 eq/mol has been found to be suitable; higher values decreased phase separation and toughness. The acrylic elastomer can be made *in situ* by mixing bisphenol A resin and the monomers with the catalyst. Also, by using a chain transfer agent the viscosity can be kept down.[121]

Acrylonitrile-acrylate copolymers with carboxylic groups are compatible with epoxy resin compositions, e.g. DEGBA resin and DDS, and have resulted in water-resistant adhesive bonds.[122] For bonding glass to glass, a photocurable adhesive from a vinyl-terminated acrylonitrile–butadiene rubber, a cycloaliphatic epoxy resin and a polyether glycol has been devised.[123] Liquid chloroprene co-polymers with reactive groups, molecular weight about 5000, impart flexibility to epoxy adhesive compositions and ratios of 1:20 to 2:1 rubber to resin have been suggested. An epoxy-resin/polyamine composition with calcium phosphate filler, thiol accelerator and up to 8% hydroxyl-terminated polybutadiene as flexibiliser has been suggested for bonding bone prostheses.[124] A composition of similar type with 2–4% silicone rubber included gave improved adhesive strength, namely an improvement of 155% at 22°C and 190% improvement at −196°C.[125]

(vi) Attempts have been made to employ the known advantages of plasticisers (e.g. benzyl butyl phthalate) or reactive diluents (e.g. phenyl glycidyl ether) in amine-cured epoxide adhesives to gain reductions in wetting angle, viscosity and/or shrinkage of the cured composition.[126,127] However, wetting angle was not found to be relevant and the highest shear strengths were obtained with compositions with the greatest shrinkage!

(vii) The diglycidyl ester of methylphosphonic acid at a concentration of 10–25 phr has been shown to double the joint strength of steel substrates when used in an epoxy–polyamine adhesive.[111]

7. CONCLUDING REMARKS

The application of epoxy resins as adhesives started 40 years ago. In the past ten years there have still been extensive investigations into new resins, compositions, applications and structure/property relationships. It has been a phase of vigorous maturity in epoxy resin technology. However, acceptance of adhesive bonding as a method of joining structural materials is still not widespread in engineering. Perhaps in the next five years or so there will be a decisive fusion of the mature technology and the acceptance of adhesive bonding on a par with other joining methods in general applications.

REFERENCES

1. ICI America Inc., US Patent 4 072 656, priority 24 May 1976.
2. Mitsui Petrochem., Japanese Patent 51 144 498, priority 6 June 1975.
3. Sumgait Petrochem., USSR Patent 513 993, priority 17 April 1974.
4. Hunston, D. L., Griffith, J. R. and Bowers, R. C., *Ind. Engng Chem., Prod. Res. Dev.*, **17** (1978) 10.
5. Griffith, J. R. and O'Rear, J. G., Silicone amine cured fluoroepoxy resins, Amer. Chem. Soc. Symp. Ser., No. 132, 1980, Paper No. 4, p. 35.
6. Daikin Kogyo Co. Ltd, European Patent 30 932, priority Japan 30 November 1979.
7. Bayer A. G., European Patent 44 422, priority W. Germany 21 July 1980.
8. CIBA–GEIGY A. G., West German Patent 2 949 385, priority US 11 December 1978.
9. CIBA–GEIGY A. G., West German Patent 2 949 401, priority US 11 December 1978.
10. Kakiuchi, H. and Okamoto, K., *Netsu Kokasei Jushi,* **3**(3) (1982) 113.
11. CIBA–GEIGY A. G., European Patent 13 258, priority US 11 December 1978.
12. CIBA–GEIGY A. G., West German Patent 2 726 824, priority Switzerland 17 June 1976.
13. CIBA–GEIGY A. G., West German Patent 2 726 846, priority Switzerland 17 June 1976.
14. Mitsui Toatsu Chemicals Inc., PCT World Patent 83 01 776, priority Japan 20 November 1981.
15. Lunak, S. and Dusek, K., *J. Polym. Sci.,* Symposium No. 53 (1975) 45.
16. Rinde, J. A., Happe, J. A. and Newey, H. A., *SPE 38th Ann. Tech. Conf.,* 1980, p. 457.
17. Bokare, U. M. and Gandhi, K. S., *J. Polym. Sci., Polym. Chem. Ed.,* **18** (1980) 857.
18. Paul, S. and Ranby, B., *J.O.C.C.A.,* **62** (1979) 153.
19. Apicella, A., Nicolais, L., Iannone, M. and Passerini, T., *J. Appl. Polym. Sci.,* **29** (1984) 2083.
20. Gupta, A., Cizmecioglu, M., Coulter, D., Liang, R. H., Yavrouian, A., Tsay, F. D. and Moacanin, J., *J. Appl. Polym. Sci.,* **28** (1983) 1011.
21. Mijovic, J., Kim, J. and Slaby, J., *J. Appl. Polym. Sci.,* **29** (1984) 1449.
22. Igarashi, T., Kondo, S. and Kurokawa, M., *Polymer,* **20** (1979) 301.
23. Shimbo, M., Ochi, M. and Shigeta, Y., *J. Appl. Polym. Sci.,* **26** (1981) 2265.
24. Noskov, A. M., *Internat. Polym. Sci. Technol.,* **8** (1981) 6.
25. Morgan, R. J., *J. Appl. Polym. Sci.,* **23** (1979) 2711.
26. Oleinik, E. F., *Pure Appl. Chem.,* **53** (1981) 1567.
27. Brewis, D. M., Comyn, J. and Fowler, J. R., *Polymer,* **20** (1979) 1548.

28. TEXACO DEV. CORP., US Patent 4 178 427, priority US 6 February 1978.
29. TEXACO DEV. CORP., US Patent 4 187 367, priority US 27 March 1978.
30. SHIMBO, M., OCHI, M. and KONISHI, Y., *Nippon Setchaku Kyokaishi,* **13** (1977) 162.
31. MUNOZ-ESCALONA, A. and OTEYZA, M., *Rev. Plast. Mod.,* **34** (1977) 835.
32. CIBA–GEIGY A. G., West German Patent 2 607 663, priority Switzerland 28 February 1975.
33. KIM, S. L., SKIBO, M. D., MANSON, J. A., HERTZBERG, R. W. and JANISZEWSKI, J., *Polym. Engng Sci.,* **18** (1978) 1093.
34. GLASGOW, D. G., NASA Contract Reports NASA CR-145022, 1976; NASA CR-145227 1977.
35. AMERICAN CYANAMID CO., US Patent 4 069 204, priority US 7 June 1976.
36. LUDECK, W., East German Patent 123 670, priority 14 January 1976.
37. CINCINNATI MILACRON INC., US Patent 4 142 034, priority 10 April 1978.
38. H. B. FULLER CO., US Patent 4 082 708, priority 4 April 1978.
39. WESTINGHOUSE ELEC. CORP., US Patent 4 113 684, priority 10 December 1976.
40. WESTINGHOUSE ELEC. CORP., US Patent 4 120 913, priority 10 December 1976.
41. SIMON, E., US Patent 4 042 544, priority 8 April 1976.
42. LOCTITE CORP., West German Patent 2 810 428, priority US 10 March 1977.
43. WEISNER, J., *Internat. Polym. Sci. Technol.,* **7** (1980) 92.
44. FEDTKE, M. and BIEROEGEL, K., *Plaste u. Kaut.,* **28** (1981) 253.
45. AMERICAN CYANAMID CO., European Patent 24 526, priority US 1 May 1980.
46. MITSUBISHI CHEM. IND., Japanese Patent 79 26000, priority 27 February 1979.
47. ASAHI CHEMICAL IND., Japanese Patent 51 100 196, priority 28 February 1975.
48. ASAHI CHEMICAL IND., Japanese Patent 51 100 197, priority 28 February 1975.
49. HITACHI, Japanese Patent 53 090 344, priority 21 January 1977.
50. LENINGRAD LENSOVET. TECH., USSR Patent 535 319, priority 6 June 1975.
51. CIBA–GEIGY A. G., West German Patent 2 800 306, priority Switzerland 7 October 1977.
52. CIBA–GEIGY A. G., West German Patent 2 800 302, priority Switzerland 7 January 1977.
53. NATIONAL STARCH and CHEM. CORP., US Patent 4 268 656, priority 16 January 1980.
54. CIBA–GEIGY A. G., European Patent 18 950, priority Switzerland 8 May 1979.
55. CIBA–GEIGY A. G., British Patent 2 070 616, priority Switzerland 29 February 1980.

56. General Electric Co., East German Patent 3 023 696, priority US 29 June 1979.
57. Kamon, T., *Setchaku,* **25**(7) (1981) 317.
58. Denki Kagaku Kogyo Co., Japanese Patent 55 104 318, priority 2 February 1979.
59. Diamond Shamrock Corp., US Patent 4 092 293, priority 24 June 1976.
60. Daw Chemical Co., US Patent 3 968 167, priority 15 September 1975.
61. Phillips Petroleum Co., US Patent 3 857 876, priority 24 October 1972.
62. CIBA–GEIGY Corp., US Patent 4 126 505, priority GB 5 November 1974.
63. Matsuda, H., *J. Appl. Polym. Sci.,* **23** (1979) 2603.
64. Matsuda, H., *J. Appl. Polym. Sci.,* **25** (1980) 1915.
65. Mitsubishi Electric Corp., Japanese Patent 52 084 248, priority 6 January 1976.
66. Matsuda, H. and Dohi, H., *J. Appl. Polym. Sci.,* **26** (1981) 1931.
67. Lord Corp., US Patent 4 002 599, priority 6 November 1975.
68. Mitsui Toatsu Chem. Inc., Japanese Patent 54 122 398, priority 21 September 1979.
69. Toa Gosei Chem. Ind. Ltd, (a) Japanese Patent 54 142 244, priority 6 November 1975; (b) Japanese Patent 54 161 649, priority 21 December 1979.
70. Gulf Research Dev. Co., US Patent 3 948 849, priority 1 November 1974.
71. Hitachi Chemical, Japanese Patent 54 096 598, priority 17 January 1978.
72. Mitsubishi Chem. Ind., Japanese Patent 54 026 000, priority 28 July 1977.
73. Micafil A. G., East German Patent 3 001 637, priority 17 January 1980.
74. Minnesota Mining Co., US Patent 4 224 422, priority 26 June 1979.
75. TRW Inc., US Patent 4 273 916, priority 13 June 1980.
76. TRW Inc., US Patent 4 283 521, priority 14 February 1979.
77. Mitsubishi Gas Chem. Ind., Japanese Patent 56 010 524, priority 9 June 1979.
78. Goldschmidt TH. A. G., West German Patent 2 853 646, priority 13 December 1978.
79. Nippon Oils and Fats, Japanese Patent 55 145 728, priority 4 May 1979.
80. Cemedine Co. Ltd, Japanese Patent 54 160 455, priority 8 June 1978.
81. Westinghouse Electric Corp., East German Patent 3 233 476, priority US 11 September 1981.
82. W. R. Grace and Co., US Patent 4 374 963, priority 2 November 1981.
83. W. R. Grace and Co., French Patent 2 536 753, priority US 29 November 1982.
84. General Electric Co., British Patent 2 065 124, priority 11 December 1979.
85. Aspbury, P. J. and Wake, W. C., *Brit. Polym. J.,* **11** (1979) 17.
86. Morgan, R. J. and O'Neal, J. E., *J. Mater. Sci.,* **12** (1977) 1966.

87. Racich, J. L., Thesis 77-14356, University of Wisconsin, Madison, USA, 1977.
88. Luettgert, K. E. and Bonart, R., *Prog. Coll. Polym. Sci.*, **64** (1978) 32.
89. Matyi, R. J. and Uhlmann, D. R., *J. Polym. Sci., Polym. Phys. Ed.*, **18** (1980) 1053.
90. Lind, A. C., *Amer. Chem. Soc., Polymer Prepr.*, **22**(2) (1981) 333.
91. Morgan, R. J., O'Neal, J. E. and Miller, D. B., *J. Mater. Sci.*, **14** (1979) 109.
92. Kochervinskii, V. V., *et al.*, *Vysokomol. Soedin., Ser. A*, **19**(8) (1977) 1697.
93. Kamon, K. and Saeki, K., *Kobunski Ronbun.*, **34**(7) (1977) 537.
94. Stevens, G. C., Champion, J. V. and Liddell, P., *J. Polym. Sci.*, **20** (1982) 327.
95. Hitachi, Japanese Patent 56 082 814, priority 10 December 1979; Japanese Patent 56 104 925, priority 25 January 1980.
96. Hitachi, Japanese Patent 56 104 979, priority 25 January 1980; Japanese Patent 56 109 220, priority 1 February 1980.
97. Matsushita Elec. Works, Japanese Patent 55 034 251, priority 31 August 1978.
98. Asahi Chem. Ind., Japanese Patent 55 048 242, priority 2 October 1978.
99. Reichold Chemicals Inc., US Patent 4 000 214, priority 29 May 1973.
100. Seitetsu Kagaku, Japanese Patent 51 125 124, priority 1 November 1976.
101. CIBA–GEIGY A. G., West German Patent 2 800 306, priority 7 January 1977.
102. Toyo Ink Mfg. Co. Ltd, Japanese Patent 59 06212, priority 2 July 1982.
103. Cemedine Co. Ltd, Japanese Patent 57 179 268, priority 27 April 1981.
104. Drake, R. and Siebert, A., *SAMPE Quarterly*, (July 1975) 11.
105. Houwink, R. and Salmon, G., *Adhesion and Adhesives*, Vol. 1, Elsevier, Amsterdam, 1965, Ch. 17.
106. Riew, C. K., Rowe, E. H. and Siebert, A. R., Amer. Chem. Soc. Div. Organic Coatings Plastics Chem., Adv. Chem. Ser., **154** (1974) 326.
107. Wang, T. T. and Zupko, H. M., *J. Appl. Polymer Sci.*, **26** (1981) 2391.
108. Brewis, D. M., Comyn, J. and Fowler, J. R., *Polymer*, **18** (1977) 951.
109. Bucknall, C. B. and Yoshii, T., *Brit. Polym. J.*, **10** (1978) 53.
110. Kunz, S. C. and Beaumont, P. W. R., *J. Mater. Sci.*, **16**(1981) 314.
111. AS USSR Kazan Org., USSR Patent 794 054, priority 4 May 1978.
112. Morgenson, O. E., Jacobsen, F. and Pethrick, R. A., *Polymer*, **20** (1979) 1034.
113. Sayre, J. A., Assink, R. A. and Lagasse, R. R., *Polymer*, **22** (1981) 87.
114. Manzione, L. T., Gillham, J. K. and McPherson, C. A., *J. Appl. Polymer. Sci.*, **26** (1981) 889.
115. Paul, N. C., Richards, D. H. and Thompson, D., *Polymer*, **18** (1977) 945.
116. UK Secretary for Defence, British Patent 1 484 797, priority 28 February 1975.

117. MURPHY, W. T. and SIEBERT, A. R., *Natl. SAMPE Symp. Proc.*, **29** (1984) 262.
118. OCHI, M. and BELL, J. P., *J. Appl. Polym. Sci.*, **29** (1984) 1381.
119. PAUL, N. C., PEARCE, P. J. and RICHARDS, D. H., *Adhesion*, **3** (1979) 65.
120. LEE, B. L., LIZAK, C. M. and RIEW, C. K. *et al.*, *SAMPE 12th Tech. Conf.*, (1980) 1116.
121. STAUFFER CHEM. CO., European Patent 11 4733, priority US 19 January 1983.
122. GOLDSCHMIDT, TH A. G., European Patent 16 247, priority 29 June 1979.
123. MINNESOTA MINING CO., US Patent 4 219 377, priority 14 March 1979.
124. COMMISSION ENERGIE ATOMIQUE, European Patent 37 759, priority 24 March 1980.
125. KOL'TSOVA T. YA., KERBER, M. L., AKUTIN, M. S., NEVEROV, A. N. and OB'EDKOV, M. L., *Plast. Massy*, (10) (1981) 40.
126. (a) DEARLOVE, T. J., *J. Appl. Polym. Sci.*, **22** (1978) 2509; (b) DEARLOVE, T. J. and ULICORY, J. C., *J. Appl. Polym. Sci.*, **22** (1978) 2523.
127. KOLESNIKOVA, YA. D., KUZNETSOVA, V. M., STAL'NOVA, I. O. and CHINILINA, N. V., *Plast. Massy*, (3) (1977) 40.

4

Bismaleimide Resins

H. Stenzenberger

Technochemie GmbH, Verfahrenstechnik, Dossenheim, Federal Republic of Germany

1. INTRODUCTION

Thanks to their high thermal resistance and outstanding mechanical and electrical properties, polyimides have attracted a great deal of interest, not only from polymer scientists but also from industry, over the last 30 years. Numerous studies have been conducted on the relationship between their properties and structure, but it has been difficult to arrive at broad generalities, this is because the properties of polyimides are significantly affected not only by their chemical structure, but also by a number of other variables such as molecular weight, nature of end groups, extent and process of imidization, processing variables such as temperature, and solvents used during synthesis and fabrication. The commercially available polyimides may be classified into three distinct groups: condensation, addition and thermoplastic. Condensation polyimides, although amongst the best in elevated-temperature performance, can only be processed via the precursor polymer route. Polyimides are by far the most successful family of high-temperature polymers because the raw materials (aromatic amines and tetracarboxylic acids) are relatively cheap and readily available. They have been converted into films, fibres, composite matrix resins and adhesives. Unfortunately the condensation-type cure and the highly polar high-boiling solvents prevented their success as matrix resins and adhesives because of void formation during cure.

From the processing point of view, real thermoplastic polyimides and thermosetting polyimides are more attractive. In recent research and development in the aerospace industry bismaleimides seem to have been accepted as the next-generation materials for composite matrix resins.

Unlike epoxies, bismaleimides satisfy the requirement for improved temperature resistance and improved hot–wet performance. It is the aim of this article to describe the state of development of bismaleimide resins and their use as matrix resins for composites and as adhesives.

2. SYNTHESIS OF BISMALEIMIDES

Bismaleimides (BMIs), the building blocks of bismaleimide resin formulations, are synthesized from maleic anhydride and a diamino compound.[1,2,3] For temperature-resistant resins aromatic diamines are reacted with maleic anhydride in an inert organic solvent at room temperature, forming the corresponding bismaleamic acid as an intermediate in almost 100% yield; this undergoes cyclodehydration at temperatures of 40–60°C, in the presence of acetic anhydride and fused sodium acetate as a catalyst, to form the bismaleimide (Fig. 1). The yield of recrystallized bismaleimide is usually in the range 65–75%. *N*-Acetyl derivatives (acetanilides) of the amine are formed as by-products.[4] Under the above experimental conditions the major by-product in the synthesis of 4,4′-bismaleimidodiphenylmethane is *N*-maleimido-*N*′-acetylaminodiphenylmethane.

Bismaleimides based on aromatic diamines are crystalline substances of high melting point (Table 1). A wide variety of bismaleimides have been synthesized with the aim of tailoring the properties of the resins and adhesives for which they are used. However, 4,4′-bismaleimidodiphenylmethane is still the most widely used BMI in commercial resin formulations due to the low price of the starting materials, 4,4′-diaminodiphenylmethane and maleic anhydride. Long-chain sulfone ether type diamines[5] have been converted to the corresponding sulfone ether type bismaleimides to provide the basis of high-temperature adhesives.[6] The molecular weight between crosslinks was varied systematically and the effect on the adhesive properties evaluated. As was to be expected, low crosslink densities (high molecular weight aromatic sulfone ether type amines) provided superior adhesive properties. An interesting bismaleimide was synthesized by Varma and

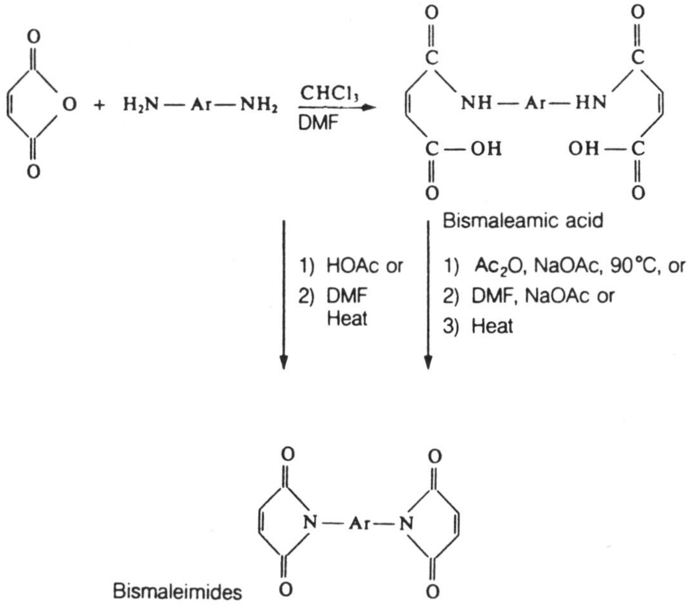

$$\text{Maleic anhydride} + H_2N—Ar—NH_2 \xrightarrow[\text{DMF}]{\text{CHCl}_3} \text{Bismaleamic acid (HOOC–CH=CH–CO–NH—Ar—HN–CO–CH=CH–COOH)}$$

Bismaleamic acid

1) HOAc or
2) DMF
Heat

1) Ac_2O, NaOAc, 90°C, or
2) DMF, NaOAc or
3) Heat

Bismaleimides (N—Ar—N)

FIG. 1. Synthesis of bismaleimides.

Parker which is based on bis(3-aminophenyl)methyl phosphine oxide. The crosslinked resin shows outstanding non-flammability characteristics. For glass fabric laminate based on this bismaleimide a LOI (Limiting Oxygen Index) of 100% was found,[7] which means that these laminates cannot be ignited, even in very high oxygen concentrations.

Long-chain maleimide-terminated poly(amide-imides) have been synthesized for use as laminating resins.[8] However, these materials show no melt transitions and are only soluble in *N*-methylpyrrolidone and therefore suffer problems of processability. Long-chain (high molecular weight) maleimide-terminated polyimides have also appeared in the patent literature.[9] Maleimide-terminated polyamides have been synthesized from maleimidobenzoic acid halides, aromatic amines and iso(tere)phthalic acid halides.[10] Polymerizable bismaleimides have further been prepared by reacting a dicarboxylic acid dichloride with *N*-(4-hydroxyphenyl)maleimide to give *N,N′*-[isophthaloylbis(oxy-*p*-phenylene)]maleimide. In principle all aromatic diamino compounds can be converted into bismaleimides, but only a

TABLE 1
PROPERTIES OF BISMALEIMIDES

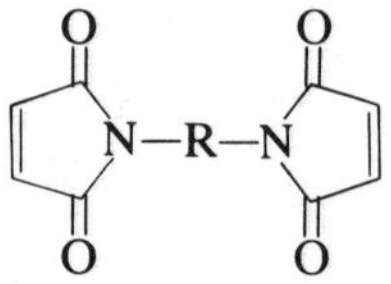

N—R—N

No.	—R—	M. pt (°C)	ΔH[a] (J/g)
1–1		250	?
1–2	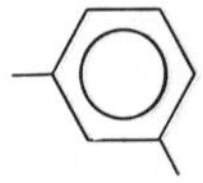	202–203	210
1–3	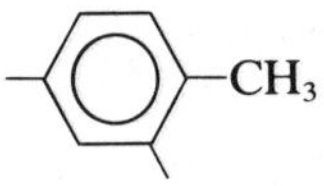 $-CH_3$	174–176	180
1–4	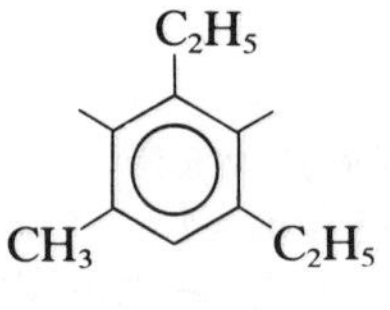 C_2H_5, CH_3, C_2H_5	146–150	160
1–5	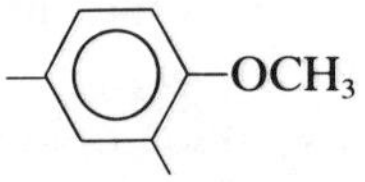 $-OCH_3$	174–175	205
2–1	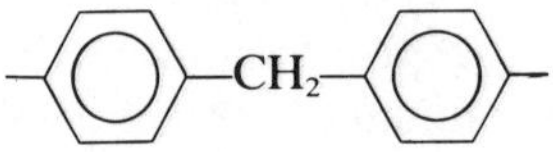 $-CH_2-$	155–157	240
2–2	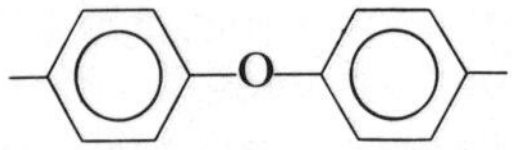 $-O-$	172–178	160
2–3	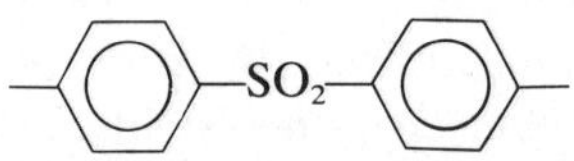 $-SO_2-$	252–255	190

TABLE 1 (*contd.*)

[Structure: bismaleimide, N—R—N]

No.	*—R—*	*M. pt* (°C)	ΔH^a (*J/g*)
2–4	SO_2-bridged diphenyl (3,3′-diphenylsulphone)	210–211	210
2–5	CH_3, C_2H_5-substituted phenyl–CH_2–phenyl (C_2H_5, CH_3)	150–154	192
2–6	C_2H_5, C_2H_5-substituted phenyl–CH_2–phenyl (C_2H_5, C_2H_5)	149–151	280
2–7	phenyl–P(=O)(CH_3)–phenyl	177–180	225

[a] ΔH, heat of polymerization.

limited number are actually used in bismaleimide matrix resin and adhesive formulations.

2.1. Difunctional Monomaleimides (AB Monomers)

In the following paragraphs the copolymerization of bismaleimides with various types of reactive monomers is described. Instead of copolymerizing bismaleimide (AA monomer) with a difunctional reactive diluent (BB monomer) it seemed of interest to synthesize

AB-type monomers of the general formula

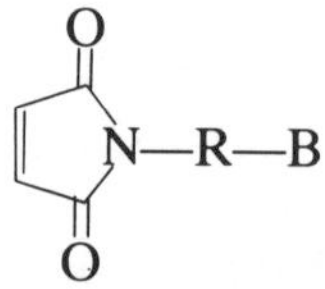

$$\text{maleimide ring (C=O, C=C, C=O)}\text{N—R—B}$$

where B represents a polymerizable group. Such AB monomers have been extensively investigated in our laboratory and some of the monomers synthesized are given in Table 2. Compounds 0–1, 0–2 and 0–3 were prepared by esterification of the corresponding aminophenylcarboxylic acid (aminobenzoic acid) with allyl alcohol and condensation of the resulting aminobenzoic acid allyl ester with maleic anhydride in the usual way. The aminobenzoic acid allyl esters have also been reacted with *m*-maleimidobenzoic acid chloride to form the corresponding maleimide-amideallyl esters (0–4, 0–5 and 0–6 in Table 2). A very attractive AB monomer is 3-ethynyl-1-maleimidobenzene (0–7 in Table 2). Because of its low melting point (130°C) and its extremely high heat of polymerization it is a versatile building block in formulating matrix resins and adhesives.

3. REACTIVITY OF BISMALEIMIDES

The reactivity of the maleimide double bond is a consequence of the electron-withdrawing nature of the two adjacent carbonyl groups creating a very electron-deficient bond (Fig. 2). When heated above their melting points, bismaleimides easily polymerize without the need of a catalyst, thus forming highly crosslinked polymers with good high-temperature properties.[11–13] Nucleophilic (electron-donating) species such as primary and secondary amines, phenols, thioalcohols, etc., undergo a so-called Michael Addition reaction,[14] but nucleophilic attack of the imide ring is also possible.

Maleimide is also a very reactive 'dienophile' and easily forms Diels Alder adducts with various dienes like bisfulvenes,[15] 2-(hydroxymethyl)-1,3-butadiene derivatives[16] and *N,N'*-furfuryl benzophenonetetracarboxylic acid diimide.[17] The most important reactions with respect to the use of bismaleimides as matrix resins for

TABLE 2
PROPERTIES OF BISMALEIMIDES

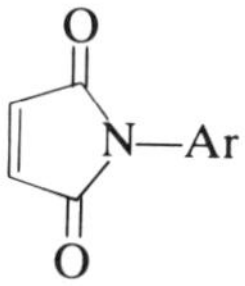

$$\text{maleimide ring (two C=O)}-N-Ar$$

No.	*—R—*[a]	*M. pt (°C)*	*ΔH*[b] *(J/g)*
0–1	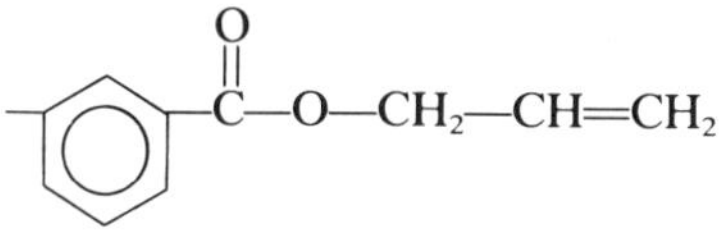 $-C_6H_4-C(=O)-O-CH_2-CH=CH_2$	116–119	240
0–2	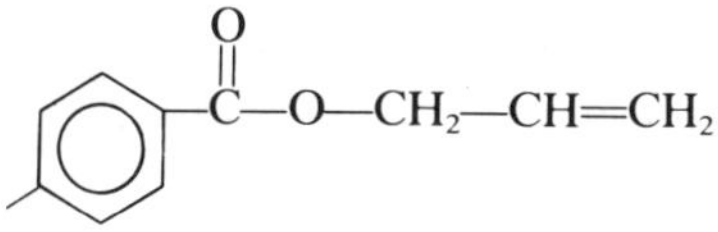 $-C_6H_4-C(=O)-O-CH_2-CH=CH_2$	Honey	228
0–3	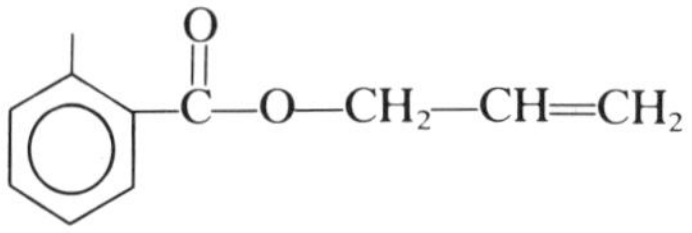 $-C_6H_3(CH_3)-C(=O)-O-CH_2-CH=CH_2$	57–59	—
0–4	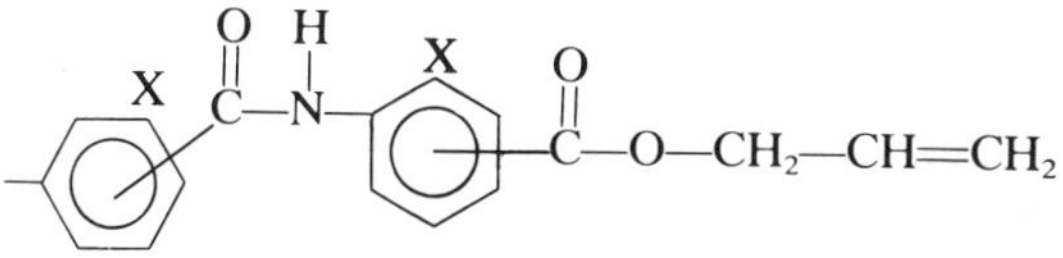 $-C_6H_3X-C(=O)-N(H)-C_6H_3X-C(=O)-O-CH_2-CH=CH_2$	136–139	226
0–5	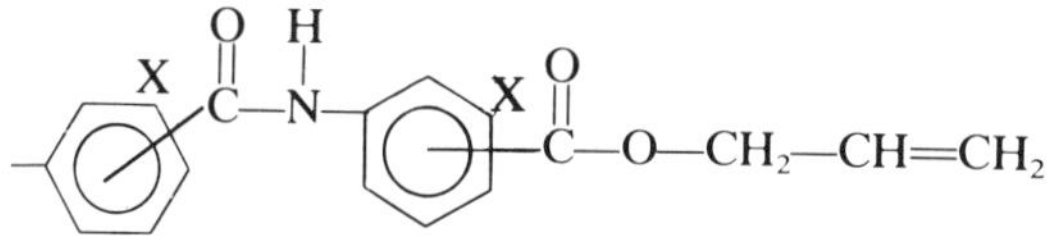 $-C_6H_3X-C(=O)-N(H)-C_6H_3X-C(=O)-O-CH_2-CH=CH_2$	99–102	120
0–6	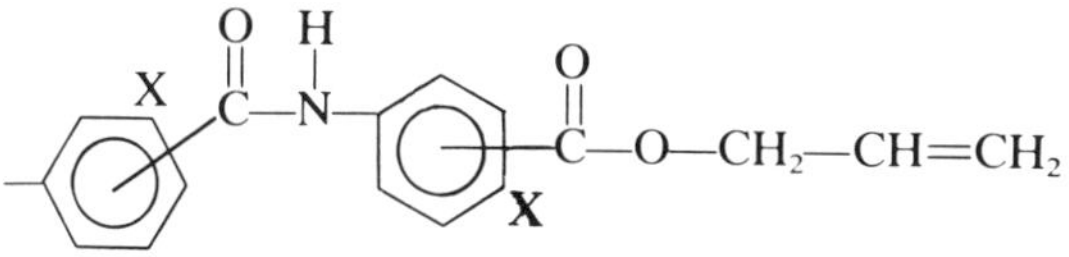 $-C_6H_3X-C(=O)-N(H)-C_6H_3X-C(=O)-O-CH_2-CH=CH_2$	145–146	—
0–7	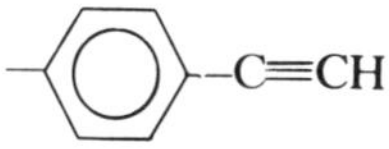 $-C_6H_4-C\equiv CH$	130	600

[a] X shows the position of substitution.
[b] ΔH, heat of polymerization.

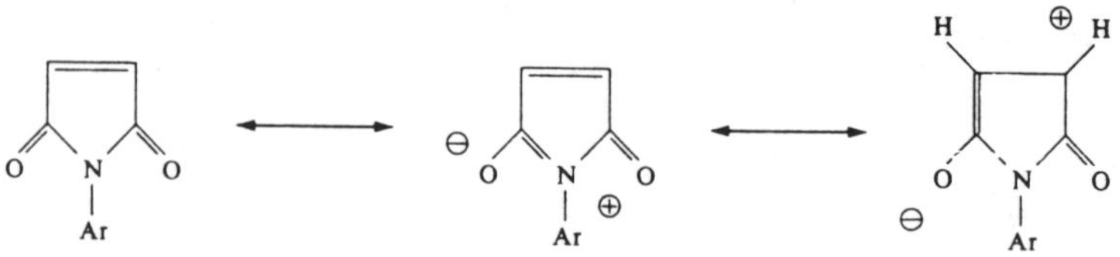

FIG. 2. Reactivity of maleimides.

fibre composites or as adhesives are:

(a) thermal polymerization;
(b) radical-type homo- and copolymerization;
(c) ionic-type homo- and copolymerization;
(d) 'diene'-type chain extension copolymerization;
(e) 'ene'-type linear chain extension copolymerization;
(f) Michael Addition copolymerization.

3.1. Thermal Homopolymerization

Bismaleimides can be involved in a radical-type self-polymerization reaction which can be simply effected by heat. The polymerization velocity at a certain temperature is a function of the chemical structure of the bridging unit between the terminating maleimide groups. The mechanism of the thermally induced curing reaction probably involves a free-radical polymerization because hydroquinone is an effective retardant for the polymerization.

Since the thermally induced homopolymerization is an exothermic reaction, differential scanning calorimetry can be employed to study the cure kinetics,[18,19] which shows that a first-order kinetic law applies for conversions up to 30%. The kinetic data published for a series of aliphatic and aromatic bismaleimides are given in Table 3, indicating that fairly high cure temperatures are necessary to complete cure in a reasonable time. Standard equipment to cure composites or adhesive joints normally operates at maximum temperatures of around 170°C; therefore catalysts are necessary for low-temperature cure of bismaleimides.

3.2. Radical-Type Homo- and Copolymerization

The homopolymerization of cyclic monomers such as maleic anhydride[20,21] and maleimide[22–24] has been investigated extensively.

TABLE 3
KINETIC DATA FOR THE BISMALEIMIDE CURING REACTION

Resin[a]	*Heat of polymerization* $\Delta H(J/g)$	*Order of reaction* n	*Reaction constant,* $K(s^{-1})$		*Activation energy,* E_A (kJ/mol)
			260°C	280°C	
I	260	1	$8{\cdot}9\times10^{-3}$	$2{\cdot}1\times10^{-2}$	103
II	193	1	$4{\cdot}0\times10^{-3}$	$1{\cdot}0\times10^{-2}$	118
III	96·5	2	—	—	124

[a] I, 4,4′-bismaleimidodiphenylmethane; II, Compimide 353; III, Kerimide 601.

N-Substituted maleimides also homo- and copolymerize with various vinyl monomers[24a] initiated by peroxides and azo compounds. Peroxides are suitable catalysts for the polymerization of bismaleimides and bismaleimide containing resin mixtures. Figure 3 shows a comparison of the DSC scans of a commercially available bismaleimide resin (Compimide 353) cured with and without peroxide (2,5-dimethyl-

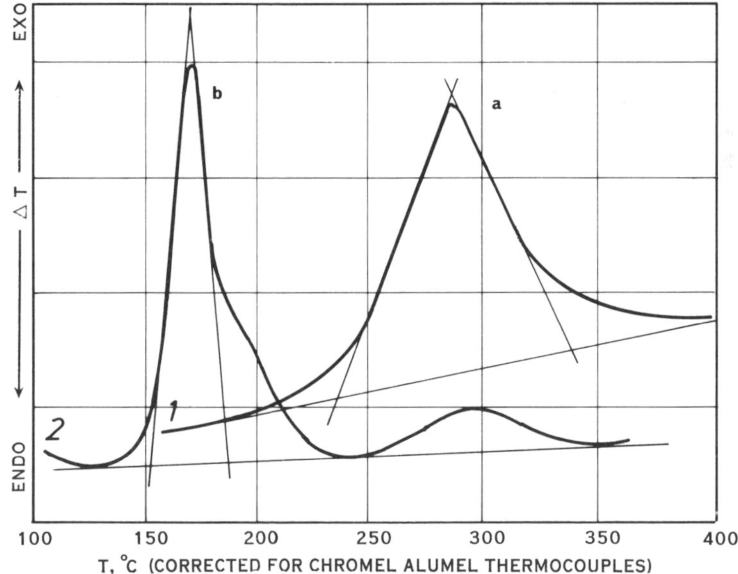

FIG. 3. Homopolymerization of bismaleimide: (a) no catalyst; (b) with catalyst (2,6-dimethylhexane-2,5-di-*t*-butyl peroxide).

hexane-2,5-di-*t*-butyl peroxide) catalyst. It is obvious that at a catalyst level of 1·5% (Fig. 3) the cure exotherm peak temperature is shifted to lower temperatures which allows the resin to be cured similarly to epoxies. Peroxide-type cure acceleration is of particular interest for rapid-curing injection-moulding compounds. Higher molecular weight phenylated free-radical curing agents, e.g. 4,4′-oxybis(triphenylmethylhydroperoxide), have been synthesized which allow the cure of bismaleimides at temperatures of 170°C within 5–10 min when used in a concentration of 3 wt%.[25] Since bismaleimide monomers melt at relatively high temperatures this initiator is attractive because it reaches its maximum rate of decomposition at a temperature near or above the melting point of the monomers. It has to be mentioned at this point that instead of high-melting bismaleimides, low-melting eutectic bismaleimide mixtures can be copolymerized which, because of their low meltability, allow a wider range of peroxide catalysts to be employed as cure accelerators.

Besides thermally and peroxide-initiated radical-type homopolymerization, a radical type of copolymerization of bismaleimide with a wide range of vinyl (styrene, divinylbenzene) or allyl (triallyl cyanurate, triallyl isocyanurate, diallyl phthalate) compounds is possible.[26,27] These liquid co-reactants are of particular interest in formulating tacky prepreg systems for low-pressure autoclave cure.[28] Again a very attractive tool to study the copolymerization behaviour with vinyl- and/or allyl-type reactive diluents or comonomers is differential scanning calorimetry. Figure 4 shows the DSC scan of a bismaleimide/triallylcyanurate mixture. The non-catalyzed BMI/TAC (73/27) mixture undergoes thermally induced radical-type copolymerization, providing a peak exotherm temperature of 278°C. However, addition of 0·5% of TBPIN (*t*-butyl-per-3,5,5-trimethylhexanoate) as a catalyst leads to a bimodal DSC trace. The peak with a maximum at 158°C presumably indicates the peroxide-induced copolymerization of bismaleimides (BMI) and triallyl cyanurate (TAC), whilst the second exotherm with a maximum at 265°C represents the thermally induced copolymerization of the remaining BMI/TAC mixture. The further increase of the catalyst to above 1 wt% leads to a single cure exotherm (copolymerization peak) at a temperature around 152°C. Similar behaviour is obtained with triallyl isocyanurate or *o*-diallyl phthalate when it is copolymerized with bismaleimides in the presence of peroxide catalysts. Obviously a great variety of peroxides, e.g. dicumyl peroxide, benzoyl peroxide, cumene hydroperoxide, di-*t*-butyl perox-

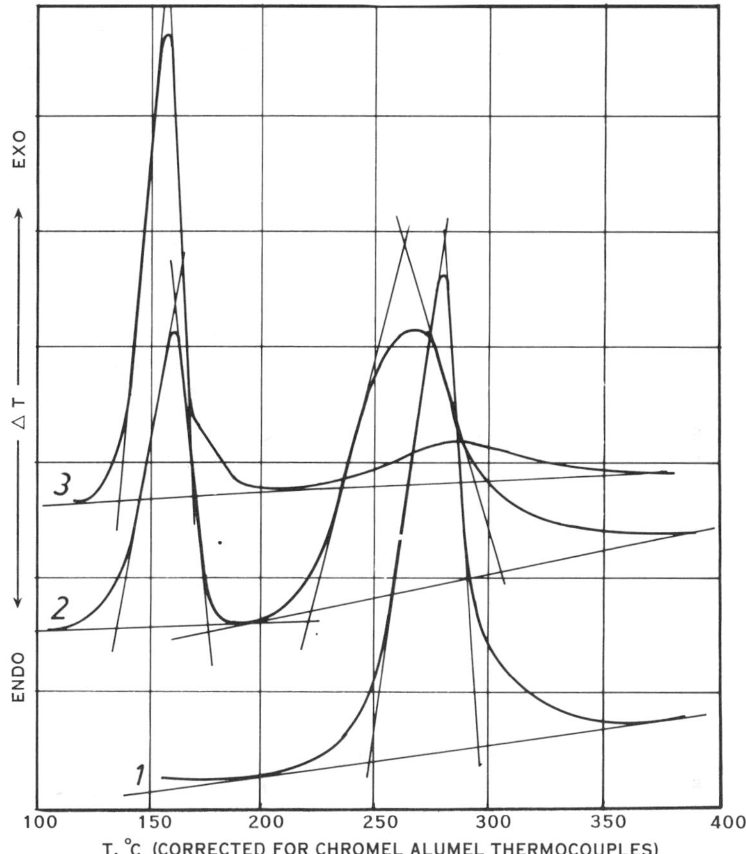

FIG. 4. Peroxide-initiated copolymerization of bismaleimide and triallyl cyanurate mixtures. Key: 1, BMI 73% + TAC 27%; 2, BMI 73% + TAC 26·5% + TBPIN 0·5%; 3, BMI 73% + TAC 26% + TBPIN 1·0% (BMI, bismaleimide; TAC, triallyl cyanurate; TBPIN, *t*-butyl-per benzoate).

ide, and azo-compounds, e.g. azobisisobutyronitrile, are useful cure accelerators for bismaleimides and bismaleimide/vinyl (allyl) comonomer mixtures.

3.3. Ionic-Type Homo- and Copolymerization

Tawney *et al.*[23] and Cubbon[24] mentioned the anionic polymerization of maleimide and *N*-*n*-butylmaleimide and the formation of red low molecular weight polymers. It was further shown by Nakayama and

Smets[30] that *N-n*-butylmaleimide undergoes ring opening together with anionic polymerization in the presence of sodium *t*-butoxide at 20°C and butyl-lithium at −40°C. Bismaleimides and formulated bismaleimide resins and adhesives can be polymerized ionically using tertiary amines or imidazoles as catalysts.

DABCO (diazabicyclo-octane) and 2-methylimidazole are recommended as cure accelerators for commercial resins.[31] The effect of these catalysts on the cure kinetics is obvious from Fig. 5. The non-accelerated bismaleimide resin shows a cure exotherm maximum of 241°C. However, the peak maximum is shifted to lower temperatures in the presence of ionic catalysts. DABCO is more effective than 2-methylimidazole. Obviously, the catalyst concentration present in the resin also influences the cure kinetics.

A knowledge of the copolymerization with reactive comonomers in the presence of ionic catalysts is very important, particularly with liquid vinyl-, allyl- and acrylate-type reactive diluents which frequently

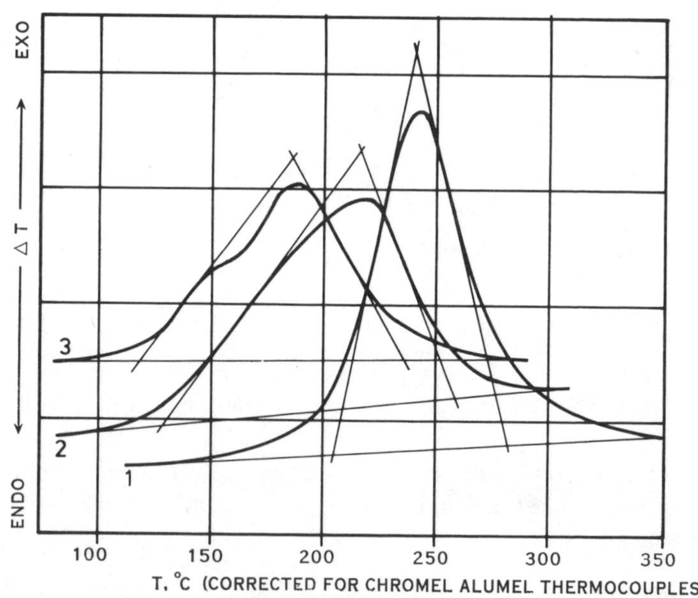

FIG. 5. Differential scanning calorimetry of bismaleimide: 1, uncatalyzed; 2, 0·5% 2-methylimidazole catalyst; 3, 0·5% diazabicyclo-octane catalyst.

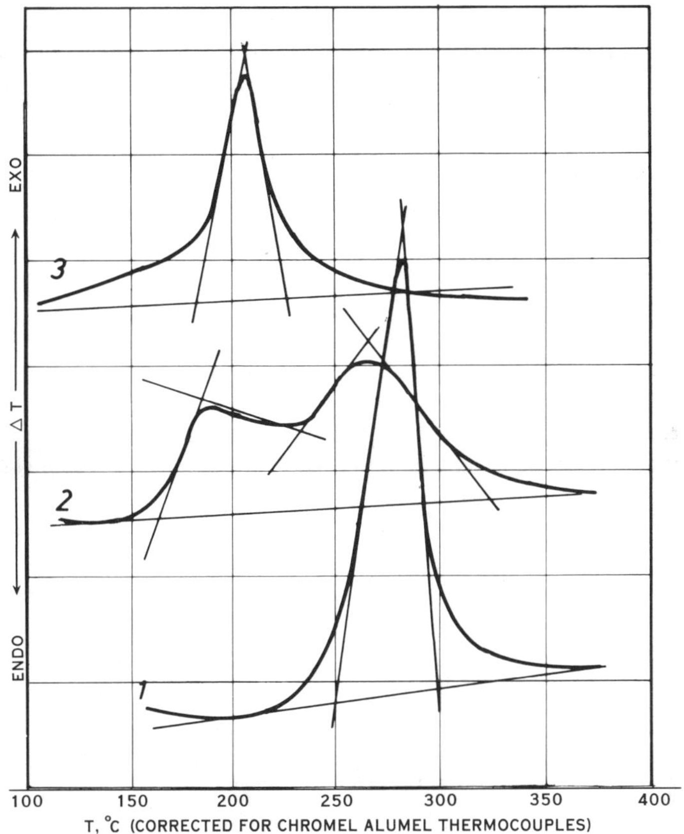

FIG. 6. Ionic-type copolymerization of bismaleimide with reactive diluents (BMI, bismaleimide; TAC, triallyl cyanurate; TMPTA, trimethylolpropane triacrylate; MI, 2-methylimidazole: 1, BMI 73% + TAC 27%; 2, BMI 73% + TAC 26·7% + DABCO 0·3%; 3, BMI 67% + TMPTA 32·7% + MI 0·3%.

are used to formulate and modify BMI resin compositions. In Fig. 6, the DSC trace of the non-catalyzed bismaleimide triallyl cyanurate mixture provides, as we already know, a peak temperature of 278°C. The addition of 0·3% of DABCO as a catalyst leads to a bimodal DSC trace which does not change significantly if the catalyst concentration is increased. The ionic-type homopolymerization of the bismaleimide is responsible for the lower-temperature peak, and the higher peak represents the thermal copolymerization of BMI and triallyl cyanurate.

Different behaviour is obtained with those reactive diluents which themselves can be activated by ionic-type accelerators. A typical example is trimethylolpropane triacrylate (TMPTA) which provides a single copolymerization peak with 2-methylimidazole as anionic accelerator. This copolymerization takes place at a fairly low temperature (205°C) and therefore could be the basis for a low-temperature curing BMI resin formulation.

3.4. Diene-Type Copolymerization

The reaction of a diene with a dienophile to yield a ring-containing compound is known as the Diels–Alder reaction. The use of this reaction for the preparation of polymers has been specially attractive to polymer chemists because of the wide variety of reactants that can undergo the reaction. Bismaleimides are highly active bis(dienophiles) and have been reacted with 2-(hydroxymethyl)-1,3-butadiene,[16] bisfulvenes,[15] etc. and pseudobis(dienes) such as cyclopentadienones,[32] pyrones[33] and thiophene dioxides,[34] which yielded some very high molecular weight Diels–Alder polymers.

An interesting Diels–Alder reaction takes place between styrene and maleic acid anhydride in the presence of polymerization inhibitors.[35–37] This is shown in Fig. 7, via a 1:1 adduct first; then further reaction with maleic acid to give a 2:1 adduct is possible. Also, the reaction of *N*-butylmaleimide with styrene takes place in a similar way. This Diels–Alder reaction sequence has recently been applied to the formulation of a modified bismaleimide resin.[38] A mixture of 1,4-divinylbenzene and 4-ethylstyrene is blended into a hot-melt-type bismaleimide resin which cures to a temperature-resistant three-dimensional network with outstanding hot–wet environmental resistance. The differential scanning calorigrams of Fig. 8 provide an insight into the copolymerization behaviour of bismaleimides and divinylbenzene. A resin mixture consisting of 73% bismaleimide, 22% triallyl cyanurate and 5 wt% of divinylbenzene leads to a bimodal DSC scan: the lower temperature peak at 180°C is only of low intensity. The gradual increase of the divinylbenzene concentration and the simultaneous decrease of the triallyl cyanurate finally provides a DSC scan with a single peak at 165°C, which is responsible for the bismaleimide–divinylbenzene copolymerization. The mixture can be cured at low temperatures because of the low-cure exotherm maximum. The

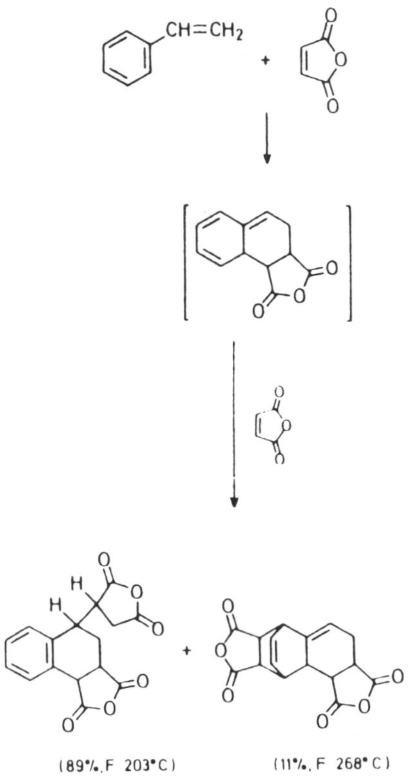

(89%, F 203°C) (11%, F 268°C)

FIG. 7. Diels–Alder addition reaction of styrene with maleic anhydride (F = melting point).

commercial resin (the product V 378 A from US Polymeric) based on the divinylbenzene concept described is discussed in detail in Section 5.

3.5. 'Ene'-Type Linear Chain Extension Copolymerization

The addition reaction between an allyl-type olefinic double bond ('ene') and a double bond (π-bond) is known as an 'ene'-reaction. Recently it has been found that allyl groups which are attached to an aromatic ring can undergo an 'ene'-type linear chain extension reaction with the double bond of a maleimide. This chemistry has been used to formulate a commercially available resin in which *o,o'*-diallylbisphenol A is reacted with a bismaleimide. This commercial resin is described in more detail in Section 4.3.

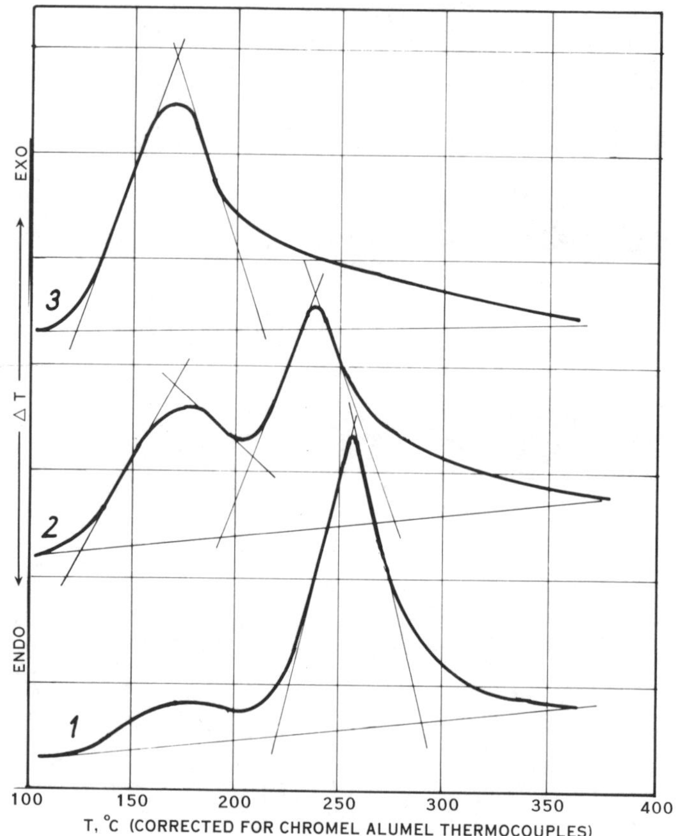

FIG. 8. DSC scan of divinylbenzene-modified bismaleimide: 1, BMI 73% + TAC 22% + DVB 5%; 2, BMI 73% + TAC 17% + DVB 10%; 3, BMI 73% + TAC 7% + DVB 20%.

3.6. Michael Addition Copolymerization

The nucleophilic addition of a C—H acidic compound onto an activated double bond is known as the Michael Addition reaction. Since the double bond of maleimides is highly activated by the two neighbouring carbonyl groups, maleimide is easily attacked by nucleophilic reactants. Sheremeteva *et al.*[39] investigated the reaction of unsaturated cyclic imides with ammonia and amines. It was shown that when an equimolecular quantity of cyclic imide (*N*-methyl-citraconimide) is reacted with methylamine in methanol solution at 0°C, the reaction proceeds in two directions: methylamine is added to the double bond, forming the methylimide of methylaspartic acid (I) and aminolysis of the imide ring takes place and dimethylcitracondiamide (II) is formed (Fig. 9). With the extension of these reactions to bifunctional compounds, it could be shown that in the absence of water, phenylenediamines are added to bismaleimides providing coloured high-melting polyphenylene aspartic acid imides, which show limited solubility in dimethylformamide and formic acid. A patent[40] describes the cure of amino-terminated elastomeric amines with bismaleimides via the Michael Addition reaction. Linear polyaspartimides have been synthesized by Crivello,[41] who reacted stoichiometric amounts of 4,4′-bismaleimidodiphenylmethane and 4,4′-diaminodiphenylmethane in cresylic acids as solvents and obtained high molecular weight polymers. Thiols and thiolate ions are very strong nucleophilic agents and have therefore been used to synthesize polyimidosulfides.[42,43] Another paper published by Crivello and Juliano reports on the synthesis of polyimidothioether–polysulfide block copolymers,[44] and the same author reports the synthesis of polyimidoethers by condensation of hydrogen sulfide and bismaleimides.[45] Alcohols, phenols and carboxylic acids are very weak nucleophiles and therefore there is only little description in the literature of their copolymerization with bismaleimides. However, Renner *et al.*[46] report the preparation of crosslinked polymers of high glass transition temperature and good high-temperature properties which are obtained by reacting bismaleimides with bisphenol A or novolacs in the presence of basic catalysts at elevated temperatures. Tertiary amines seem to have the right pK_b value to maximize the yield of the adduct. It is, however, important to know that the catalyst also activates the

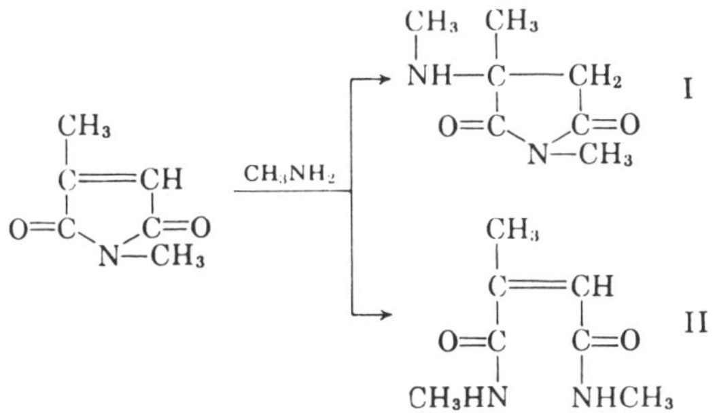

FIG. 9. Reaction sequences of cyclic imide with methylamine.

ionic homopolymerization of the maleimide and at higher temperatures degrades the Michael adduct.

The most important resins are obtained by a non-stoichiometric reaction between bismaleimides and aromatic diprimary amines.[47] Usually the bismaleimide is employed in excess and the reaction performed in melt to give low molecular weight crosslinkable prepolymers. The Michael Addition reaction also takes place between acid hydrazides and bismaleimides; this concept was used to prepare processable bismaleimide resins.[48]

Many other nucleophilic compounds, such as dimethylhydantoin, barbituric acid and compounds containing an activated methylene group, can undergo an addition reaction with the bismaleimides. At this point it should be noted that acetone can, in the presence of a basic catalyst, undergo a Michael Addition reaction. This is of particular interest, because acetone and methyl ethyl ketone are frequently used as prepregging solvents for bismaleimide resins. If amines, hydrazides or other strong bases are used as chain extenders in BMI resins, this can cause a problem because of uncontrolled co-reaction. Furthermore, it should be noted that the Michael adducts are of lower thermal stability than the polybismaleimides. In Fig. 10 the comparative thermal-gravimetric analysis of polybismaleimidodiphenylmethane and a polyaspartimide is provided, showing a much

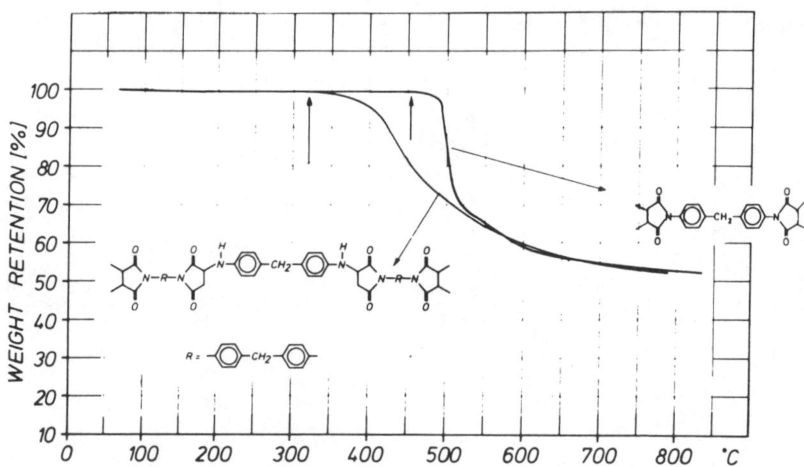

FIG. 10. Comparative TGA thermograms of polybismaleimide and polyaminobismaleimide (Michael adduct). Conditions: $dT/dt = 10°C/min$, nitrogen.

higher polymer decomposition temperature for the unmodified poly(BMI).

3.7. Other Co-monomers

The patent literature outlines various other co-reactants for bismaleimides, e.g. cyanates,[49] isocyanates,[50,51] azomethanes[52] and epoxies.[53,54] In most of these patents no proved reaction sequences or product structures are described.

4. COMMERCIAL BISMALEIMIDES AND BISMALEIMIDE RESINS

Bismaleimides are a relatively new class of thermosetting resins. The requirement for improved temperature resistance combined with easy processing made them candidates for the next generation of matrix resins and adhesives. Looking back over the last 20 years, epoxies have been the class of resin chosen for advanced composites and adhesives. However, their temperature capability is limited to 180°C in the dry, and to 110°C in a wet atmosphere. Totally aromatic linear polyimides for use at 250–300°C have been available for 20 years, but their condensation-type cure and the extremely high cure temperature required never allowed their development as matrix resins for advanced composites. In the previous sections the properties and reactivity of bismaleimides were outlined. Bismaleimide model compounds (Table 1) are crystalline substances of high melt transition and are not at all useful as matrices or adhesives as such. In order to fulfil the processing requirements, they have to be formulated into products which enable their use as highly concentrated solutions or as hot melts. Beside bismaleimide building blocks, some semi-formulated resins and finished products are available from resin manufacturers and fully formulated finished products in the form of prepreg or adhesive film are also available. 4,4′-Bismaleimidodiphenylmethane is the most widely used building block for BMI resins. It is available from several sources at relatively low cost but only experienced formulators can readily handle the material. Formulation, i.e. pre-reaction with co-monomers and/or blending with reactive monomers, additives, elastomers and other processing aids, is necessary to provide the required handling and use properties. From the prepregger's point of view, processability means good solubility in low boiling point solvents like acetone, methyl ethyl ketone, methylene chloride, etc., or meltability at temperatures not in excess of 70–100°C, in order to perform the mixing and pre-reaction operations.

4.1. Compimide 353

Compimide 353 is a nearly eutectic mixture of bismaleimides of extremely low melting point. Typical resin data are given in Table 4. Upon heating, the resin softens between 40 and 70°C and forms a very low viscosity melt at 110°C. Extensive heating at 110°C does not advance (prepolymerize) the resins and therefore provides a very wide processing window at 90–100°C, at which temperature blending and formulation can easily be performed. The vendor (Boots–Technochemie) describes the material as an unformulated basic BMI, with the processing potential for a hot-melt-type composite matrix resin or an adhesive.

4.2. Michael Addition Resins

The route by far most widely accepted to convert bismaleimide building blocks into resins is via Michael Addition with difunctional nucleophilic reactants.

The most popular and frequently used BMI resins are available from Rhone–Poulenc, France. Their polyaminobismaleimides are prepared by reacting a bismaleimide with an aromatic diamine either in the melt

TABLE 4
COMPIMIDE 353 RESIN DATA

Property	*Test method*[a]	*Value*	*Comment*
Physical form	—	—	Resolidified melt.
Appearance	—	—	Yellow–red brown transparent mass.
Gel time	DIN 16945	35–65 min	
Viscosity	DIN 16945	400–1400 mPa s	
Differential scanning calorigram	TC–TM 14	T_B 193 ± 10°C, T_{MAX} 275 ± 15°C	Heating rate 20°C/min.
Polymerization energy	TC–TM 14	ΔH 220 ± 40 J/g	
Composition	TC–TM 25	Comparison with standard	HPLC

[a] DIN = German industry norm; TC–TM = Technochemie test method.

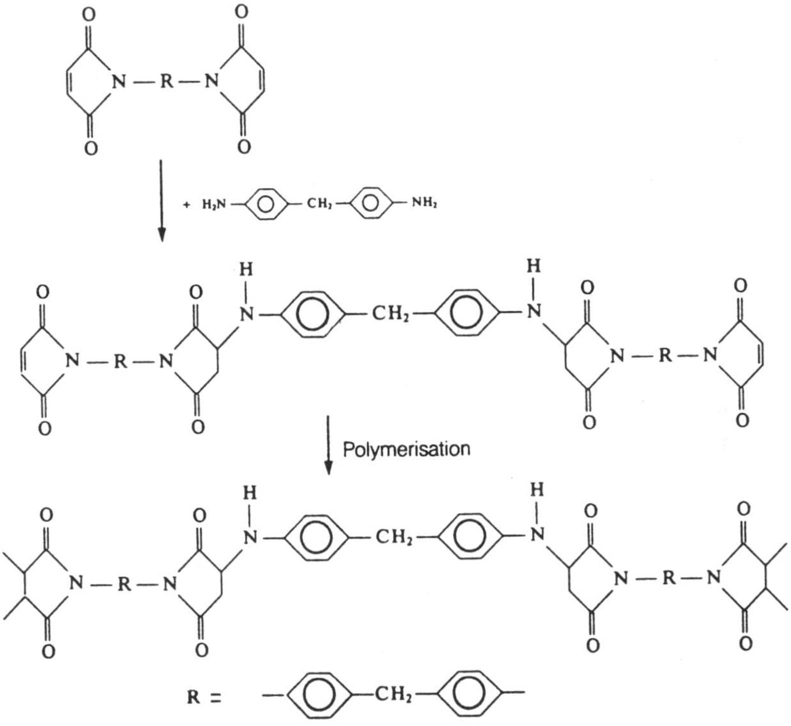

FIG. 11. Polyaminobismaleimide chemistry.

or in solution (Fig. 11). They can be converted into the polyimides by simply heating the resin under pressure up to temperatures of 160–250°C.[47] The addition of nucleophilic difunctional reactants such as diamines (Michael Addition reaction) takes place at temperatures between 40 and 150°C. The resulting macro-bismaleimide is meltable and soluble and therefore can be processed.

However, the reaction product of 4,4′-bismaleimidodiphenylmethane and 4,4′-diaminodiphenylmethane, known as Kerimide 601, is prepolymerized to such an extent that the resulting prepolymer is only soluble in *N*-methylpyrrolidone as a solvent and therefore it is almost impossible to prepare prepregs with a sufficiently low residual solvent content. The problem associated with non-volatile solvents for the manufacture of prepregs and laminates is that they cause void formation during cure and therefore decrease the overall laminate

TABLE 5
THERMAL AGEING OF KERIMIDE 601 GLASS FABRIC LAMINATES

Ageing time at 200°C (h)	*Flexural strength (MPa/ksi)*		*Wt. loss at 200°C (%)*
	25°C	*200°C*	
0	489/71·0	391/56·8	—
1000	450/65·3	381/55·3	0·4
2000	450/65·3	372/53·9	0·5
5000	391/56·8	372/53·8	1·1
8000	313/45·4	254/36·9	1·9
10000	274/39·7	260/29·8	2·7

performance. Kerimide 601 is used mainly in glass fabric laminates for PCBs (printed circuit boards). Laminates based on K 601 show a half-life at 200°C of approximately 12 000 h as shown in Table 5.

Another approach for processable bismaleimide resins is based on the Michael adducts obtained by reacting a bismaleimide, or a mixture of bismaleimides, with aminobenzoic acid hydrazide in the molten state, so providing a resin which shows solubility in low-boiling solvents. The molar ratio of the bismaleimide and the amino acid hydrazide is such that the resulting mixture of products is soluble and the melt viscosity of the resin is low enough to allow its application in hot-melt formulations. The reaction sequence of the synthesis is given in Fig. 12. In the first step Michael Addition between the BMI and the amino group of the aminobenzoic acid hydrazide takes place, but at a later stage the hydrazide group is involved again in a Michael Addition reaction. The resin as supplied contains mainly Michael adduct (III, Fig. 12) besides unreacted bismaleimide. Two different resins based on this chemical concept are available from Boots–Technochemie under the general trademark Compimide 795 and Compimide 183.

Compimide 795 is tailored to meet the processing requirements for hot-melt prepregging. Various derivatives have been developed (Compimide 795E, Compimide 800) which can be moulded via low-pressure autoclave techniques to composites.[55] Compimide 800 is a material which finds extensive use as a carbon fibre matrix resin. Typical properties of unidirectional laminates are given in Table 6. Besides its high-temperature capability, the resin also shows superior hot–wet

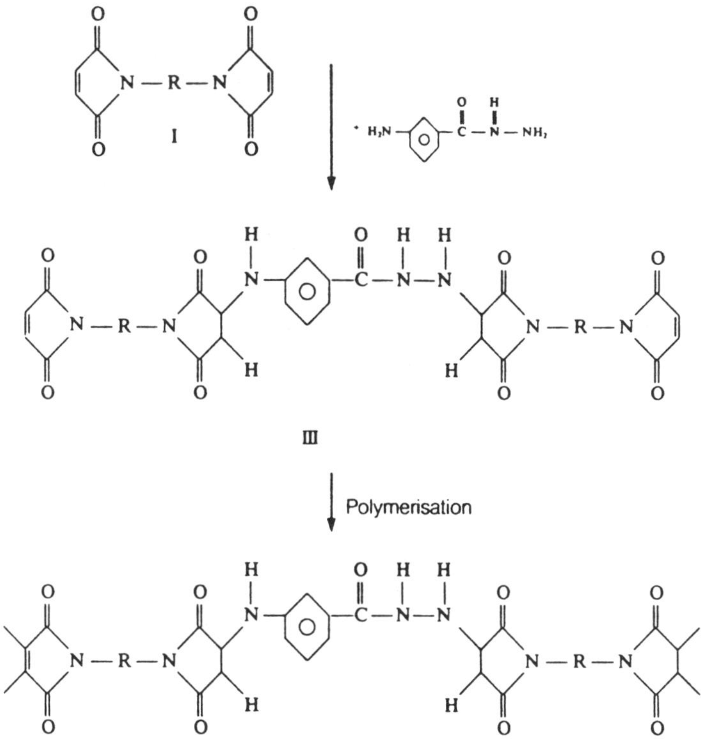

FIG. 12. Amino acid hydrazide–bismaleimide Michael Addition.

environmental stability over epoxies. Mechanical property retention after ageing in wet conditions (94% relative humidity) at 70°C is summarized in Table 7.

One of the advantages of many bismaleimide resins is their high glass transition temperature after hot–wet ageing, which is required for military aerospace applications. Compimide 183 is a bismaleimide resin tailored for the processing via solution techniques to prepregs. The most favourable solvent is dimethylformamide, which can also be used in combination with other diluents such as methylglycol acetate, or ketones such as acetone or methyl ethyl ketone. Of major importance with respect to the proper processing of Compimide 183 is the use of catalysts such as tertiary amines or imidazoles in order to adjust the cure kinetics, so that the laminates can be processed at 170°C by high-pressure techniques. The resin is recommended for use

TABLE 6

PROPERTIES OF COMPIMIDE 800 CARBON FIBRE LAMINATES

Property	*Temp.* (°C)	*Fibre direction* (*deg*)	*T 300/6000*	*Celion 6000, PI finish*[a]	*IM-6*	*Besfight HTA-7*	*Commercial prepreg: epoxy-sized Besfight HTA-7*
Fibre content (vol %)			63	61	63	65·7	64
Flexural strength (MPa)	20	0	1892	1884	1839	2253	2069
	250	0	1432	1250	1150	1374	1360
Flexural modulus (GPa)	20	0	121	118	142	127	121
	250	0	127	120	144	130	121
Flexural strength (MPa)	20	90	85	80	56	73	48
	250	90	—	35	—	33	—
Flexural modulus (GPa)	20	90	8·26	7·92	7·17	8·24	8·05
	250	90	—	5·4	—	5·28	—
ILSS (MPa)	20	0	113	108	92	98	99
	250	0	53	53	46	—	45
	20	0 ± 45	63	58	43	50	50
	250	0 ± 45	37	47	35	38	36
G_{Ic} (J/m^2)	20	0	*ca* 90–100	254	374	—	225

[a] Reproduced with permission from Ref. 77. Finish based on NR 150 polyimide resin.

TABLE 7

HYDROTHERMAL AGEING OF CELION 6000/COMPIMIDE 800 UNIDIRECTIONAL LAMINATES (AGEING CONDITIONS: TEMPERATURE 70°C, RELATIVE HUMIDITY 94%)

Property	*Test temp.* (°C)	*Property value after ageing*			
		0*h*	100*h*	500*h*	1000*h*
Fibre content (vol%)	—	58·7	—	—	—
Moisture absorption (%)	—	—	1·19	1·49	1·66
Flexural strength (N/mm^2)	20	1917	1749	1777	1756
	150	1515	1327	1303	1319
Flexural modulus (kN/mm^2)	20	111	112	111	112
	150	112	114	112	112
Horizontal shear, 5:1 (N/mm^2)	20	90	76	84	82
	150	74	56	54	52

in PCBs, for thermal insulation boards and as face sheets for low FST (Fire Smoke Toxicity) Nomex honeycomb sandwich panels.[56]

Typical properties of a Compimide 183 glass fabric laminate are given in Table 8. Postcured laminates (postcure temperature 200°C) show outstanding properties up to 250°C. A very interesting observation was made in that the moisture absorption of US-2116-style glass fabric laminates is a function of the postcure temperature. Postcure temperatures up to 230°C provided laminates which absorb 0·5–0·8% water, whilst postcure at 250°C for 4 h pushes the moisture absorption up to 1·6%. It is consistent with the free volume theory that cure and postcure at high temperatures create a high free volume in the resin, resulting in a high moisture absorption. This is a dilemma to a certain extent because, on the one hand, high glass transition temperatures in thermosets are only achieved if high postcure temperatures are employed, but simultaneously the free volume of the resin is so dramatically increased that the resulting high moisture absorption reduces the hot–wet performance to the level of low-T_g resins.

There is only a little published information about the mechanical properties of neat resins. For Compimide 795 and Compimide 183 flexural properties at ambient and high temperatures are compiled in Table 9. Interestingly, Compimide 183, which was cured in the presence of diazabicyclo-octane as a catalyst at lower temperatures, shows a much lower elastic modulus and a much higher fracture

TABLE 8
MECHANICAL PROPERTIES OF COMPIMIDE 183 GLASS FABRIC LAMINATES[a]

Property	*Unit*	*Test temperature (°C)*			
		23	150	200	250
Fibre content	% by weight	59	—	—	—
Flexural strength[b]	MPa	474	424	361	288
	ksi	68·7	61·5	52·4	41·8
Flexural modulus[b]	GPa	17·1	17·6	16·2	15·1
	10^6 psi	2·48	2·55	2·35	2·19
Flexural elongation	%	2·84	2·56	2·39	2·15
Flexural strength[c]	MPa	481	—	372	299
	ksi	69·3	—	54·0	43·4
Flexural modulus[c]	GPa	19·6	—	16·2	15·3
	10^6 psi	2·84	—	2·35	2·22

[a] Glass fabric, US-style 2116, size A1100; prepreg resin content, 42 wt%; solvent for prepregging, methyl glycol acetate; catalyst concentration, 0·35% (DABCO).
[b] No postcure: green laminates.
[c] Postcure: 2h at 200°C, $T_g > 250$°C.

energy than Compimide 795, which was cured without a catalyst. Ageing for 500 h at 250°C does not degrade the room-temperature mechanical properties of either resin, but does increase the high-temperature elastic moduli due to further crosslinking.

The thermal-oxidative stability of glass fabric laminates is of interest for high-temperature applications such as thermal insulation boards.

TABLE 9
MECHANICAL PROPERTIES OF COMPIMIDE RESINS C 795 AND C183

Property	*Unit*	*Compimide* 183		*Compimide* 795	
		23°C	*250°C*	*23°C*	*250°C*
Density	g/cm^3	1·31	—	1·32	—
Flexural strength	MPa	110–125	58–74	90–110	60–65
Flexural modulus	GPa	4·10–4·59	2·3–2·9	5·4	3·30
Flexural strength[a]	MPa	106	54	100	40
Flexural modulus[a]	GPa	4·37	3·54	5·5	4·2
Fracture energy, G_{Ic}	J/m^2	150–180	—	40	—

[a] Flexural specimens aged in air for 500h at 250°C.

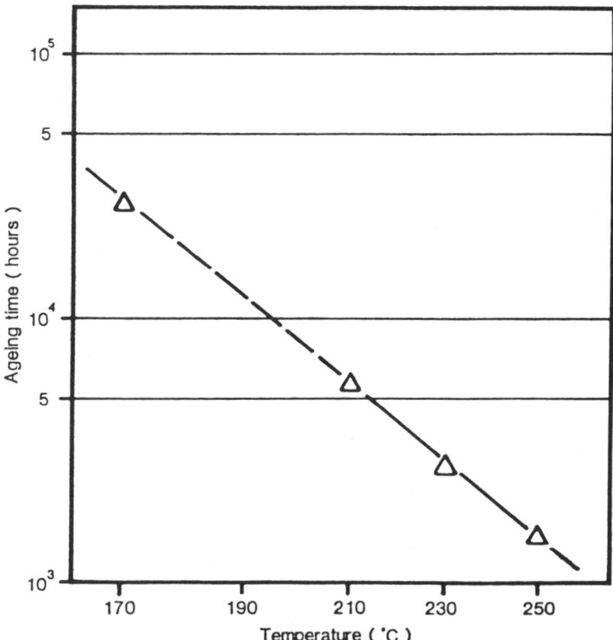

FIG. 13. Arrhenius plot for the thermal-oxidative stability of Compimide 183 glass fibre laminates.

Figure 13 shows the Arrhenius plot for the half-life of Compimide 183 glass fabric laminates, indicating a 250°C capability for 1 350 h and a 25 000 h long-term stability at 175°C.

The aim of looking for new BMI resin concepts is to improve their fracture resistance. This can be achieved by simply increasing the molecular weight between crosslinks, but without decreasing the temperature capability. This problem was confronted by attempting to approach the targets via the chemical concept outlined in Fig. 14. Maleimidobenzoic acid chloride (I) is reacted with 4,4′-diaminodiphenylmethane (II), preferably in methylene chloride as a solvent, in the presence of a hydrogen chloride acceptor in such a way that the molar ratio of the difunctional aromatic amine and the acid halide is between 1:2 and 1:1. The resulting polymaleimide prepolymer consists, depending on the molar proportions of the reactants, of a mixture of the bismaleimide (IV) and the amino-terminated monomaleimide (III). Prepolymerization of the resin takes place via the

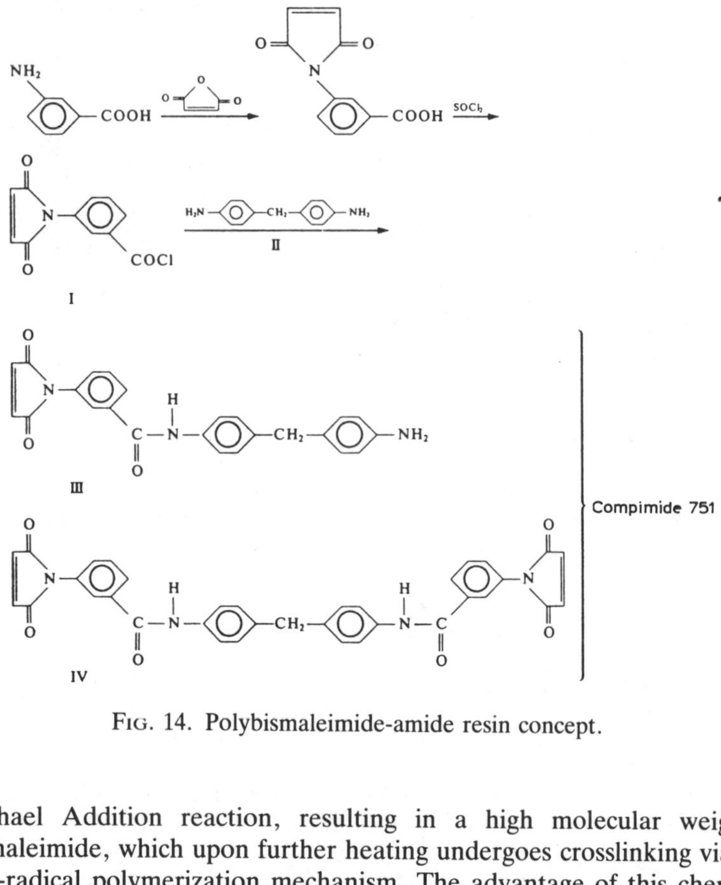

I

II

III

Compimide 751

IV

FIG. 14. Polybismaleimide-amide resin concept.

Michael Addition reaction, resulting in a high molecular weight bismaleimide, which upon further heating undergoes crosslinking via a free-radical polymerization mechanism. The advantage of this chemical concept is that resins with controlled prepolymer molecular weight can be synthesized by simply adjusting the molar composition of m-MIC (I) and diaminodiphenylmethane. The commercially available resin Compimide 751 has an average molecular weight of 1140, providing after cure a resin with a very high extensibility ($\varepsilon = 4{\cdot}8\%$). Two other resins with a lower molecular weight have been synthesized and were cured under the same conditions as Compimide 751. The mechanical properties of these resins are presented in Table 10. As was to be expected, the lower the molecular weight between the crosslinks, the lower the extensibility (elongation) of the cured resin. The chemical structure of the backbone seems to be responsible for the elastic modulus, which is almost the same in all three resins. From

TABLE 10
MECHANICAL PROPERTIES OF COMPIMIDE 751 NEAT RESINS

Resin no.	*Formula mol. wt.*	*Flexural strength* (N/mm^2)	*Flexural modulus* (kN/mm^2)	ε (%)
C751-1	1140	220	4.38	4.81
C751-2	790	180	4.65	3.71
C751-3	670	121	4.28	2.68

the patent literature it can be seen that many BMI resins which are chain-extended with different nucleophilic difunctional organic materials are possible. *p*-Aminophenol,[57] diphenols such as hydroquinone and bisphenol A,[58] dicarboxylic acids,[59] 5,5′-dimethylhydantoin,[60] barbituric acid derivatives,[61] dimercaptans,[62] secondary amines[63] and many other compounds are described as effective chain-extenders for bismaleimides. These chain-extenders work very well in the presence of tertiary amines as catalysts. However, 4,4′-diaminodiphenylmethane is preferred as a co-chain-extender because it is basic enough to catalyze the Michael Addition of the aforementioned co-monomers and is also co-reacted into the polymer backbone. To the author's knowledge, no commercial product is based on one of these concepts.

4.3. The XU 292 Resin Concept

Recently Ciba–Geigy launched a new two-component bismaleimide resin coded XU 292. The XU 292 A-part is 4,4′-bismaleimidodiphenylmethane and the XU 292 B-part is *o,o*′-diallylbisphenol A. The chemistry, i.e. the cure mechanism of the system, is outlined in Fig. 15. The inventors[64,65] claim that initially the allyl function and the maleimide undergo an 'ene'-type linear chain extension reaction and the resulting 'substituted styrene' undergoes a Diels–Alder reaction with another maleimide residue. A very high fracture energy is reported for the neat resin (Table 11), which is an indication of the proposed linear 'ene'-type chain extension during the initial phase of the cure reaction. A typical DSC scan for the cure of the resin is given in Fig. 16, indicating bimodal characteristics: the first exotherm correlates with the 'ene' addition reaction, whilst the second exotherm is likely to be a combination of the Diels–Alder addition reaction and homopolymerization of the bismaleimide. The cured resin as compared with tetraglycidylmethylenedianiline/diaminodiphenyl sulfone

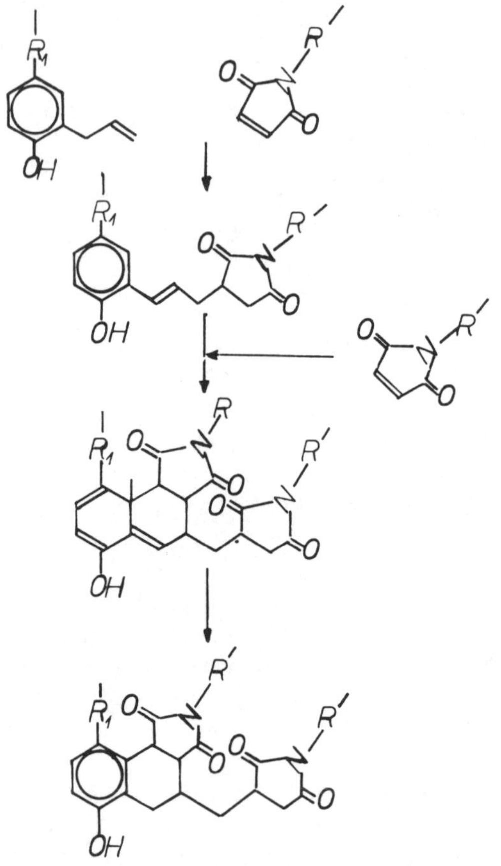

FIG. 15. The XU 292 resin concept (bismaleimide + *o,o'*-diallylbisphenol A).

high-temperature epoxy shows improved strength, stiffness, fracture toughness and glass transition temperature.

4.4. The Bismaleimide/Cyanate Resin Concept

Based on the old triazine resin technology from Bayer,[66] Mitsubishi Gas Oil Chemicals developed a new family of resins which are mixtures of difunctional cyanuric acid derivatives or cyanuric acid prepolymers and bismaleimides (BT resins; B stands for bismaleimide, T triazine). The proposed cured structure is given in Fig. 17. Since the cyanurate prepolymer homopolymerizes in the presence of tertiary

TABLE 11
COMPARISON OF XU 292 WITH MY 720/HT 976 EPOXY RESIN SYSTEM

Property[a]	*Neat resin XU 292*		*MY 720/HT 976 (100/44 pbw)*
	System I[b]	*System II*[b]	
RT Tensile:			
Strength (ksi)	11·9	13·6	8·5
Modulus (ksi)	620	564	542
Elongation (%)	2·3	3·0	1·8
RT Flexural:			
Strength (ksi)	24·2	26·8	13·3
Modulus (ksi)	589	580	499
RT Compression:			
Yield strength (ksi)	29·9	30·4	29·2
Modulus (ksi)	348	360	284
300°F (149°C) Tensile:			
Strength (ksi)	7·4	10·1	6·5
Modulus (ksi)	354	412	378
Elongation (%)	2·6	3·05	1·9
HDT (°F/°C)	523/273	545/285	460/238
Glass transition temp. (TMA) T_g (°F/°C)	523/273	540/282	480/249
Fracture energy, G_{Ic} (compact tension)			
(lb/in^2)	0·97	1·2	0·3
(J/m^2)	170	210	53

Reproduced with permission from Ref. 65.
[a] RT, measured at room temperature.
[b] XU 292 is a two component system consisting of a bismaleimide (XU292A) and *o′,o′*-diallylbisphenol A (XU292B). The ratios of XU292A to XU292B can vary. Systems I and II are different with respect to the ratio XU292A/XU292B.

amines and zinc octoate as a catalyst, it is possible to add almost any amount of bismaleimide, which either copolymerizes with the cyanurate, or homopolymerizes when present in excess. Obviously the higher the concentration of bismaleimides in the mixture, the higher is the glass transition temperature of the cured resin. BT resin can also be used in combination with epoxy resins to achieve epoxy-like handling

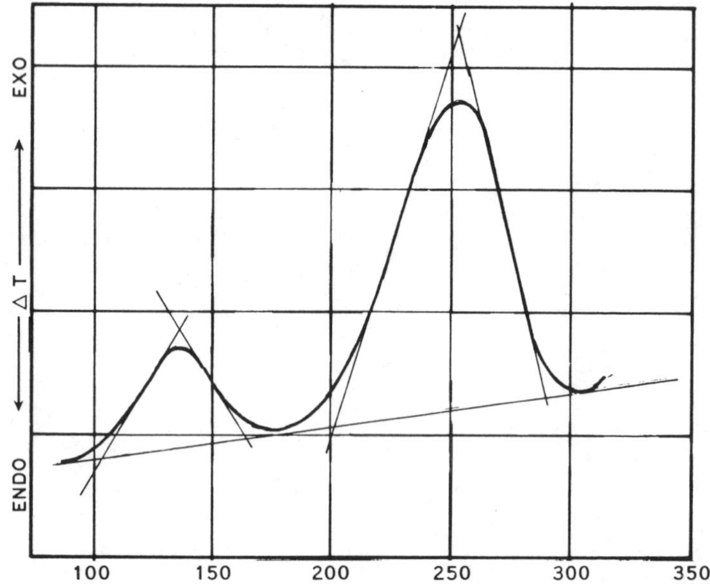

FIG. 16. DSC scan of XU 292 bismaleimide resin.

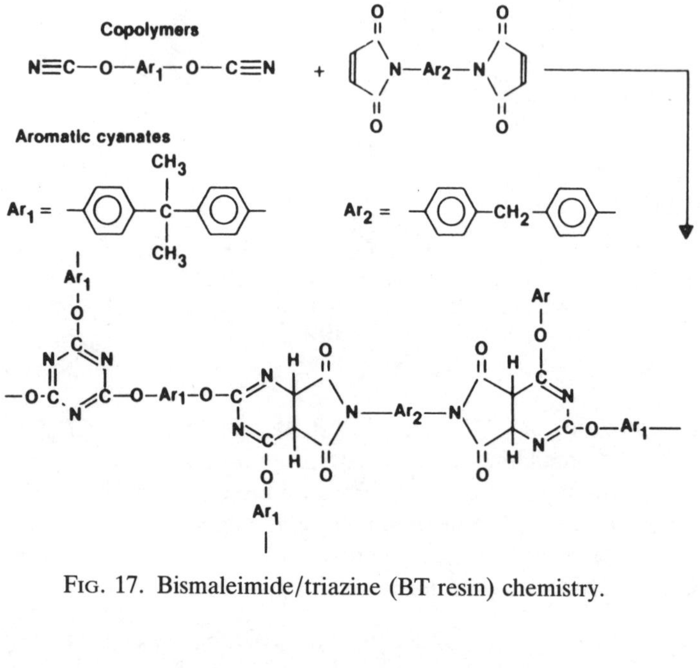

Copolymers

N≡C—O—Ar$_1$—O—C≡N

Aromatic cyanates

Ar$_1$ =

Ar$_2$ =

FIG. 17. Bismaleimide/triazine (BT resin) chemistry.

performance and improved peel strength against copper film. BT resin is mainly used for PCBs and multilayer printed wiring boards.

5. COMMERCIAL PREPREG SYSTEMS

In Section 4, commercially available bismaleimides and bismaleimide systems were described. Prepreggers who convert resins to prepreg in most cases further formulate these resins with the aim of tailoring the processing properties for high- or low-pressure autoclave moulding and adjust the performance properties according to the desired end-use. In most cases, the exact chemical composition is kept proprietary. Almost every prepregger has a modified bismaleimide material in his product line and some of the established materials are described below.

By far the most widely used modified bismaleimide prepreg resin is US Polymeric's V378A. The material is the result of an extensive research and development programme[67] between General Dynamics and US Polymeric. Chemically, the resin system is based on a low-melting eutectic mixture of bismaleimides which is modified with 1,4-divinylbenzene as a reactive diluent and small amounts of phenoxy resin and polysulfone.[68,69] V378A can be cured under standard low autoclave pressure of 7 bar and temperatures of 350°F (177°C) similar to current 350°F curing epoxy systems. However, unrestrained post-cure at temperatures of 475–500°F (246–260°C) are required to maximize elevated-temperature strength retention. At the time of its development, V378A was the only material that met the criterion of resistance to matrix cracking caused by moisture and thermal transients. The only complaints from industry have been regarding the odour problem associated with divinylbenzene. V378A–bismaleimide resin shows unique flow properties in that the viscosity is much higher than that of most of the normally used high-temperature epoxies like 5208 (Narmco) or 3501 (Hercules). Figure 18 illustrates the different behaviour and indicates a short flow period for V378A and a much higher minimum viscosity when compared with Narmco's 5208 epoxy resin. The narrow processing window is related to the reaction kinetics of the Diels–Alder copolymerization between the bismaleimide and the divinylbenzene (see Section 3.4). Nevertheless, if the resin is used in 'net resin' prepreg systems, thick laminates can be moulded consistently without voids and porosity.

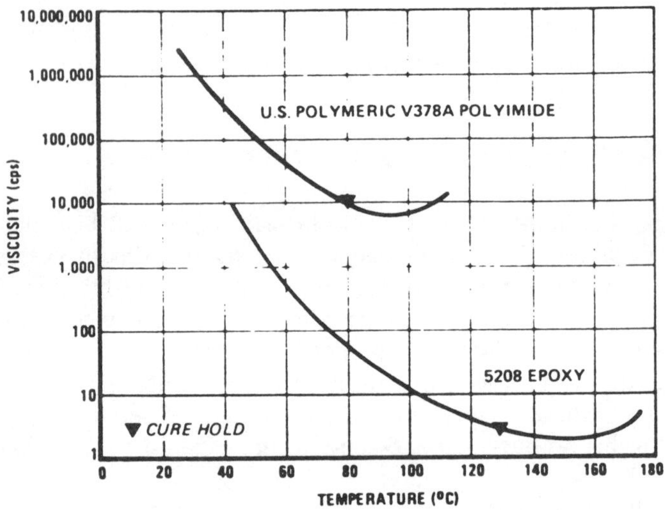

FIG. 18. Viscosity comparison for V378A and 5208 resins.

Prepreg based on V378A/T300 have been used for the manufacture of the F-16 XL fighter aircraft cranked-arrow wing skins. This successful application of a bismaleimide resin was responsible for the rapidly growing interest in bismaleimide technology for military applications. Glass and carbon fabric prepregs based on V378A are also in use for the AV-8B vertical take-off combat aircraft.

Typical mechanical and physical properties of cured V378A/T300 6K tape composites are summarized in Table 12. In short, V378A offers easy epoxy-like processing characteristics with no volatile condensation products given off during cure, together with significantly superior high-temperature strength retention after severe humidity ageing. However, V378A has to be considered as a first-generation bismaleimide resin with respect to its toughness.

Cycom® 3100, recently launched by American Cyanamid, is described as a tough modified-bismaleimide system that exhibits structural performance in all respects equal to or better than high-performance 350°F cure epoxies, but with temperature performance extended to 350°F under hot–wet conditions. Nothing is published about the chemistry of this prepreg resin, but neat-resin mechanical properties published by American Cyanamid indicate superior strain

TABLE 12
TYPICAL MECHANICAL AND PHYSICAL PROPERTIES OF CURED V 378A/T 300 6K TAPE COMPOSITES

Test	*Temp (°F)*	*'As is'*	*4 wk, 160°F, 98% RH*	*8 wk, 160°F, 98% RH*
0° Flexure, ultimate, (ksi)	RT	265.0	263.0 (1.73)[b]	254.0 (1.78)[b]
	350	197.5	157.0	135.0
	450	179.0	—	—
	550	122.0	—	—
	600	107.0	—	—
0° Flexural modulus (10^6 psi)	RT	19.8	20.7	18.4
	350[a]	20.7	18.3	15.3
	450	19.2	—	—
	550	18.4	—	—
	600	17.6	—	—
0° Horizontal shear (ksi)	RT	18.3	16.1 (1.85)[b]	13.7 (1.91)[b]
	350	10.9	8.0	7.4
	450	9.2	—	—
	550	6.4	—	—
	600	5.8	—	—
0° Tensile, ultimate, (ksi)	RT	9.0	4.3 (1.74)[b]	4.6 (1.87)[b]
	350	5.9	1.40	1.58
0° Tensile modulus (10^6 psi)	RT	1.40	1.60	1.40
	350	1.00	0.70	0.85
0° Tensile strain, (10^{-6}in/in)	RT	6 570	2 400	3 100
	350	5 900	2 200	1 800
±45° Tensile, ultimate, (ksi)	RT	23.0	22.6 (1.60)[b]	22.3 (1.59)[b]
	350	15.7	15.8	15.5
±45° Tensile modulus (10^6 psi)	RT	2.43	3.00	2.50
	350	1.63	1.35	1.34
±45° Tensile strain (10^{-6}in/in)	RT	22 500	22 600	22 300
	350	29 280	30 000	29 000
° Tensile, ultimate, (ksi)	RT	228.9	—	—
° Tensile modulus (10^6 psi)	RT	21.8	—	—
° Tensile strain (10^{-6} in/in)	RT	10 470	—	—

	0° *Tensile*	0° *Flexure*	±45° *Tensile*	90° *Tensile*
Specific gravity (g/cm^3)	1·60	1·60	1·57	1·60
Void content (%)	0·5	−0·9	0·9	−0·9
Fibre volume (%)	68·3	64·8	63·6	64·8
Resin solids (wt %)	25·2	28·6	29·0	28·6

Reproduced with permission from Ref. 64.
[a] Soak for 5s at test temperature.
[b] Weight gain % of moisture listed in parentheses.

TABLE 13
CYCOM® 3100 NEAR RESIN FLEXURAL PROPERTIES

Condition and property	Result
RT, dry	
Modulus, 10^6 psi (GPa)	0·683 (4·71)
Strength, ksi (MPa)	25·7 (177·2)
Strain, %	4·1
350°F (177°C), dry after 2 weeks at 350°F (177°C)	
Modulus, 10^6 psi (GPa)	0·531 (3·66)
Strength, ksi (MPa)	13·1 (90·3)
Strain, %	2·5
350°F (177°C), wet after 2 weeks immersed at 156°F (71°C)	
Modulus, 10^6 psi (GPa)	0·329 (2·27)
Strength, ksi (MPa)	11·6 (80.0)
Strain, %	4·0
450°F (232°C) after 30 h at 450°F (232°C)	
Modulus, 10^6 psi (GPa)	0·464 (3·20)
Strength, ksi (MPa)	16·7 (115·1)
Strain, %	4·1

Reproduced by courtesy of American Cyanamide, USA.

capability, high-temperature property retention and hot–wet resistance over the state-of-the-art epoxy resin systems (Table 13). The edge delamination resistance (280 N/mm^2 on T 300 fibres, 180 N/mm^2 on IM 6 intermediate modulus fibres) is superior to that of the first-generation bismaleimide resins.

Another prepreg system (R6451) is available from Ciba–Geigy's Composite Materials Department. The graphite prepreg exhibits excellent handling ability with no odour, good tack and drape at normal ambient conditions.[70] R6451 resin matrix is a modified bismaleimide system formulated with commercially available resin materials. Its viscosity characteristics are shown in Fig. 19, indicating very low viscosity above 210°F (99°C), assuring good fibre wet-out during both prepreg fabrication and laminate cure. The gel time of the resin is 15 min at 177°F (81°C), indicating a very wide processing window. A typical cure cycle requires a dwell at 175°F (79°C) for 2h plus a 4h cure at 350°F under a pressure of 4–6 bar, followed by a 4 h oven postcure at 475°F (246°C). It is claimed that the material is resistant to micro-cracking after thermal ageing for 25h at 350°F plus

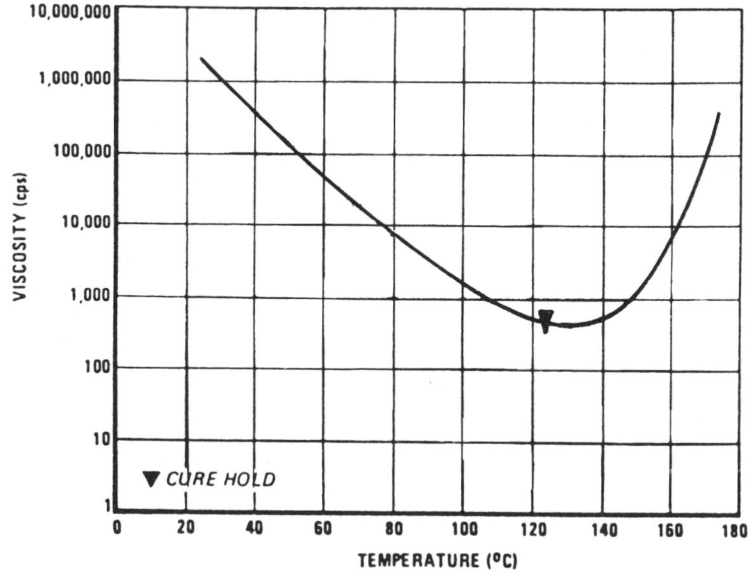

FIG. 19. Viscosity characteristics of Ciba–Geigy's R6451 resin.

TABLE 14
HEAT AGEING OF CIBA–GEIGY R6451 GRAPHITE LAMINATES

Ageing temperature (°F)	*Ageing time (h)*	*0° Flexural modulus (ksi/msi)*		*S.B.S.*[a] *(ksi)*		*Wt loss (%)*
		RT	*350°F*	*RT*	*350°F*	
Initial value		264/19·1	190/20·0	16·9	11·2	0
350	1000	274/18·9	219/18·4	12·6	10·3	1·4
	2000	256/18·2	194/17·8	14·5	9·9	2·2
400	1000	271/18·6	235/18·7	12·1	9·6	1·5
	2000	254/19·1	201/18·3	13·3	9·9	2·4
450	1000	262/20·2	219/17·7	13·1	7·4	3·2
	2000	217/19·2	179/18·4	12·9	9·0	6·3
500	500	280/18·0	256/17·0	14·1	10·0	3·6

Reproduced with permission from Ref. 70.
[a] S.B.S. = short beam shear strength.

624h at 275°F (135°C) (fibre orientation 0, +45/−45, 0). The long-term heat ageing study on unidirectional laminates shows outstanding high-temperature resistance with a 500h performance at 500°F (Table 14).

Other prepreggers like Avco, Narmco and Hysol–Dexter also recently launched bismaleimide-based prepreg systems. A recent patent[53] claims that epoxy/bismaleimide blends and the prepregs designated Avcomp 151 A and Avcomp 130 from Avco are based on the concept described there. The Hysol Division (the Dexter Corporation and Hysol Grafil) recently published a paper[71] in which they claim to have developed a tough resin with a neat resin fracture energy of

TABLE 15

RESIN AND CURED RESIN PROPERTIES OF FIBERITE X86

Resin properties	
Gel time at 200°C (min)	10–14
Viscosity (P)	
Minimum at 160°C, temp. rise 2°C/min	15–20
Isothermal at 80°C	500
DSC, exotherm profile	
Onset temp. (°C)	202
Maximum peak temp. (°C)	259
Cured resin properties	
Cure cycle (h/°C); postcure (h/°C)	4/180; 6/240
T_g[a] (°C)	280–290
Specific gravity at 23 ± 2°C (g/cm^3)	1·22
Gardener impact (ft lb/in)	3·4 (N = 14)
Flexural strength[b] (ksi)	
RT, dry	18·9 ± 6·8
350°F (177°C), dry	12·3 ± 2·5
350°F (177°C), wet[c]	6·7 ± 1·2
450°F (232°C), dry	10·5 ± 2·0
Flexural Modulus[b] (10^6 psi)	
RT, dry	0·664 ± 0·020
350°F (177°C), dry	0·441 ± 0·020
350°F (177°C), wet[c]	0·254 ± 0·020
450°F (232°C), dry	0·380 ± 0·020
Water boil equilibrium wt gain[d] (%)	4·44

Reproduced by courtesy of Fiberite.

[a] TMA, expansion probe, Perkin–Elmer.

[b] Lot #49173, Method ASTM-D-790.

[c] Wet-7 Day Immersion at 165°F.

[d] Modified ASTM-D-570.

TABLE 16

TYPICAL CARBON FABRIC 12-PLY LAMINATE PROPERTIES FOR FIBERITE X86: WARP DIRECTION[a]

Tensile strength, ksi (MPa)		
RT, dry	91·4 ± 1·6	(630 ± 11)
375°F (190°C), dry	82·6 ± 7·0	(569 ± 48)
Tensile modulus, msi (GPa)		
RT, dry	9·34 ± 0·10	(64·4 ± 0·7)
375°F (190°C), dry	9·39 ± 0·02	(64·7 ± 0·1)
Tensile strain		
RT, dry	10 100 ± 400	
375°F (190°C), dry	8 900 ± 1 040	
Compression strength, ksi (MPa)		
RT, dry	101·0 ± 6·5	(696·5 ± 44·8)
375°F (190°C), dry	70·0 ± 10	(482 ± 69)
375°F (190°C), wet[b]	53.4 ± 2·1	(368 ± 14·5)
Flexural strength, ksi, (MPa)		
RT, dry	114 ± 6	(786·2 ± 41·3)
375°F (190°C), wet[b]	89·8 ± 3	(619·3 ± 20·7)
Flexural modulus, msi (GPa)		
RT, dry	8·7 ± 0·2	(60·0 ± 1·3)
375°F (190°C), wet[b]	8·9 ± 0·1	(61·3 ± 0·7)
Interlaminar shear strength, ksi (MPa)		
RT, dry	8·7 ± 0·4	(60·0 ± 2·7)
375°F (190°C), wet[b]	5·5 ± 0·1	(37·9 ± 0·7)
420°F (215°C), wet[b]	4·7 ± 0·1	(32·4 ± 0·7)
450°F (232°C), dry	6·5 ± 0·4	(44·8 ± 2·7)

	Tensile and flexural	*Compression and shear*
Resin solids, %	36·2	36·0
Fibre volume, %	56·1	56·3
Specific gravity	1·53	1·53
Cured ply thickness, 10^3	8·0	8·0
Prepreg lot No.	48898-1	48898-2

Reproduced by courtesy of Fiberite.

[a] Fabric style W-322, plain weave; U.C. T-300 untwisted yarn, $10{\cdot}0 \times 10^{-3}$in (0·25 mm), wt 192 g/m^3 (5·71 oz/yd^2).

[b] Wet: 95% RH at 60°C to 1% water wt gain.

220 J/m^2. The published edge-delamination strength (207 MPa) for intermediate-modulus fibres indicates translation of the toughness into carbon fibre composites.

A new high-temperature laminating resin, X 86, has recently been launched by Fiberite. The resin and the cured resin properties of this

TABLE 17
TYPICAL PROPERTIES OF COMPOSITE CI 1020[a]

Property	*Units*	*Room temp.*	*250°C*
Density	g/cm^3		
Flexural strength	N/mm^2	1850	1100
Young's modulus (tension)	kN/mm^2	130	130
Young's modulus (flexure)	kN/mm^2	120	120
Tensile strength	N/mm^2	1750	1500
Interlaminar shear strength τ (0°)	N/mm^2	90–100	40
Glass transition temp. T_g, dry	°C		287
Fibre content	% by vol.	60	60

Reproduced by courtesy of Sigri.
[a] Determined in accordance with DIN 29.971.

modified bismaleimide are given in Table 15. The resin shows a very high glass transition temperature and good hot–wet stability at 350°F. Typical laminate properties for carbon fabric are given in Table 16. The advantages of this resin system are good mechanical performance at temperatures up to 450°F (232°C), good drape and tack, low toxicity and good toughness on intermediate-modulus fibres, namely a G_{Ic} (interlaminar) value of 440 J/m^2.

All the commercial prepreg systems described so far are designed for the US market. However, in Europe, prepreggers are becoming more interested in high-T_g resins, and bismaleimides are possible candidates. Cyanamid Forthergill, UK, have launched a product called Cycom X 780. The resin is based on Compimide 795 (Boots–Technochemie), a bismaleimide resin which is chain-extended with m-aminobenzoic acid hydrazide.

Compimide 800 is the main constituent of CI 1020, the bismaleimide resin/carbon fibre prepreg, which is available from Sigri/Meitingen, FRG. Typical laminate properties are given in Table 17. The material is used in several programmes by German aerospace companies and has found acceptance because of its outstanding handling characteristics and good environmental (hot–wet) stability.

6. ADHESIVES BASED ON BISMALEIMIDE RESINS

High-temperature adhesives, which can be used above 200°C, are based on linear condensation polymers such as polyimides and

TABLE 18

TENSILE SINGLE LAP-SHEAR PROPERTIES OF AMERICAN CYANAMID FM 32 (BISMALEIMIDE) ADHESIVE[a]

Test temperature (°F)/(°C)	*Control with no conditioning (psi)*	*After 72 h water–boil*		*After 100 h at 400°F*		*After 500 h at 400°F*	
		(psi)	*(% of control)*	*(psi)*	*(% of control)*	*(psi)*	*(% of control)*
RT	2690	2950	110	2320	86	2075	77
350/177	3050	2900	95	3120	102	2675	88
400/204	2930	2750	94	3270	112	3025	103
450/232	3070	2425	79	3200	104	2725	89
500/260	2860	1150	54	1700	59	1550	54

[a] Adhesive film—0·096 psf on 112 1-617 glass cloth, 0·4% volatile; substrate metal—2024 T3 aluminium alloy, surface pretreated in a chromic–sulphuric acid (Fph) etch; cure—press rate to 350°F, 4h at 350°F, 40 psi; 4h postcure at 400°F, no pressure.

TABLE 19

TENSILE SINGLE LAP-SHEAR PROPERTIES OF HYSOL DEXTER EA 9655 AND EA 9673

Temperature (°F)	*Substrate: aluminium alloy*		*Substrate: titanium alloy (6A 14 VTi)*		*Substrate: BMI composite (V 378A Graphite)*	
	EA 9655	*EA 9673*	*EA 9655*	*EA 9673*	*EA 9655*	*EA 9673*
75	2700	2000	—	—	2100	1800
75 (after 30 days at 120°F/100% RH)	3700	—	—	—	—	—
350	—	—	—	—	1800–3300	—
450	2900	1900	—	—	—	2200
450 (aged 700h at 500°F)	1700	1800	—	—	—	—
450 (aged 3000h at 450°F)	1200	—	—	—	—	—
450 (after 30 days at 120°F/100% RH)	1800	—	—	—	—	—
500	1800	2200	1400–2600	—	1900–2400	1800
550	900	1900	—	—	—	1800
600	—	500	—	—	—	—

Cure: 1h at 350°F/25 psi + 2h at 450–475°F; all values in psi.

polyquinoxalines.[72] PMR-type polyimides (Larc 13) have also been tried for high-temperature adhesive applications.[73] Bismaleimide adhesives are commercially available only from two sources. American Cyanamid has recently introduced its first adhesive system based on bismaleimide chemistry. FM 32 adhesive offers 450°F dry and 350°F wet performance with epoxy-like handling and processing. Typical lap shear data are provided in Table 18.

The Hysol Division introduced two bismaleimide resins for adhesive applications. EA 9655, which is based on the conventional bismaleimide technology,[74] and a product of an unique technology coded EA 9673: both formulations are available as film adhesives/pastes and syntactic paste adhesive and have also been formulated into primers. Typical single lap-shear data for both adhesives for aluminium substrate are given in Table 19. As can be seen, EA 9673 offers substantial strength at 550°F, when cured for 1h at 350°F plus 2h at 450–475°F. Lap-shear data on other substrates are also given in Table 19 and are self-explanatory.

6.1. New Toughened Bismaleimide Adhesives

Kinloch and Shaw[75,76] recently reported that bismaleimides can be toughened with carboxyl-terminated acrylonitrile–butadiene rubbers.[75] Interestingly, the cured system forms a microphase-separated structure similar to epoxies. Toughening of epoxies is normally performed in such a way that the rubber is employed in a pre-reacted form (epoxy adduct) and is blended with the epoxy resin at ambient temperature to form a compatible system at room temperature, which during cure undergoes a microphase separation, forming a rubbery particulate structure (see Chapter 5). The rubber particle yield and size are finally responsible for the improvement in fracture toughness. In contrast to epoxy resins, CTBN rubber (carboxyl-terminated acrylonitrile–butadiene rubber, Hycar 1300 × 8, from BF–Goodrich) is not compatible with bismaleimide (Compimide 353) at room temperature and, even at elevated temperature (110°C), the rubber is not soluble in the BMI resin. However, during cure the rubber chemically reacts with the BMI resin (via the double bonds present in the rubber backbone) and thus compatibilization takes place, to such an extent that finally a microvoid structure similar to the CTBN-rubber-modified epoxies is achieved. The comparison between the differential scanning calorigram of the virgin Compimide 353 and the CTBN-rubber-modified Compimide 353 indicates a lower cure-onset temperature for the

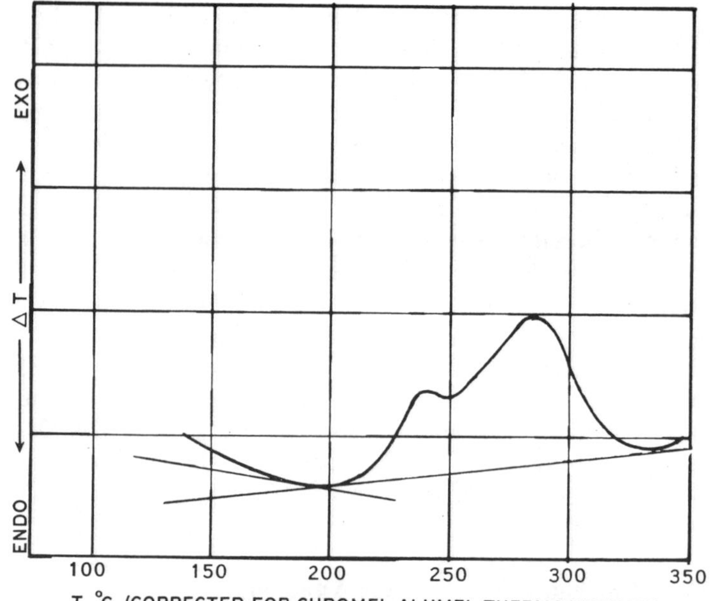

FIG. 20. DSC scan of CTBN-rubber-modified bismaleimide (100 parts Compimide 353, 100 parts CTBN—Hycar 1300 × 8).

rubber-modified version, but the total heat of polymerization of the mixture is very similar to that of the basic resin (Fig. 20). If there were no reactive double bonds in the CTBN rubber backbone available for the copolymerization, then the polymerization energy would be lower.

The influence of the rubber concentration on the fracture energy G_{Ic} is given in Fig. 21. Obviously, a significant improvement is achieved when rubber concentrations in excess of 50 phr are used. The unmodified basic resin has only a very low fracture toughness, but 50–100 parts of rubber provide significant improvements. However, the elastic moduli of the modified resins are dramatically reduced. The high-temperature moduli (Table 20) also indicate a significant reduction of the glass transition temperature. Nevertheless, the improvement in fracture toughness, together with the reduction of the elastic modulus, makes these systems interesting for adhesive applications.

Lap-shear data on steel substrate as a function of rubber concentration is given in Fig. 22.[76] Above a 50 phr level no further increase in lap shear is obtained. Finally, it has to be mentioned that a reduction

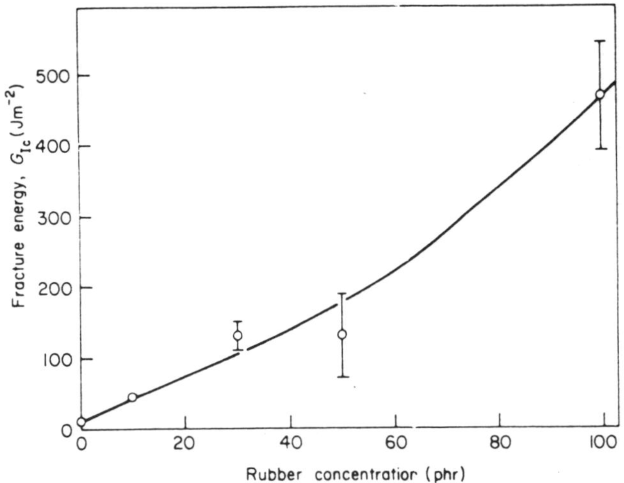

FIG. 21. Effect of rubber concentration on fracture energy, G_{Ic}. (Reproduced with permission from Ref. 76.)

in the thermal oxidative stability, compared with that of non-modified bismaleimide, is to be expected. The comparison between the kinematic TGA curves (10°C/min) indicates a much lower degradation temperature for the rubber-modified bismaleimide (Fig. 23). The effect of long-term ageing and the influence on the adhesive properties (lap-shear strength) has not yet been reported.

TABLE 20

FRACTURE TOUGHNESS OF MODIFIED BISMALEIMIDES

Composition	*Properties*				
	Flex. str. (MPa)		*Flex. mod. (GPa)*		G_{Ic} (J/m^2)
	23°C	250°C	23°C	250°C	23°C
C353	50–70	40–60	5·5	3·4	30
C353 33%–CTBN	87	30	3·26	1·42	536
C353 50%–CTBN	99	26	2·87	0·52	776
C353 43%–CTBN	83	7	2·55	0·21	1190

Key: C = Compimide; CTBN = carboxy terminated acrylonitrile butadiene rubber.

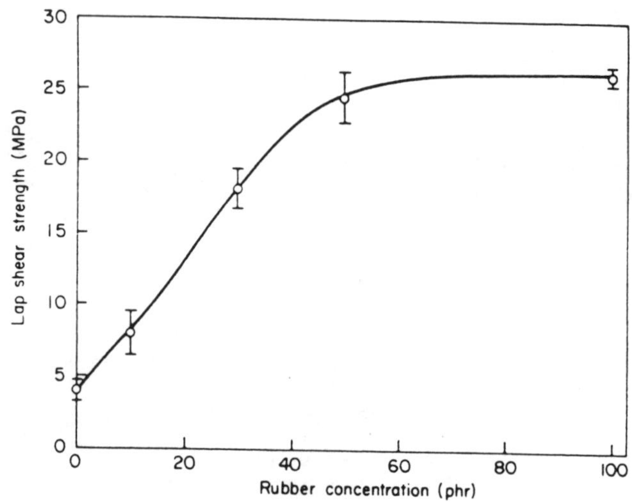

FIG. 22. Effect of rubber concentration on tensile lap-shear strength (steel substrates). (Reproduced with permission from Ref. 76.)

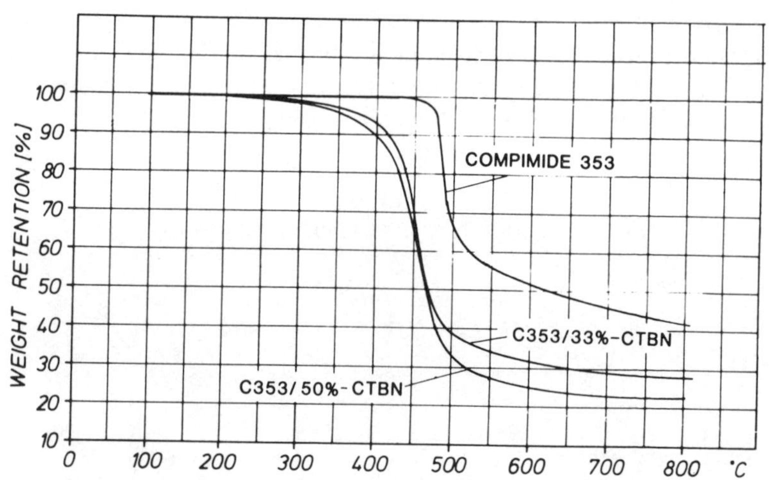

FIG. 23. TGA thermogram of CTBN-rubber-modified bismaleimide ($dT/dt = 10°C/min$).

7. FUTURE BISMALEIMIDE DEVELOPMENT REQUIREMENTS

The past five years have seen a major step towards the application of bismaleimides as matrix resins for composites and as adhesives. Improvements are necessary with respect to their temperature capability, their toughness and also their stability in hot–wet environments. The strain to failure has to be improved to make BMIs compatible with new high-strain high-modulus fibres. Also of major concern is the fibre–resin interface. Fibre surface treatments (physical and chemical) have to be tailored for the new resin chemistry. Research and development effort has to be concentrated in the following areas.

(a) Synthesis of new bismaleimide building blocks.
(b) New bismaleimide co-monomer chemistry for linear chain extension polymerization.
(c) Synthesis of new co-monomers for bismaleimides.
(d) Study of the fibre surface chemistry for improved fibre–resin adhesion.
(e) Bismaleimide-reactive elastomer copolymerization for improved adhesive systems.

Materials research groups are aware of these requirements and are trying to achieve these ambitious targets.

REFERENCES

1. Cole, N. D. and Gruber, W. F., US Patent 3127414, 1964.
2. Searle, N. E., US Patent, 2444536, 1948.
3. Kovacic, P. and Hein, R. W., *J. Amer. Chem. Soc.*, **81** (1959) 1187.
4. Sauers, C. K., *J. Org. Chem.*, **34**(8) (1969) 2275.
5. Kwiatkowski, G. T. and Brode, G. L., US Patent 3839287, 1974.
6. Kwiatkowski, G. T., Robenson, L. M., Brode, G. L. and Bedwin, A. W., *J. Polym. Sci., Polym. Chem. Ed.*, **13** (1975) 961.
7. Varma, I. K., Fohlen, G. and Parker, J. A., US Patent 4276344 1981.
8. Kauh, D. and Jablonski, R. J., *J. Polym. Sci., Polym. Chem. Ed.*, **17** (1979) 1945.
9. D'Alelio, G. F., US Patent 3929713, 1975.
10. Hiedo, K. and Tateki, M., Japanese Patent 74 35379, 1974.
11. Grundschober, F., US Patent 3533996, 1967.
12. Stenzenberger, H., *Appl. Polym. Symp.*, **31** (1977) 91.
13. Hummel, D. O., Heinen, K. U., Stenzenberger, H. D. and Siesler, H., *J. Appl. Polym. Sci.*, **18** (1974) 2015.

14. Sheremeteva, T. V., Larina, G. N., Zhenenvskaya, M. G. and Gusinskaya, V. A., *J. Polym. Sci. Part C,* **16** (1967) 1631.
15. Reeder, J. A., US Patent 3334071, 1967.
16. Bailey, W. J., Economy, J. and Hermes, M. E., *J. Org. Chem.,* **27** (1962) 3295.
17. Jones, R. J., O'Rell, M. K., Sheppard, C. H. and Vaughan, R. W., *SAMPE Quart.,* **8**(3) (1977) 18.
18. Stenzenberger, H. D., *Appl. Polym. Symp.,* **22** (1973) 77.
19. Barton, J. M. and Knight, G. J., RAE Technical Report 75082.
20. Joshi, R. M., *Makromol. Chem.,* **53** (1962) 33.
21. Lang, J. L., Pavelich, W. A. and Clarey, H. D., *J. Polym. Sci.,* **55** (1961) 532.
22. Joshi, R. M., *Makromol. Chem.,* **62** (1963) 140.
23. Tawney, P. O., Snyder, R. H., Conger, R. P., Liebbrand, K. A., Stiteler, G. H. and Williams, A. R., *J. Org. Chem.,* **26** (1961) 15.
24. Cubbon, R. C. P., *Polymer,* **6** (1965) 419.
24a. Coleman, L. E., Jr, and Conrady, J. A., *J. Polym. Sci.,* **38** (1959) 241.
25. Iwata, K. and Stille, J. K., *J. Polym. Sci., Polym. Chem. Ed.,* **14** (1976) 2841.
26. de Koenig, A., European Patent Appl. 108461, 1984.
27. Nash, H. C., Paranski, C. F., Jr, and Ting, R. J., *Amer. Chem. Soc. Org. Coatings Plast. Chem.,* **40** (1979) 877.,
28. Stenzenberger, H. D., Herzog, M., Römer, W., Scheiblich, R., Reeves, N. and Pierce, S., *29th Nat. SAMPE Symp.,* **29** (1984) 1043.
29. Stenzenberger, H. D. and Herzog, M., unpublished results.
30. Nakayama, Y. and Smets, G., *J. Polym. Sci., Part A-1,* **5** (1967) 1619.
31. Stenzenberger, H. D., Herzog, M., Römer, W., Scheiblich, R., Pierce, S. and Canning, M., *30th Nat. SAMPE Symp.* **30** (1985) 1568
32. Kraiman, E. A., US Patent 2890206, 1959; *Chem. Abstr.,* **53** (1959) 17572.
33. Kraiman, E. A., US Patent 2890207, 1959; *Chem. Abstr.,* **53** (1959) 17572.
34. Chow, S. W. and Whelan, J. M., Jr, US Patent 2971944, 1961; *Chem. Abstr.,* **55** (1961) 12941.
35. Brückner, V., *Ber. Dtsch. Chem. Ges.,* **75** (1942) 2043.
36. Brückner, V. and Kovacs, J., *J. Org. Chem.,* **13** (1948) 641.
37. Hudson, B. J. F. and Robinson, R., *J. Chem. Soc.,* (1941) 715.
38. Street, S., US Patent 4351932, 1982.
39. Sheremeteva, T. V., Larina, G. N., Zhenevskaya, M. G. and Gusinskaya, V. A., *J. Polym. Sci., Part C,* **16** (1967) 1631.
40. Kovacic, P., US Patent 2818405, 1957.
41. Crivello, J. V., *J. Polym. Sci., Polym. Chem. Ed.,* **11** (1973) 1185.
42. Smyth, P. G., Nagamatsu, A. and Truton, J. S., *J. Amer. Chem. Soc.,* **82** (1960) 4600.
43. Sinon, S. and Konigsberg, W., *Proc. Nat. Acad. Sci. (US),* **56** (1966) 749.
44. Crivello, J. V. and Juliano, P. C., *Amer. Chem. Soc. Prepr., Div. Polym. Sci.,* **14**(2) (1973) 1220.

45. Crivello, J. V., *Amer. Chem. Soc. Prepr. Div. Polym. Chem.*, **13**(2) (1972) 924.
46. Renner, A., Forgo, I., Hoffmann, W. and Ramsteiner, K., *Helv. Chim. Acta,* **61** (1978) 1443.
47. Bergain, M., Combet, A. and Grosjean, P., British Patent 1190718, 1970.
48. Stenzenberger, H. D., US Patent 4211860, 1980; US Patent 4303799, 1981.
49. Ayano, S., *Chem. Economy Eng. Rev.*, **10**(3) (1978) 25.
50. Locatelli, J. L. and Robin, J., US Patent 4342860, 1982; US Patent 4302572, 1981.
51. Stenzenberger, H. D., US Patent 4520145, 1985.
52. Darmory, F. P. and Di Benedetto, M., US Patent 3944525, 1976.
53. Loszewski, R. C. and Alverez, R. T., British Patent Application 2139628, 1984.
54. Akiyama, K. and Makimo, K., US Patent 3730948, 1973.
55. Stenzenberger, H. D., Herzog, M., Römer, W., Scheiblich, R., Reeves, N. J. and Pierce, S., *29th Nat. SAMPE Symp.*, **29** (1984) 1043–59.
56. Stenzenberger, H. D., Herzog, M., Römer, W., Pierce, S., Canning, M. and Fear, K., *Proc. ANTEC 1985,* (1985) 1246–9.
57. Kiyoje, M. and Tsutomu, O., Japanese Patent 77 51499, 1977.
58. Forgo, I., Schreiber, B., Renner, A. and Haug, Th., German Offen. 2459925, 1975.
59. Forgo, I., Renner, A. and Schmitter, A., German Offen. 2459938, 1974.
60. Akira, F. and Yashushi, Y., Japanese Patent 77 141900, 1977.
61. Anon, Japanese Patent Appl. 48-055994, 1971.
62. White, J. E., *Amer. Chem. Soc. Polymer Prepr.*, **26**(1) (1985) 132.
63. Asahara, T., Yoda, N. and Minami, N., US Patent 3669930, 1972.
64. Zahir, S. A. and Renner, A., US Patent 4100140, 1978.
65. King, J. J., Chaudhari, M. and Zahir, S., *29th Nat. SAMPE Symp.*, **29** (1984) 392.
66. Grigat, E. and Putter, R. (Bayer), West German Patent 1195764, 1963.
67. McKague, L., in *Composites for Extreme Environments,* ASTM STP 768, 1982, pp 20–32.
68. Street, S., in *Polyimides,* Ed. K. L. Mittal, Plenum Press, New York and London, 1984, p. 77.
69. Street, S., *25th Nat. SAMPE Symp.*, **25** (1980) 366.
70. Ho, V., *29th Nat. SAMPE Symp.*, **29** (1984) 409.
71. Steiner, P. A., Blair, M., Gong, G., McKillen, J. and Brown, J., *6th Internat. SAMPE Europ. Chapter Conf.*, (1985) 297.
72. Hergenrother, P. M. and Pagar, D. J., *Adhesive Age,* **20**(12) (1977) 38–43.
73. Pogar, D. J., *11th Nat. SAMPE Techn. Conf.*, (1979) 746.
74. Darmory, F. P., in *New Industrial Polymers,* Amer. Chem. Soc. Symp. Series **29** (1974) 124–44.

75. Kinloch, A. J. and Shaw, S. J., *Amer. Chem. Soc. Polym. Mater. Sci. Engng*, **49** (1983) 307.
76. Shaw, S. J. and Kinloch, A. J., *Int. J. Adhesion Adhesives*, **5**(3) (1985) 123.
77. Segal, C. L., Stenzenberger, H. D., Herzog, M., Römer, W., Pierce, S. and Canning, M. S., *7th National SAMPE Technical Conference Proceedings*, **17** (1985). 147.

5

Rubber-Toughened Thermosetting Polymers

A. J. KINLOCH

Department of Mechanical Engineering, Imperial College of Science and Technology, London, UK

NOTATION

E	modulus
G_{Ic}	fracture energy
ΔG_m	free energy of mixing
ΔH_m	enthalpy of mixing
$\mathbf{R}$	gas constant
ΔS_m	entropy of mixing
T	temperature
T_g	glass transition temperature
T_β	relaxation temperature of epoxy phase
V	molar volume
a_T	time–temperature superposition shift factor
n	mole fraction
t_f	time-of-failure
v_f	volume fraction of dispersed rubbery phase
δ	solubility parameter
Φ	volume fraction

1. INTRODUCTION

The use of adhesives and continuous-fibre-reinforced plastics in structural engineering applications has increased markedly over the last few years and for both technical and economic reasons this trend will certainly continue. At present these materials employ thermosetting polymers almost exclusively, either as the basis for the structural-adhesive compositions or as the matrix for glass-, polyamide- and carbon-fibre composites.

This dominance of thermosetting polymers may be ascribed to two main aspects. Firstly, typical precursors are low molar mass resins which contain relatively polar chemical groups, and such resins are usually polymerised via the addition of a curing agent or catalyst and/or the application of an external stimulus such as heat or light. The resin/hardener mixture may therefore be formulated so that it exhibits a relatively low viscosity at some stage during the polymerisation, particularly at elevated temperatures and/or upon solvent dilution. Obviously a low viscosity and the presence of polar chemical groups are important requirements for attaining good wetting and high intrinsic adhesion at the adhesive/substrate or matrix/fibre interface.[1] Secondly, when polymerised, thermosetting polymers are amorphous, highly crosslinked polymers. This microstructure results in many useful properties for engineering applications, such as high modulus and failure strength, low creep and good performance at elevated temperatures.

However, the structure of thermosetting polymers also leads to one highly undesirable property: they are relatively brittle materials, having a very poor resistance to crack initiation and growth. To increase the crack resistance, but without seriously impairing other important properties, two methods have been reported[2] for modifying the structure of such polymers. They are both based upon attaining a dispersion of a particulate second phase in the thermosetting polymer. In one method the second phase consists of brittle rigid particles, such as alumina, silica or glass spheres, whilst in the other the second phase is rubbery in character. The former rigid-particulate materials are low-cost fillers and are often added for reasons of economy or to control the rheological behaviour of the polymers. They may also significantly increase the toughness of brittle thermosetting polymers, as discussed in Chapter 6. However, the incorporation of a second phase of dispersed rubbery particles in the thermosetting polymer has resulted in the greatest improvements in toughness.

Various types of thermosetting polymers have been modified by the incorporation of a second rubbery phase. They include epoxies, phenolics, polyimides and polyesters; but of these, rubber-toughened

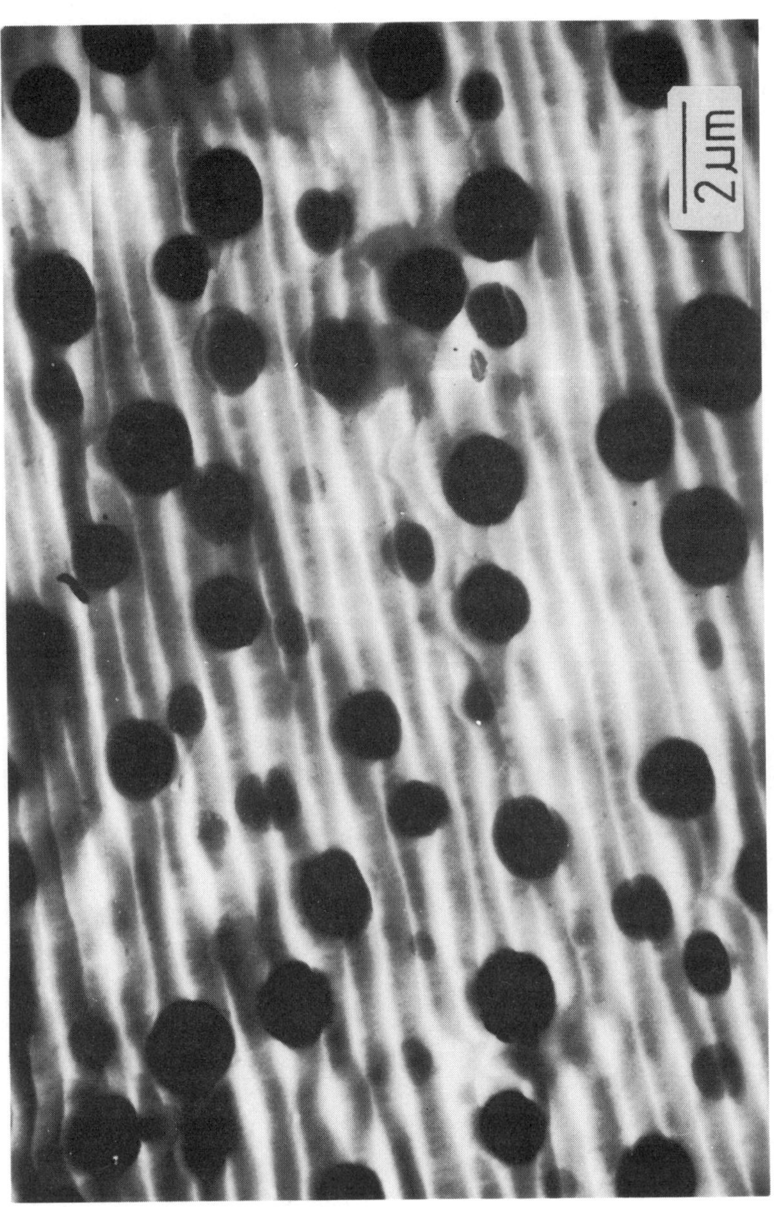

FIG. 1. Transmission electron micrograph of osmium-tetroxide-stained section of a rubber-modified epoxy polymer.

epoxies have been the most extensively studied polymers, reflecting their wide use as structural adhesives and increasing use in fibre composites.

The multiphase microstructure of a rubber-modified epoxy may be clearly seen in the transmission electron micrograph of a thin section shown in Fig. 1. The rubber employed was a carboxyl-terminated butadiene–acrylonitrile copolymer and the section was exposed to osmium tetroxide prior to examination. The selective reactions of reagent with the unsaturated butadiene group increases the contrast between the particles and the epoxy matrix.[3,4] The data given in Table 1 clearly reveal that this microstructure leads to major improvements in toughness, without significant losses in other important properties. For example, the fracture energy, G_{Ic}, has been increased by a factor of 20–30 whilst the modulus has been decreased by about only 6% and the upper glass transition temperature of the toughened polymer is unchanged.

TABLE 1
PROPERTIES OF AN UNMODIFIED AND A RUBBER-TOUGHENED EPOXY

Property	*Unmodified epoxy*	*Rubber-toughened epoxy*
Molar mass between crosslinks in epoxy phase	~500	~500
T_g of epoxy phase (°C)	100	100
T_β of epoxy phase (°C)	−74	−74
Volume fraction of dispersed rubbery phase		0·18
Mean diameter of dispersed rubbery phase (μm)		1·6
T_g of rubbery phase (°C)		−55
Tensile fracture stress (MPa)	63	58
Tensile fracture strain (%)	5	9
Young's modulus (GPa)	3·2	2·8
Fracture energy, G_{Ic} (kJ/m^2)	0·2	4–6
Izod impact strength (J/m of notch)	0·7	3–5

The present chapter will therefore review recent developments in our understanding of the relationships between the chemistry, resulting microstructure and observed mechanical behaviour, particularly the toughness, of rubber-toughened thermosetting polymers.

2. CHEMISTRY/MICROSTRUCTURE RELATIONSHIPS

2.1. Introduction

The microstructure of a rubber-modified thermosetting polymer is obviously dependent upon the detailed polymerisation chemistry of the starting products, namely resin/hardener/rubber. The microstructure, in turn, controls the mechanical behaviour of the cured polymer.

For reasons such as the ease of manufacture, stability of formulated material and uniformity of end-product the reactants are usually selected to give an initially homogeneous solution of resin/hardener/rubber. Polymerisation, or cure, generally involves the sequential processes of *in situ* phase separation, gelation and vitrification. During cure, phase separation of the rubber results from increases in the molar masses of the rubber and resin. The molar mass increase of the rubber often occurs via copolymerisation of the rubber with resin monomer or oligomers. The resulting rubbery polymer is probably similar in structure to an ABA block copolymer, where A represents the resin units and B is the chain of a rubber molecule. This copolymerisation may result in the volume fraction of dispersed rubbery particles actually being greater than the volume fraction of added rubber. It may also lead to the rubbery phase not simply consisting of homogeneous particles, but the particles themselves possessing a complex multiphase morphology. On the other hand, if the rubber is too compatible with the resin, then no phase separation, or only limited phase separation, may occur. Such a single-phase microstructure may lead to a small increase in toughness but also invariably causes major decreases in the strength, modulus and elevated-temperature performance.

The main factors which affect the microstructure are the chemical and physical properties of the rubber and resin/hardener and the cure conditions, e.g. temperature and time of cure.

2.2. Chemical and Physical Properties of the Rubber

2.2.1. Molecular Structure

The optimum molecular structure of the rubber is determined by two main requirements: the rubber must be in the form of a well-dispersed separate phase in the cured polymer and the rubber particles need to have good adhesion to the thermoset matrix.

The first important requirement is that the rubber forms a separate phase when the resin is cured. Since the rubber is usually initially dissolved in the resin, this must normally involve an *in situ* phase separation. The problem of obtaining a rubber which is initially miscible with the resin, but which will phase-separate during the cure process, is met by using polar, low molar mass (liquid) rubbers which possess reactive end-groups. A typical such rubber is a carboxyl-terminated butadiene–acrylonitrile rubber (CTBN) represented by:

$$\mathrm{HOOC}\left[(\mathrm{CH_2{-}CH{=}CH{-}CH_2})_x{-}(\mathrm{CH_2{-}\underset{\substack{|\\ \mathrm{CN}}}{CH}})_y\right]_m\mathrm{COOH}$$

where typically $m = 10$ and the molar mass is about 3500g/mol. The values of x and y will determine the acrylonitrile content and hence the polarity and compatibility of the rubber.

Now miscibility and phase separation are governed by the thermodynamics of the system. When the free energy, ΔG_m, of isothermal mixing is negative then miscibility will result, but when ΔG_m is positive then phase separation occurs.

The value of ΔG_m is given by:

$$\Delta G_m = \Delta H_m - T\Delta S_m \tag{1}$$

where ΔH_m and ΔS_m are the molar enthalpy and entropy of mixing and T is the temperature. Now ΔH_m and ΔS_m may be expressed[5,6] by:

$$\Delta H_m = V(\delta_1 - \delta_2)^2\Phi_1\Phi_2 \tag{2}$$

and

$$\Delta S_m = -\mathbf{R}(n_1 \ln \Phi_1 + n_2 \ln \Phi_2) \tag{3}$$

Thus:

$$\Delta G_m = V(\delta_1 - \delta_2)^2\Phi_1\Phi_2 + \mathbf{R}T(n_1 \ln \Phi_1 + n_2 \ln \Phi_2) \tag{4}$$

where n_1 and n_2 are mole fractions of the liquid rubber and the resin system, Φ_1 and Φ_2 are volume fractions, δ_1 and δ_2 are the solubility parameters, $\mathbf{R}$ is the gas constant and V is the molar volume of the liquid mixture. Considering eqn (4), then the first term represents ΔH_m and is always predicted to be positive, i.e. mixing is assumed endothermic. The second term is, however, usually negative. This represents $-T\Delta S_m$ and the entropy, ΔS_m, is usually positive since mixing involves an increase in the disorder of the system. Thus, the two terms in eqn (4) tend to counterbalance each other and the factor which decides whether ΔG_m is negative or positive is the relative magnitude of the two terms. This is where the molar mass plays an important role since, as the molar mass of the components increases, the first term in eqn (4), i.e. ΔH_m, becomes increasingly positive (due to V increasing), whilst the second term, i.e. $-T\Delta S_m$, is not significantly affected. However, it should be recognised that the approach outlined above is essentially a simplistic treatment which does not, for example, admit the possibility of exothermic mixing. Nevertheless, the treatment is useful in explaining, at least qualitatively, many of the chemistry/microstructure relationships in rubber-toughened thermosets, and the reader is referred elsewhere for more detailed reviews[7–9] of the thermodynamics of polymer mixing.

Considering the resin/hardener/liquid rubber systems, then the aim is to match the respective values of δ_{resin} and δ_{rubber} fairly closely so that ΔG_m is initially negative and the rubber dissolves in the resin. However, the values of δ_{resin} and δ_{rubber} need to be sufficiently different so that, upon the rubber and resin polymerising and increasing their molar masses, the first term in eqn (4) becomes sufficiently positive to make ΔG_m positive—and phase separation results. The preferred liquid rubbers for toughening thermoset resins usually have a solubility parameter lower than the resin by about $1(J/cm^3)^{1/2}$. This difference meets the above requirements without the rubber possessing an excessively polar structure; a polar structure leads to high interchain attractions and a high T_g (rubbery phase) which impairs toughness. This is illustrated[10,11] in Fig. 2, which shows the relationship between solubility parameter and acrylonitrile content in a series of carboxyl-terminated butadiene–acrylonitrile copolymer rubbers. An exact match of solubility parameters with the diglycidyl ether of bisphenol A epoxy (DGEBA) resin occurs at an acrylonitrile content of 28·5 wt%. However, the preferred rubbers have an acrylonitrile content of about half this level when the value of $(\delta_{resin} - \delta_{rubber})$ is about 1 $(J/cm^3)^{1/2}$. But even at this acrylonitrile level the T_g of the rubbery phase is some 20°C higher than for a polybutadiene rubber.

The second important requirement is that the rubbery particles need to be well bonded to the thermoset matrix; weak bonding across the particle/matrix interface leads to poor toughness, as shown later. To ensure that there are intrinsically strong chemical bonds across the interface demands using a rubber which possesses end-groups which

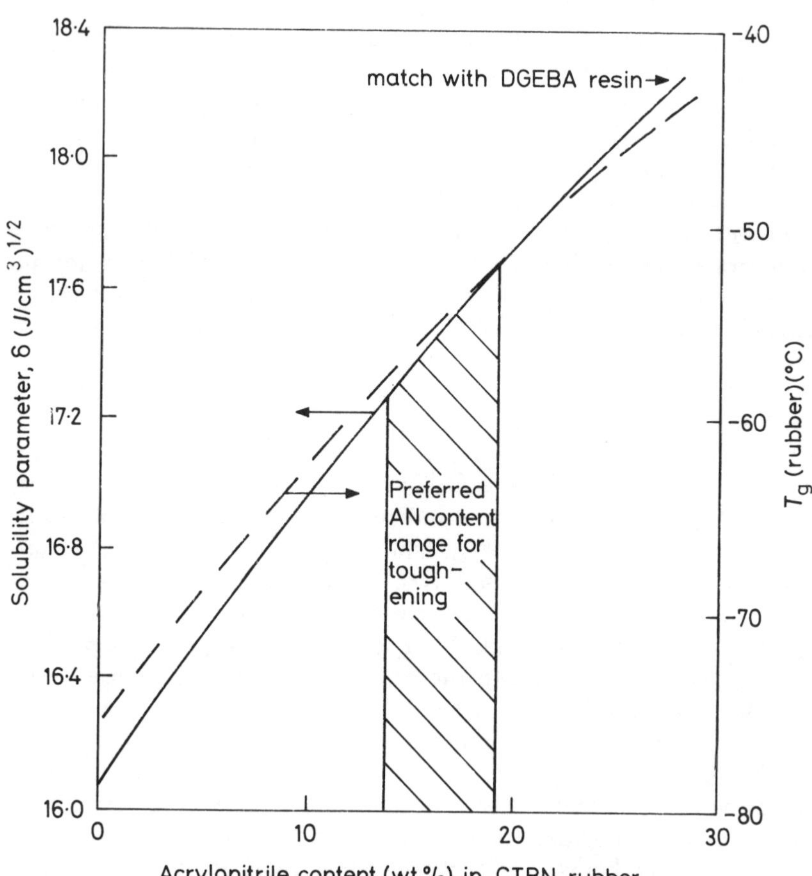

FIG. 2. Relationships between composition and properties of carboxyl-terminated butadiene–acrylonitrile rubbers.[10]

are reactive towards the resin. However, the nature of the 'interphase' between the rubbery particles and thermoset matrix has not been investigated in great detail. As may be seen from Fig. 1, the boundaries of the dispersed phase of CTBN-based particles appear sharp. Indeed, energy-dispersive X-ray analysis for the concentration of osmium across the particle/matrix interface has confirmed[12] that the interfacial boundary is sharply defined and any interphase region where segmental mixing has occurred is less than 50 nm in width, i.e. below the spatial resolution of the electron beam. If an amine-terminated butadiene–acrylonitrile (ATBN) rubber is used, instead of CTBN, then the interfacial boundary from similar transmission electron micrographs appears less well defined, often appearing somewhat diffuse.[13,14] However, ^{1}H and ^{13}C nuclear magnetic resonance spectroscopy results, together with differential scanning calorimetry data, has indicated[14] that for both types of rubber no significant segmental mixing occurs and a sharp interfacial boundary exists. The diffuse appearance of the interface between the ATBN-based particles and the epoxy matrix has been attributed to an irregular particle shape.

From the above comments it is evident that the requirements of both phase separation and strong interfacial bonding demand that a reactive rubber be employed. However, it does not necessarily have to be a carboxyl-terminated rubber. Amongst the butadiene–acrylonitrile copolymers that have been used,[15] terminal groups have included carboxyl (—COOH), phenol (—C_6H_5OH), epoxy (—$\overset{\overset{O}{\diagup\;\diagdown}}{CH—CH_2}$), hydroxyl (—$CH_2OH$) and mercaptan (—$CH_2SH$). The above list is in order of decreasing selectivity but increasing reactivity towards the epoxy resin monomer, and in order of decreasing efficiency of phase separation. This is reflected by the fracture energies of the resulting toughened epoxy polymers decreasing down the series. As Bucknall[6] has commented, the same considerations apply to other rubbers: carboxyl-terminated polycaprolactone and polypropylene oxide rubbers also toughen epoxy resins, provided the molar mass of the rubber is sufficiently high to ensure phase separation.[16] In the case of polyester resins, vinyl- and carboxyl-terminated butadiene–acrylonitrile rubbers have been employed.[17,18] An alternative approach, which avoids the problem of *in situ* phase separation, is to add a high molar mass solid polymer to the resin: both an acrylonitrile butadiene–styrene terpolymer and a polybutadiene-based rubber (molar mass $2{\cdot}6 \times 10^5$) have been employed in this type of formulation.[19,20] However, problems may arise in their incorporation and the viscosity of the resin/rubber system may be somewhat high, but this approach has yet to be fully explored and should be a fruitful area for further study.

Finally, it is important to note that to ensure 'end-capping' of the reactive liquid rubber by resin monomer units, the rubber and resin are often first pre-reacted. In some instances a catalyst is used, triphenylphosphine for example, to reduce the reaction temperature and time. The rubber/resin adduct is then added to the resin/hardener mixture to form the homogeneous mixture of reactants.

2.2.2. Concentration of Rubber

As discussed previously, there is no simple relation between the concentration of initially added rubber and the volume fraction, v_f, of precipitated rubbery particles.

If the rubber remains completely in solution in the thermosetting polymer, possibly as a random copolymer or as a block copolymer with very short rubber blocks, then no phase separation occurs and v_f is obviously zero.

On the other hand, the end-capped rubber molecules may selectively react mainly with each other to give comparatively high molar mass material, and under these conditions phase separation is likely to be efficient and the value of v_f will be high. Also, since the precipitated rubbery phase usually contains copolymerised resin, the value of v_f may even be appreciably greater than the volume fraction of initially added rubber. The copolymerised resin may be in solution in the rubbery particle, and this is likely to be the case when the molar mass of the resin units or blocks is low. Alternatively, if the resin blocks are relatively long then they may phase-separate to form inclusions within the rubbery particle. An example of both homogeneous particles and phase-separated particles occurring in the same material is shown in Fig. 3 where the smaller rubber particles ($<1\ \mu m$) are homogeneous whilst the larger particles ($>1\ \mu m$) are two-phase. It should be noted that whether phase separation has occurred inside a particle cannot always be detected using transmission electron microscopy. If the inclusions of resin are less than the thickness of the section (~ 100 nm) then they are difficult to observe. However, a knowledge of the volume of initially added rubber, the value of v_f from electron microscopy and the glass transition temperature of the rubbery phase (from, for example, dynamic mechanical analysis) can resolve the internal morphology of the particle.[13]

Finally, it is usually found that the maximum volume fraction of rubbery particles that can be obtained using the technology described above is about 0·2–0·3. Attempts to produce higher values of v_f have usually resulted in phase inversion, where the rubber forms the continuous matrix phase with dispersed particles of thermosetting polymer.[22] Obviously such materials are rubbery in character and their low modulus and strength makes them unacceptable for structural applications.

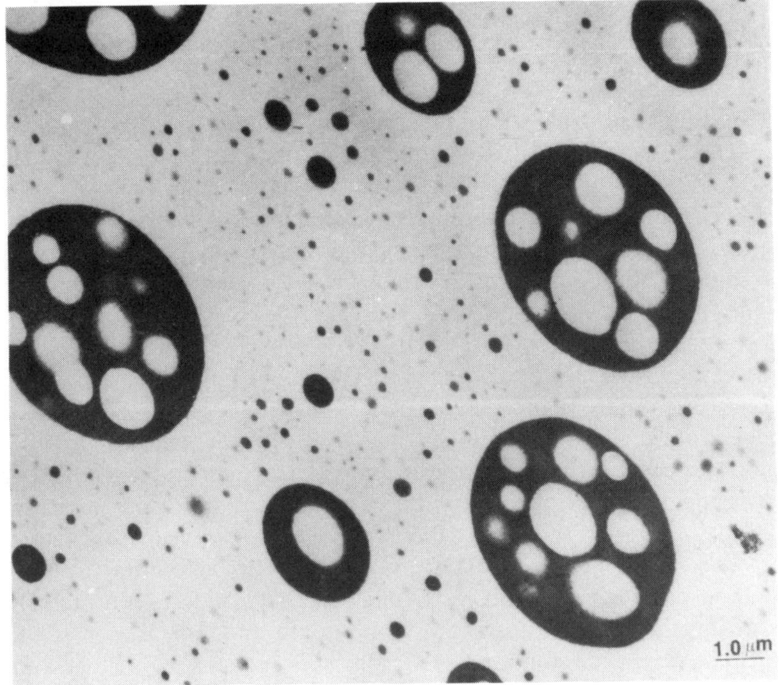

FIG. 3. Transmission electron micrograph of osmium-tetroxide-stained section of a rubber-modified epoxy polymer. The chemistry has been selected to give a more complex microstructure than that shown in Fig. 1.[21]

2.2.3. Miscellaneous Properties

Thermosetting polyimides have extremely good high-temperature properties, as discussed in Chapter 4. In an effort to maintain this aspect of their performance, but increase their toughness, work has been reported[23] on using silicone rubbers to obtain multiphase rubber-toughened polyimides; silicone rubbers possess outstanding resistance to thermal degradation. However, only relatively small increases in toughness were recorded and more recent work[24,25] has indicated that CTBN rubbers are far more effective in increasing the fracture energy and strain capability of polyimides. Further, the presence of CTBN-based particles appears, at least from the initial results, not to affect the thermal stability of the polymer to a major degree.

2.3. Chemical and Physical Properties of Resin/Hardener

The chemical and physical properties of the resin/hardener obviously have a major effect on such factors as the overall compatibility and extent of phase separation, the type of reactive groups the rubber needs to possess, etc. However, the choice of resin/hardener is usually determined by the end-application and consideration of such factors as cost, pot-life, cure temperature, desired thermal resistance, etc. Hence, it is the chemical and physical properties of the rubber which are usually tailored in order to achieve a multiphase microstructure.

The most detailed study of the effects of hardener type and concentration are those of Bucknall and Yoshii,[26] who examined the microstructure of an epoxy resin/CTBN rubber system when it was cured using different amines as shown in Fig. 4. At any given hardener concentration the largest rubbery particles are obtained using piperidine, the least reactive of the four amine hardeners employed.[10] With increasing hardener reactivity, the particle size falls and increasing hardener content also reduces the particle size. It is clear from Fig. 4b that the volume fraction, v_f, of rubbery particles follows the same trend as particle size: with increasing reactivity and concentration of hardener there is a substantial decrease in the amount of rubber precipitated from solution. At high concentrations of the more reactive hardeners most of the rubber remains dissolved in the resin matrix phase. Thus, the different hardeners and the different concentrations of hardener used affect the rates of the competing reactions to different degrees; they determine, for example, whether the rubber molecules will preferentially react with each other and whether this reaction will be completed prior to gelation of the resin. As discussed previously, the former reaction will lead to phase separation, assuming it is largely completed prior to gelation of the resin.

Apart from the diglycidyl ether of bisphenol A epoxy resin, cycloaliphatic epoxy,[16,19,27] epoxy–novolac[19] and tetraglycidyl 4,4′-diaminodiphenyl epoxy[28] resins have been toughened by generating a multiphase microstructure, together with polyester[17,18] and polyimide[23–25] resins.

Although it is strictly not part of the resin/hardener system it is convenient to discuss in the present section the effects of adding bisphenol A to an epoxy resin/hardener/rubber system. Several workers[15,21,26] have reported that addition of bisphenol A [typically about 24 parts per hundred of resin (phr)] aids separation of the

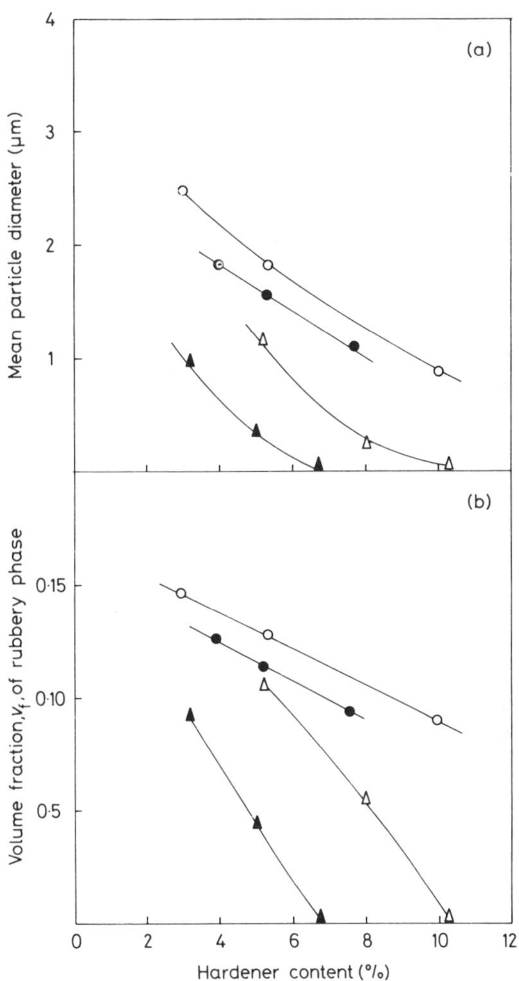

FIG. 4. Effects of type and concentration of amine hardener on the microstructure of rubber-toughened epoxy polymers.[26] Resin: diglycidyl ether of bisphenol A (DGEBA). Rubber: 8·7 wt% CTBN. Hardeners: ○, piperidine; ●, tri-(2-ethylhexanoic acid) salt of 2,4,6-tris(dimethylaminomethyl) phenol; △, triethyleneamine; ▲, 2,4,6-tris(dimethylaminomethyl)phenol.

rubbery phase and results in a bimodal distribution of particle size with one family around 0·1 μm in diameter and a second family around 1–5 μm in diameter. The detailed chemistry behind this observation is not clearly established but a reaction sequence has been suggested by Riew *et al.*[15]

2.4. Time and Temperature of Cure

The curing of the resin/hardener/rubber mixture generally involves the sequential processes of phase separation, gelation and vitrification. The time and temperature employed for the overall cure will affect the kinetics of each of these processes differently and the extent to which these processes occur. Thus, the cure conditions have a profound effect on the microstructure.

The processes of phase separation, gelation and vitrification in an epoxy resin/amine hardener/CTBN rubber system are illustrated in Fig. 5 in the form of a time–temperature–transformation (TTT) isothermal cure diagram.[13] Such diagrams have been considered in

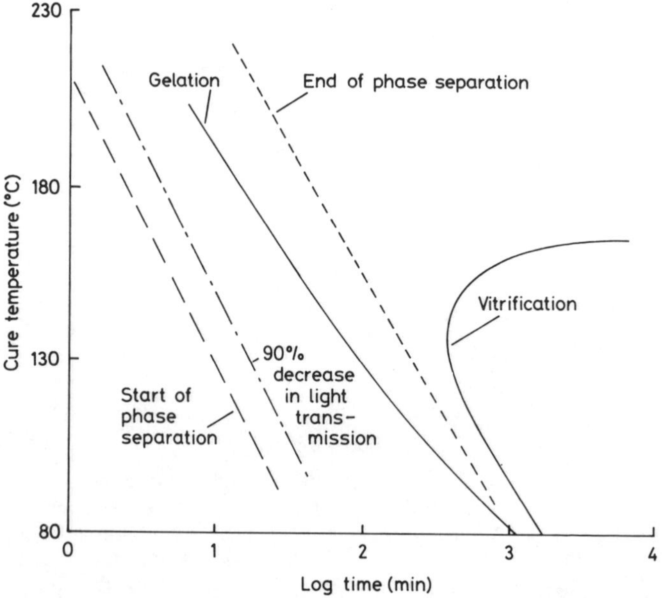

FIG. 5. Time–temperature–transformation cure diagram for an epoxy/CTBN/amine hardener system. The extent of phase separation was determined using light transmission.[13]

TABLE 2
EFFECT OF CURE CONDITIONS ON THE MICROSTRUCTURE[a 13]

Cure conditions	*T_g (epoxy matrix) (°C)*	*T_g (rubbery phase) (°C)*	*Mean diameter of rubbery particles (μm)*	*v_f of rubbery phase*
100°C/40 h	122	−50	0·6	0·11
200°C/3 h	161	−50	2·9	0·14

[a] Resin, DGEBA; rubber, CTBN (15 phr); hardener, trimethyleneglycol di-*p*-aminobenzoate (TMAB) (51 phr).

detail in Chapter 1 by Gillham and only the essential features will be discussed here. By consideration of a particular isothermal cure temperature this diagram indicates, for example, the times at which phase separation initiates and is completed, and when gelation and vitrification occur. As observed by other workers[29,30] the process of phase separation is virtually completed before gelation of the epoxy phase. If this were not the case, then the rapid increase in viscosity which accompanies gelation would inhibit any subsequent phase separation and result in a low volume fraction of dispersed rubbery phase. For the system illustrated the data in Table 2 reveal the effects on the microstructure of curing at 100°C for 40 h, as opposed to 200°C for 3 h. Further, whilst postcuring the former material for, say, 5 h at 170°C will increase the value of T_g (epoxy matrix) to 163°C, it will not of course change the volume fraction of dispersed rubbery phase. Thus, the cure cycle employed for rubber-modified-resin systems is of crucial importance in determining the microstructure, and hence mechanical properties, of the cured product.

3. MICROSTRUCTURE/MECHANICAL PROPERTY RELATIONSHIPS

3.1. Introduction

The microstructural features of rubber-toughened thermosetting polymers which may affect the mechanical properties are as follows.

(*a*) *Thermosetting matrix*
  (i) Crosslink density and T_g of matrix phase.
  (ii) Concentration of non-phase-separated rubber.

(*b*) *Rubbery dispersed phase*
- (i) Volume fraction.
- (ii) Particle size.
- (iii) Distribution of particle size.
- (iv) Intrinsic adhesion acrosss particle/matrix interface.
- (v) Morphology of dispersed phase.
- (vi) Glass transition temperature of dispersed phase.

However, before considering the influence of the above features on the mechanical properties it is of interest to discuss the mechanisms whereby the presence of rubbery particles increases the toughness of a thermosetting polymer, since toughness is the property of most interest.

3.2. Toughening Mechanisms

3.2.1. Introduction

Many different mechanisms have been proposed to explain the greatly improved fracture energy that may result when a thermosetting polymer possesses a multiphase microstructure of dispersed rubbery particles. Much of the dispute has surrounded the issues of whether the rubbery particles[31,32] or the thermoset matrix[20–22,26,33–39]absorbs most of the energy and, if the latter, whether the deformation mechanism involves the formation of crazes.[15,26,35] The reader is referred elsewhere[2,33] for detailed reviews of these mechanisms but recent work by Kinloch *et al.*,[33,38] and independently by Yee and Pearson,[21,39] has clearly established that plastic shear-yielding in the matrix is the main source of energy dissipation and increased toughness. Such plastic deformation occurs to a far greater extent in the matrix of rubber-toughened thermosets, compared with the unmodified material, due to interactions between the stress field ahead of the crack and the rubbery particles. Hence, the first step towards understanding the role of the rubbery particle is to consider the stress field around a particle.

3.2.2. Stress Fields Around Rubbery Particles

Goodier[40] has derived equations for the stresses around an isolated elastic spherical particle embedded in an isotropic elastic matrix which is subjected to an applied uniaxial tensile stress remote from the particle. His equations reveal that for a rubbery particle, which typically possesses a considerably lower shear modulus than the matrix, the maximum stress concentration occurs at the equator of the particle and has a value of about 1·9. Furthermore, assuming the particle is well bonded to the matrix, this stress is triaxial tension. This arises essentially because of the volume constraint represented by the bulk modulus of the rubbery particle, which is comparable with that of the matrix. This feature also means that, in contrast to a hole which would produce a stress concentration similar in magnitude, the rubbery particle can fully bear its share of the load at the crack front. This, together with the high shear deformations which rubbers can withstand, largely explains why rubbery particles are particularly effective in producing large increases in the toughness of brittle plastics.

More recently Broutman and Panizza[41] have developed finite-element stress analyses to obtain the stress concentration around rubbery particles when the volume fraction of particles is sufficiently high so that the stress fields of the particles interact, i.e. for values of v_f greater than about 0·09. In such cases the maximum stress concentration at the equator of the particle may be appreciably higher than for the isolated case.

Finally, it is of interest to note that Goodier's analysis also indicates that the stress concentration at the particle equator will still be present even at temperatures below the glass transition temperature of the rubber: under such circumstances the shear modulus of the rubbery particle would still be expected to be slightly lower than that of the highly crosslinked thermoset matrix, which is even more below its T_g. However, as the shear moduli will now be much closer in value, the extent of the stress concentration will be decreased somewhat. This aspect is important since it readily explains why rubber-modified thermosets are still a fair amount tougher than the unmodified polymer at temperatures well below T_g(rubbery phase), as shown later in Fig. 14 where T_g(rubbery phase) is about −55°C.

3.2.3. *Matrix Shear-Yielding and Particle Cavitation*

The stress field associated with the rubbery particles leads to the initiation of two important processes which can strongly interact, especially at high test temperatures or slow test rates.

The first process that occurs is the initiation and growth of shear-yield deformations in the matrix. The stress concentrations around the rubbery particles act as the initiation sites for the plastic

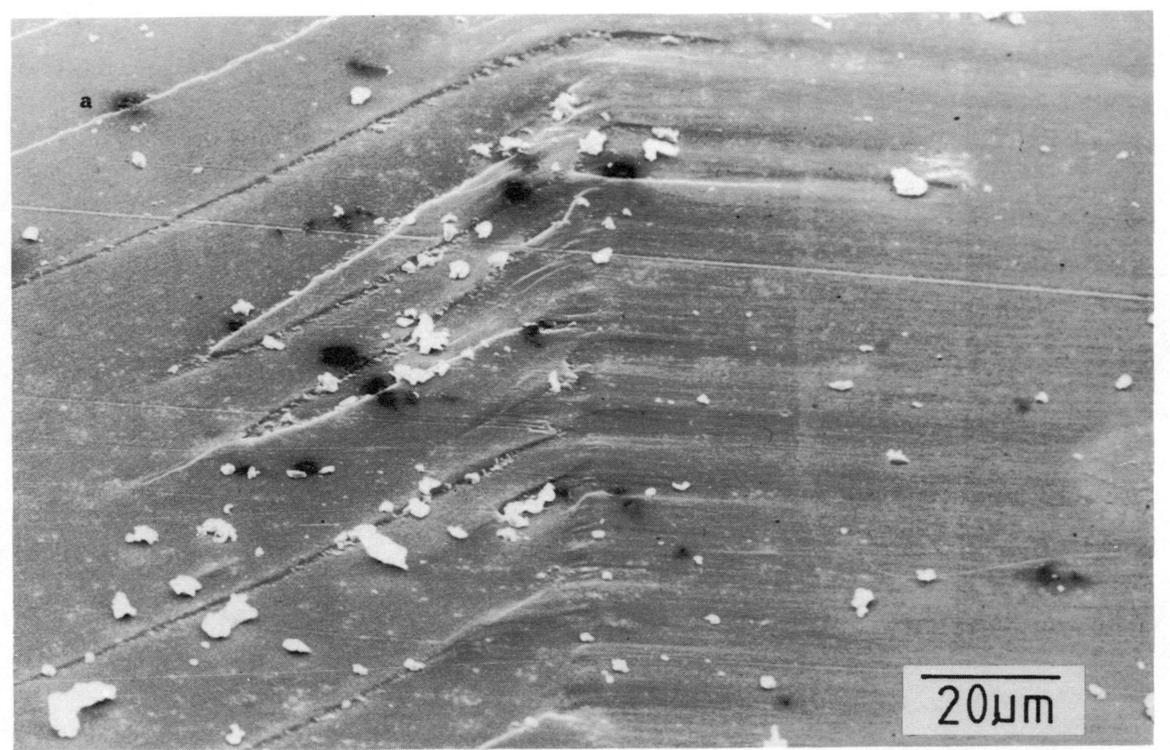

FIG. 6. Scanning electron micrographs of fracture surfaces (test temperature, 23°C).[42] (a) Unmodified epoxy polymer.

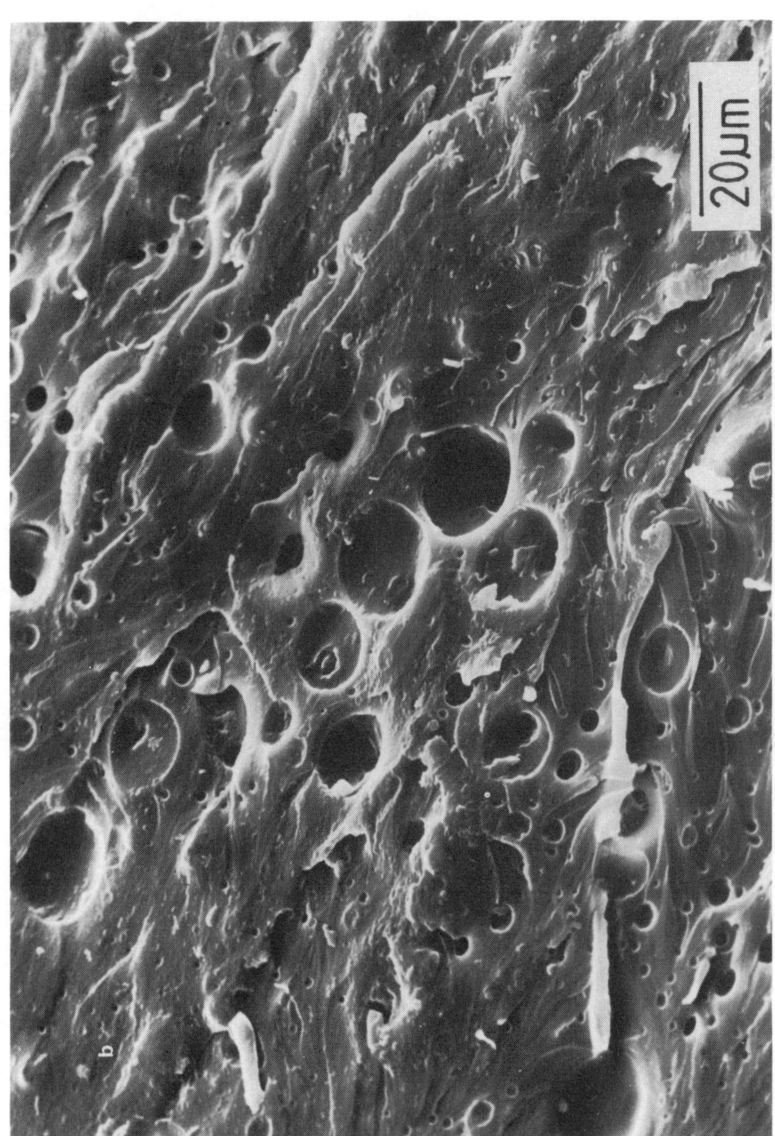

FIG. 6—*contd.* (b) Rubber-toughened epoxy polymer [v_f(rubber phase) = 0·14].

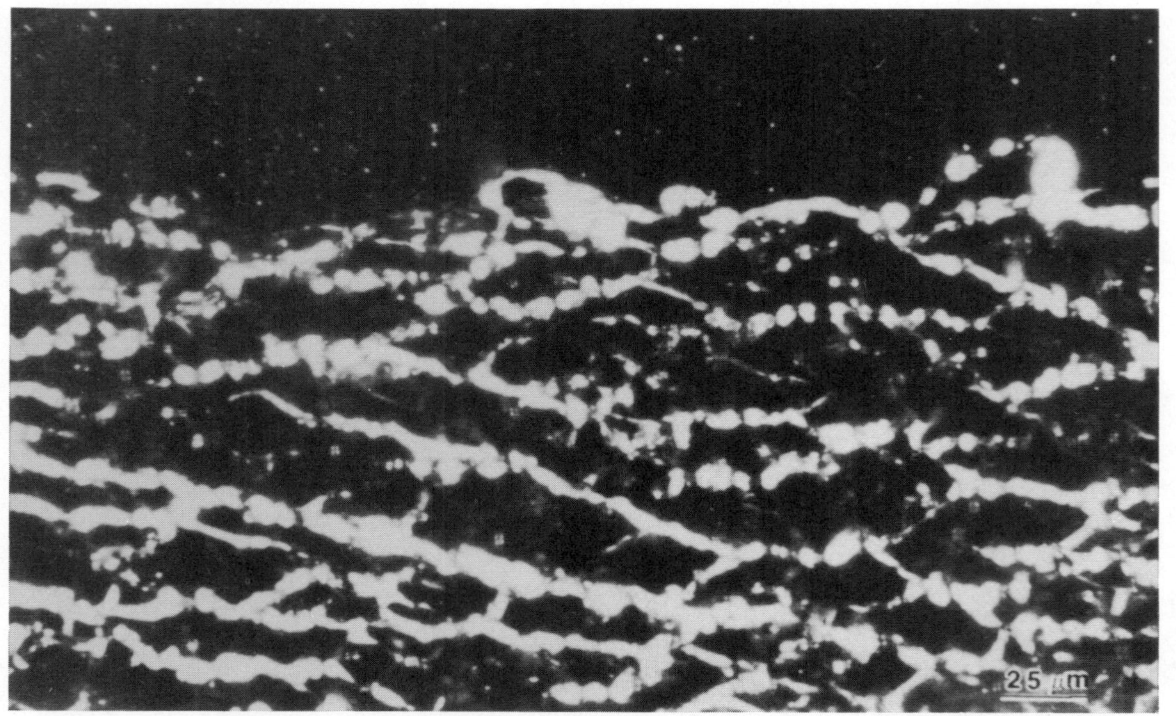

FIG. 7. Optical transmission micrograph, taken using cross-polarisers, of fracture region of a rubber-toughened epoxy polymer.[21]

shear deformation and, since there are many particles, there is considerably more plastic energy dissipation in the multiphase material than in the unmodified polymer. However, the fact that the shear deformations initiate at one particle but probably terminate at another tends to keep them localised. The enhanced plastic deformation that occurs in the multiphase material is clearly evident in Fig. 6 and the localised nature of the plastic shear bands, running at angles of approximately 45° to the principal tensile stress, i.e. in the direction of maximum shear stress, is illustrated in Fig. 7.

In considering the second it is necessary to recall that a triaxial stress state usually exists ahead of a crack.[2] This produces dilatation which, combined with the stresses that are induced in the particle by cooling after cure, causes failure and void formation either in the particle or at the particle/matrix interface. One feature of this dilatation and cavitation process is that, once formed, the voids grow and so dissipate energy. However, a more important aspect is that such voids can now greatly enhance shear-yielding.[2] Essentially, they lower the extent of triaxiality of the stress in the adjacent matrix. Since the yield stress increases with the degree of constraint, the cavitation of particles

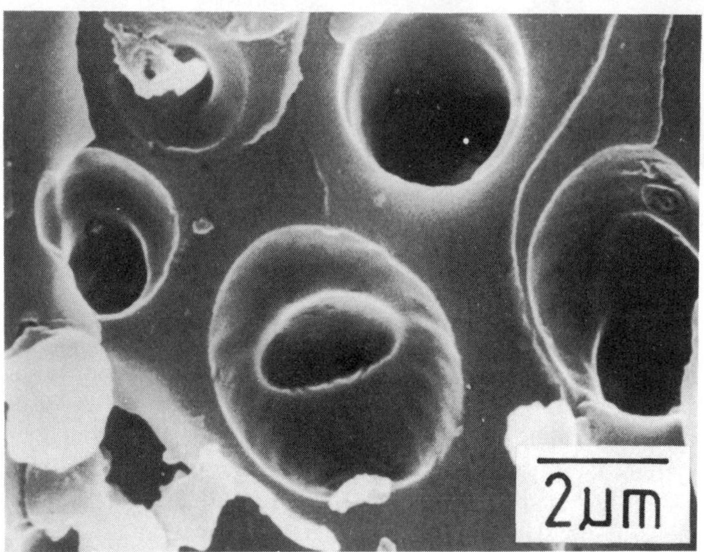

FIG. 8. Scanning electron micrograph of fracture surface of rubber-toughened epoxy polymer showing cavitated rubbery particles.

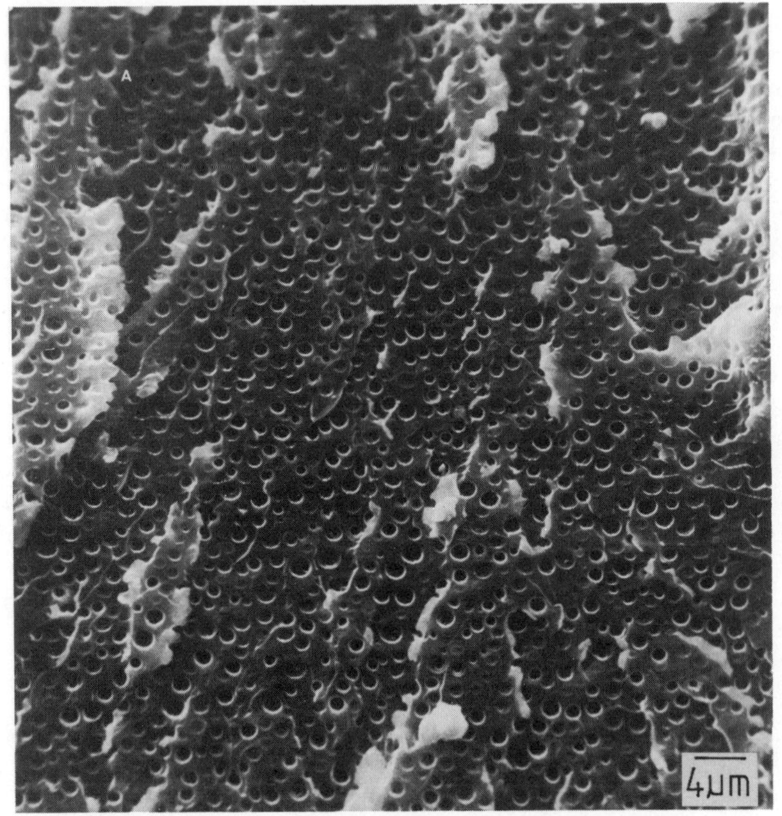

FIG. 9. Scanning electron micrographs of fracture surfaces of a rubber-toughened epoxy polymer.[33] (A) Before exposure to a solvent.

lowers the stress required for shear-yielding and so promotes even more extensive plastic shear deformations in the matrix. These mechanisms are responsible for the apparent 'holes' in the scanning electron fractographs and such features are particularly pronounced at high test temperatures when the rubber particles more readily cavitate[2] and plastic deformation of the matrix can occur more easily; see for example Fig. 6b. However, the higher magnification used in Fig. 8 does confirm that the 'holes' are in fact filled with rubber and that, in some instances, an internal void or tear can be clearly seen in the rubbery particle. Further, other experiments have revealed that such

FIG. 9—*contd.* (B) After exposure to a solvent which swells the rubbery phase.

'holes' become 'hillocks' after swelling the rubbery phase with a solvent (Fig. 9). Both pieces of evidence clearly indicate that the cavitated rubber is still in the 'holes' as a lining. Finally, it is this initiation and growth of voids in the rubbery particles which give rise to the stress-whitening that is often observed ahead of the crack tip and on the fracture surfaces.

Having discussed the mechanisms of toughening, the microstructural aspects which influence these mechanisms, and so control the measured fracture energy and other mechanical properties, will be reviewed.

3.3. Microstructural Features of the Matrix Phase

3.3.1. Introduction

The toughening mechanism outlined above highlights the role of the inherent ductility of the matrix in influencing the toughness of the multiphase thermosetting polymer. For example, a decrease in the stress needed for localised shear-yielding should obviously assist in increasing the toughness, all other important microstructural features being kept unchanged. However, an increase in the inherent ductility can usually only be achieved at the expense of other important properties.

3.3.2. Molar Mass Between Crosslinks and T_g of the Matrix Phase

Pearson and Yee[39] have studied the toughness of a series of CTBN epoxy polymers cured with 4,4′-diaminodiphenyl sulphone where the molar mass, M_c, between crosslinks was varied by using epoxy resins of different epoxy equivalent masses: the higher the epoxy equivalent mass, the higher is the value of M_c. Their results are shown in Table 3 and, as may be seen, increasing M_c produces a small increase in the value of the fracture energy, G_{Ic}, for the unmodified polymers but a dramatic increase in that for the rubber-toughened polymers. These results may be ascribed to an increase in the inherent ductility of the matrix as M_c increases, leading to greater plastic energy dissipation and an increase in G_{Ic}. A greater effect is observed for the rubber-toughened polymer since the rubbery particles initiate a large number of such energy-dissipating shear zones, as discussed above.

TABLE 3
EFFECT OF MOLAR MASS, M_c, BETWEEN CROSSLINKS ON THE FRACTURE ENERGY, G_{Ic}[a]

Epoxy equivalent mass	G_{Ic} *(kJ/m²)*	
	Unmodified polymer	*Rubber-toughened polymer*
172–176	0·16	0·22
475–575	0·20	2·41
1600–2000	0·33	11·5

[a] Resin, epoxy; hardener, 4,4′-diaminodiphenyl sulphone; rubber, CTBN.

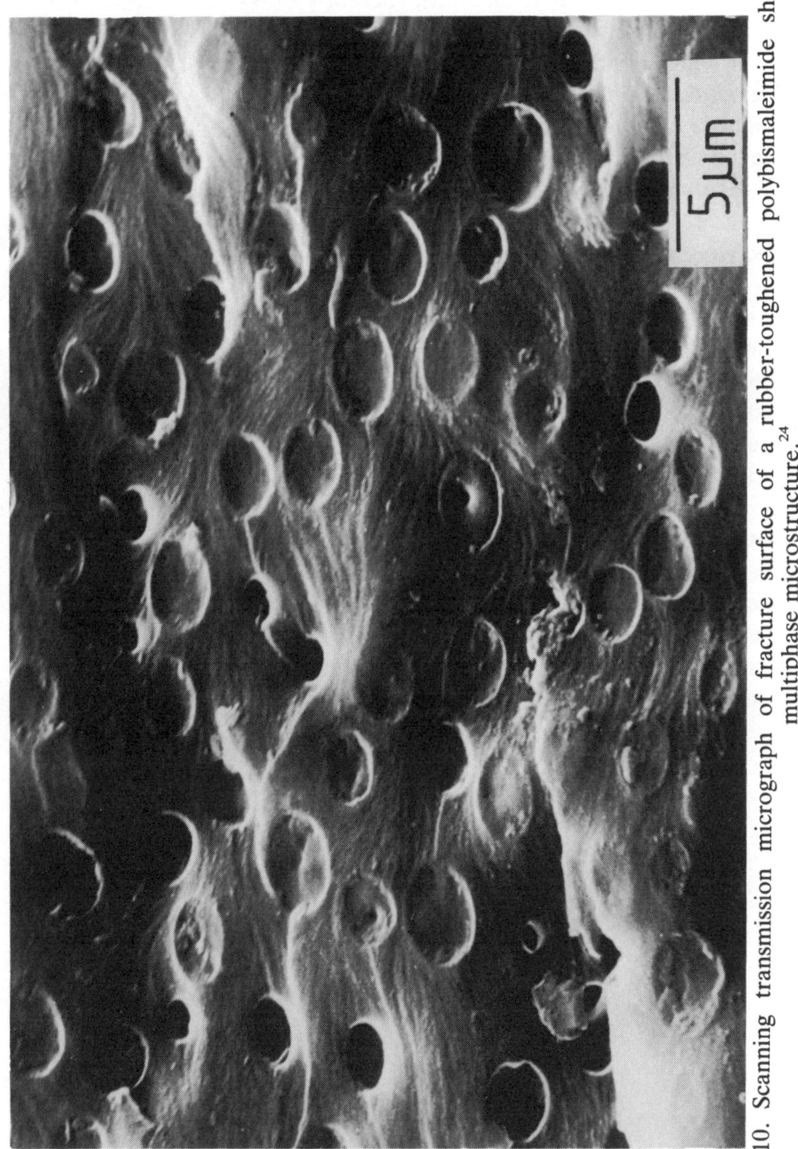

FIG. 10. Scanning transmission micrograph of fracture surface of a rubber-toughened polybismaleimide showing multiphase microstructure.[24]

The glass transition, T_g, of the matrix phase is obviously affected by the value of M_c, but it is also influenced by other parameters such as the degree of interchain attractions. As might be expected, those factors which lead to an increase in matrix ductility, e.g. high M_c values and low interchain attractive forces, also lead to low values of T_g (matrix). Thus, increased ductility and toughness are often only achieved at the expense of elevated-temperature properties.

A corollary to the above argument is that thermosetting polymers which are extremely tightly crosslinked and possess high interchain attractive forces, and hence have a high T_g, are difficult to toughen to a high absolute level. This is evident from work on rubber-toughened epoxies,[42,43] but even more dramatically illustrated by recent studies[24,25] on rubber-toughened bismaleimide polymers. In this work addition-polymerised bismaleimides, possessing glass transition temperatures of over 300°C, have been toughened using CTBN liquid rubbers. The typical microstructure obtained is shown in Fig. 10 and a two-phase structure is clearly present, with dispersed particles about 1–2·5 μm in diameter. As with other thermosets, the evidence suggests that the dispersed phase is a copolymer of CTBN and bismaleimide resin. Increasing the added rubber concentration produces a decrease in modulus (Fig. 11), as expected, but even with a large addition of rubber the modulus is still over 3 GPa. The fracture energy, G_{Ic}, is shown as a function of rubber concentration added to the bismaleimide in Fig. 12, and the fracture energy steadily increases as the

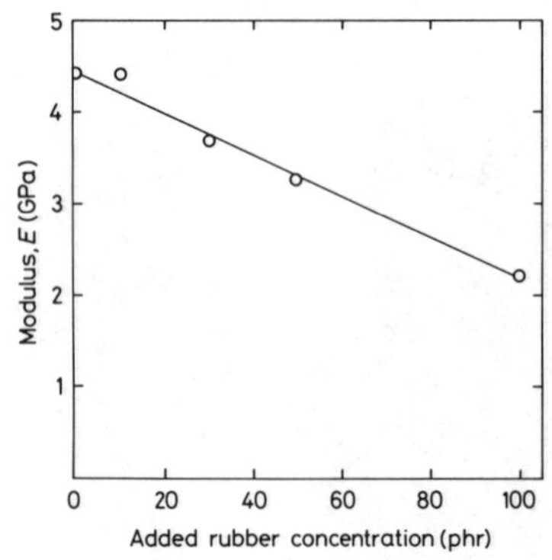

FIG. 11. Effect of rubber addition on the modulus (at 23°C) of a rubber-toughened polybismaleimide.[24]

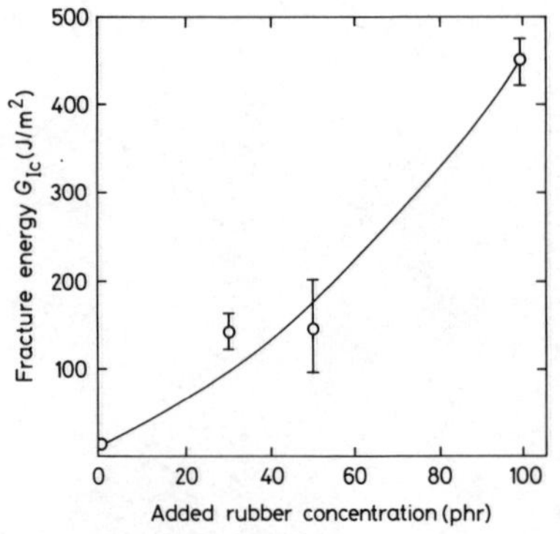

FIG. 12. Effect of rubber addition on the fracture energy, G_{Ic}, of a rubber-toughened polybismaleimide.[24]

rubber concentration increases. At a rubber concentration of 50 phr the value of G_{Ic} is about an order of magnitude higher than for the unmodified polybismaleimide, and the T_g(bismaleimide matrix) has not been significantly decreased. Nevertheless, the tightly crosslinked structure of the polybismaleimide leads to relatively low *absolute* values of G_{Ic} for both the unmodified and rubber-toughened polymers. Comparison of the values of G_{Ic} with those for a rubber-toughened epoxy shown in Table 1, with a T_g(epoxy matrix) of 100°C, clearly reveals the much lower values for the high-T_g polyimides.

The need to obtain very tough thermosets but without sacrificing the very high glass transition temperatures that they may possess is self-evident, and presents an exciting and rewarding challenge.

3.3.3. *Test Temperature and Rate*

Although they are obviously not microstructural features, it is convenient to consider the effects of test temperature and rate at this point since their main influence appears to be on the inherent ductility of the matrix. This is illustrated in Fig. 13, which shows the compressive yield stress as a function of test temperature for different

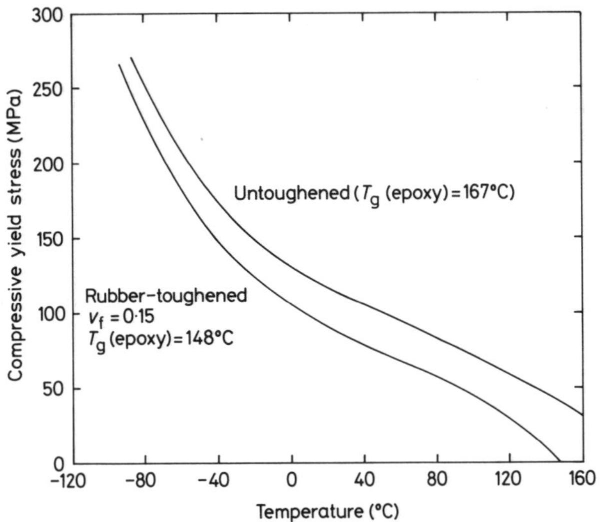

FIG. 13. Uniaxial compressive yield stress of an unmodified epoxy and rubber-toughened epoxy polymers as a function of test temperature.[42] Resin: DGEBA. Hardener: TMAB. Rubber: CTBN.

epoxy polymers. (The compressive yield stress is measured because the materials fracture before yielding in a standard uniaxial tensile experiment.) At any particular temperature, the presence of the rubbery phase decreases the yield stress a little, but the dominating parameter is the test temperature: the higher the test temperature, the lower is the yield stress and the easier becomes the generation of plastic deformation. This is reflected in the increase of fracture energy, G_{Ic}, with test temperature as shown in Fig. 14.

Various studies[2,33,45,46] have also revealed that, as is often observed, decreasing the rate of test is equivalent to increasing the temperature. Indeed, Hunston and co-workers[45–47] have found that rate and temperature effects may be interrelated by constructing a master curve of G_{Ic} versus reduced time-of-test, t_f/a_T, *below* the T_g of the epoxy matrix. The values of the time–temperature shift factor, a_T, were ascertained experimentally by superimposing yield, stress relaxation and dynamic mechanical relaxation data. Thus, the effects of rate and temperature on the value of G_{Ic} may be simply combined, as shown in Fig. 15. This approach not only enables different polymers to be objectively compared using an intrinsic reference point, but also permits the value of G_{Ic} to be predicted for any given test temperature/rate combination.

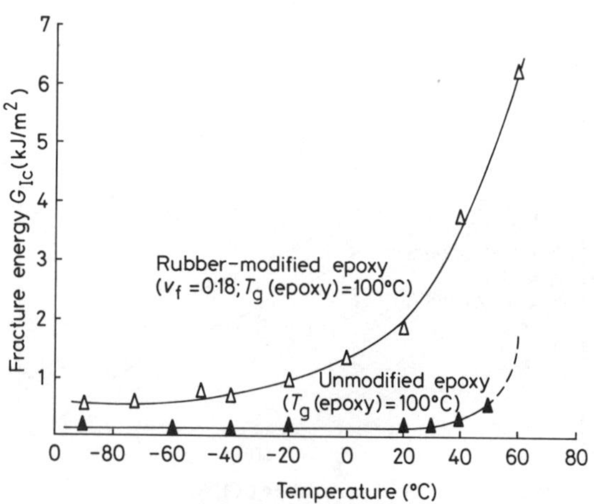

FIG. 14. Fracture energy, G_{Ic}, as a function of test temperature for an unmodified and a rubber-toughened epoxy polymer. Resin: DGEBA. Hardener: piperidine. Rubber: CTBN.

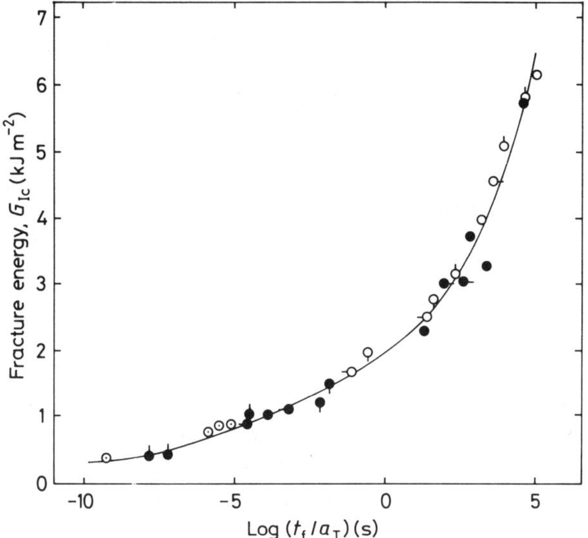

FIG. 15. Fracture energy, G_{Ic}, for an unmodified and a rubber-toughened epoxy polymer as a function of $\log(t_f/a_T)$. The shift factor, a_T, is determined from yield, stress relaxation and dynamic mechanical relaxation data.[47]

○, 60°C; ⚲, 37°C; ●—, 15°C; ●, −40°C;

○, 50°C; —○, 30°C; ●, 0°C; ⊙, −60°C;

○—, 40°C; ●, 23°C; —●, 20°C.

Reference temperature, $T_g - 80°C = 20°C$.

3.4. Microstructural Features of the Rubbery Phase

3.4.1. *Introduction*

Although the microstructural features of the rubbery dispersed phase have a major effect on the toughness of the multiphase polymer, there have been few definitive studies of the influence of such features as volume fraction, particle size, etc. In many studies crucial features like the value of v_f have not been directly ascertained, or several features have been changed simultaneously. However, an attempt is made below to establish correlations between mechanical behaviour and microstructural features of the rubbery phase.

3.4.2. Volume Fraction

The toughness of a multiphase thermoset generally increases as the volume fraction, v_f, of dispersed rubbery phase is increased, but the modulus and yield strength will usually decrease slightly.[14,22,26,31,33,38,46]

Bucknall and Yoshii[26] found a linear relation between G_{Ic} and v_f for a series of different resin/hardener systems using a CTBN rubber, as shown in Fig. 16. However, Kunz *et al.*[14] found, for both ATBN- and CTBN-modified epoxies, that once a v_f of about 0·1 had been achieved in the polymer then further increases in volume fraction resulted in only comparatively minor increases in the value of G_{Ic}. Whilst this apparent disagreement remains to be resolved, recent work[42,46,47] does indicate that the apparently unique relation for all epoxy resin/hardener/rubber systems shown in Fig. 16 may be somewhat fortuitous. It is clearly revealed that the exact dependence of G_{Ic} upon v_f is a function of the test temperature and rate, as illustrated in Fig. 17. This strong dependence of the fracture energy upon temperature largely arises from the inherent ductility of the matrix being temperature-dependent, as discussed in Section 3.3.3. To

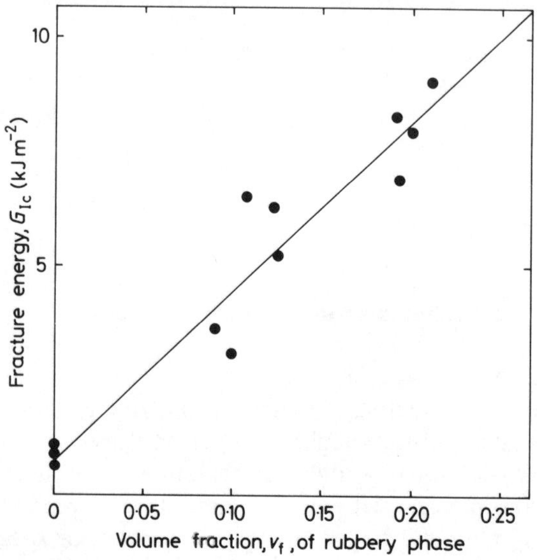

FIG. 16. Relationship between fracture energy, G_{Ic}, and volume fraction, v_f, of rubbery phase for CTBN-toughened epoxy polymers.[26] Test temperature, 23°C.

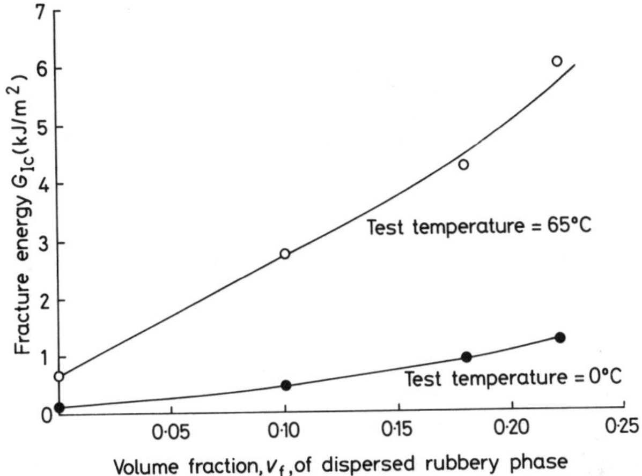

FIG. 17. Relationship between fracture energy, G_{Ic}, and volume fraction, v_f, of rubbery phase for CTBN-toughened epoxy polymers as a function of test temperature. T_g(epoxy matrix) = 100°C ± 2°C for all polymers; $t_f = 1$ s.[47]

allow for such matrix effects, the values of G_{Ic} for different toughened thermosets having different values of T_g(matrix phase) need to be compared with reference to an intrinsic reference temperature (e.g. T_g or $T_g - X°$), rather than just compared at some arbitrary test temperature. For some materials the arbitrarily selected test temperature might be well below T_g(matrix phase) whilst for others it might be very close to T_g(matrix phase).

As discussed previously, it is usually found that the maximum value of v_f that can be achieved is about 0·2–0·3, and attempts to produce higher fractions result in phase inversion and a loss of mechanical properties.

To summarise, the volume fraction of dispersed rubbery phase is generally considered to have a major influence on the toughness of the multiphase thermoset. However, the relationship between v_f and G_{Ic} is not yet conclusively established.

3.4.3. *Particle Size*

Several studies[35,37] have suggested that particle size may influence the extent and type of toughening mechanism, and so affect the measured toughness. For example, Sultan and McGarry[35] proposed that large particles (0·1–1 μm in diameter) cause crazing and are five times more effective in toughening the thermoset than small particles (~0·01 μm) which initiate shear-yielding.

However, the evidence for crazing was not convincing and this mechanism has now been discounted in such thermosets. Furthermore, in these studies the value of v_f was not determined: in order to elucidate the effect of particle size then the value of v_f must be measured, and ideally held constant.

In the few cases[26,42,47] where this has been achieved, the particle size does not appear to have a major influence on the mechanical properties. However, the average particle diameter was only varied between about 0·5 and 5 μm and no directly comparable materials have yet been prepared which contain solely very small particles (i.e. <0·1 μm). Other workers[14] have prepared toughened epoxy polymers where small particles (<0·1 μm) were present, but since these had a bimodal distribution of particle sizes their results are discussed below.

3.4.4. Particle Size Distribution

Again, the evidence concerning the influence of particle size distribution upon the measured toughness is somewhat contradictory. The above-mentioned workers[14] prepared a series of toughened-epoxy polymers which contained a bimodal distribution of very small (<0·1 μm) and larger (0·2–1 μm) particles. However, they had to contend with the problem that both the distribution and volume fraction of rubber were changing together: as the distribution was changing towards a relatively greater number of larger particles the value of v_f was also increasing. Now they found that the value of G_{Ic} did not increase with increasing value of v_f as much as had been expected (Section 3.4.2.). These observations led to the suggestions that the larger particles were not as effective as the smaller ones in improving the toughness and that the distribution of rubbery particle sizes is a more critical parameter than the overall volume fraction of particles.

However, the work of Hunston and Kinloch[47] indicates that the effect of particle size distribution is more complex. They have found that, at a fixed value of v_f, a bimodal distribution of particle sizes (centred upon maxima of 0·1 μm and 1·3 μm) can indeed result in higher values of G_{Ic} compared to a unimodal distribution (mean diameter 1·2 μm). However, this enhancement in the case of the bimodal distribution is only observed over a certain range of test temperatures and rates, and matrix effects may also play a role.

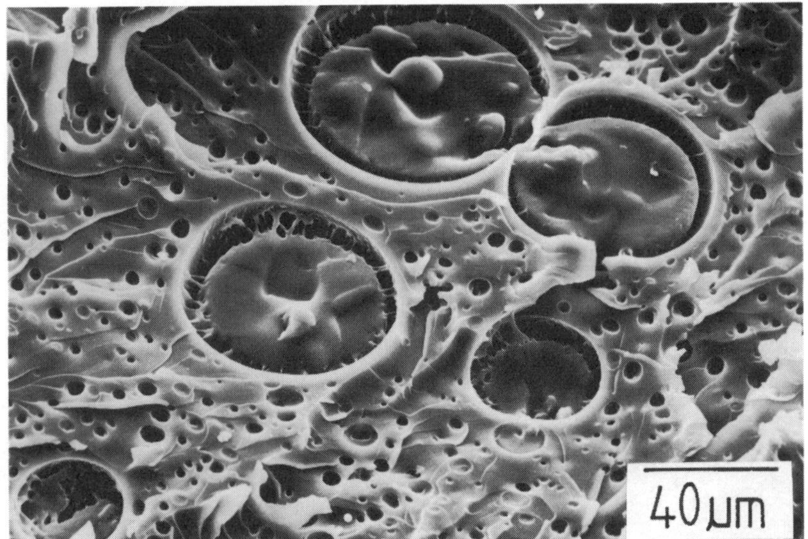

FIG. 18. Scanning electron micrograph of a fracture surface of a rubber-modified epoxy polymer where poor adhesion was especially engineered across the particle/matrix interface.[42]

3.4.5. *Interfacial Adhesion*

Nearly all of the studies reported in the literature have been concerned with well bonded rubbery particles, as a consequence of the chemical reactivity of the rubber. But poor bonding across the particle/matrix interface has been specifically engineered, by using an unreactive liquid rubber, as is evident from the fracture surface shown in Fig. 18. As might be expected, this multiphase epoxy polymer showed virtually no increase in toughness compared with the unmodified material.[42] Thus, the intrinsic adhesion across the particle/matrix interface needs to be sufficiently high to prevent premature particle debonding. Strong interfacial chemical bonds are usually present, of course, when a reactive rubber is used (Section 2.2.1).

3.4.6. *Morphology of Dispersed Phase*

In only a few studies[21,26] has the presence of a phase-separated morphology in the rubbery particles been reported; such a microstructure of resin inclusions in the particles has been illustrated in Fig. 3 above. However, there is no evidence to indicate whether a complex morphology of the particle is desirable or not—but at the very least it should ensure that a relatively high value of v_f is attained.

3.4.7. Glass Transition Temperature of the Dispersed Phase
As discussed earlier (Section 3.2.2) it is not necessary for the particles to be truly rubbery for a multiphase thermoset to be appreciably tougher than the unmodified material. Figures 14 and 15 show quite clearly that the multiphase epoxy polymer is significantly tougher even when the test temperature is well below T_g(rubbery phase), i.e. below about −55°C. Nevertheless, the most substantial increases in G_{Ic} are observed at temperatures above T_g(rubbery phase). The detailed reasons for this have been discussed in Sections 3.2 and 3.3.3.

4. CONCLUDING REMARKS

The development of multiphase thermosetting polymers based upon the inclusion of a second rubbery phase has led to major improvements in the toughness of highly crosslinked thermosets. However, it is only very recently that quantitative studies to ascertain the relationships between the chemistry, resulting microstructure and observed mechanical properties have been undertaken. Until such data are firmly established, it is difficult to define structure/property relationships in any detailed quantitative manner and this obviously inhibits attempts to define the microstructure needed to attain specific fracture properties for a given application. Nevertheless, the necessary data will undoubtedly be forthcoming and so enable the science and technology of rubber-toughened thermosets to advance even further, thereby ensuring that such materials will increasingly be employed in demanding structural engineering applications.

ACKNOWLEDGEMENTS

The author would like to thank present and former colleagues for many stimulating discussions on rubber-toughened thermosetting polymers, especially Dr D. L. Hunston, Dr S. J. Shaw, Professor J. K. Gillham and Professor J. G. Williams.

REFERENCES

1. KINLOCH, A. J., *J. Mater. Sci.*, **15** (1980) 241.
2. KINLOCH, A. J. and YOUNG, R. J., *Fracture Behaviour of Polymers*, Applied Science, London, 1983.
3. RIEW, C. K. and SMITH, R. W., *J. Polym. Sci., A1*, **9** (1971) 2739.
4. KATO, K., *J. Electron Microsc.*, **14** (1965) 220.
5. FLORY, P. J., *Principles of Polymer Chemistry*, Cornell University Press, Cornell, 1953.
6. BUCKNALL, C. B., *Toughened Plastics*, Applied Science, London, 1977.
7. KONINGSVELD, R., CHERMIN, H. A. G. and GORDON, M., *Proc. Roy. Soc.*, **A319** (1970) 331.
8. MCMASTER, L. P. (Ed.), *Aspects of Liquid–Liquid Phase Transition Phenomena in Multicomponent Polymeric Systems*, Amer. Chem. Soc. Adv. Chem. Series, **142** (1975).
9. KWEI, T. K. and WANG, T. T., in *Polymer Blends*, Vol. 1, Ed. D. R. Paul and S. Newman, Academic Press, New York, 1978, p. 141.
10. ROWE, E. H., SIEBERT, A. R. and DRAKE, R. S., *Modern Plastics*, **49** (Aug. 1970) 110.
11. DRAKE, R. and SIEBERT, A., *SAMPE Quart.*, **6**(4) (1975) 11.
12. SAYRE, J. A., ASSINK, R. A. and LAGASSE, R. R., *Polymer*, **22** (1981) 87.
13. CHAN, L. C., GILLHAM, J. C., KINLOCH, A. J. and Shaw, S. J., in *Rubber-Modified Thermoset Resins*, Ed. C. K. Riew and J. C. Gillham, Amer. Chem. Soc. Adv. Chem. Series, **208** (1984) 235.
14. KUNZ, S. C., SAYRE, J. A. and ASSINK, R. A., *Polymer*, **23** (1982) 1897.
15. RIEW, C. K., ROWE, E. H. and SIEBERT, A. R., in *Toughness and Brittleness of Plastics*, Ed. R. Deanin and A. Crugnola, Amer. Chem. Soc. Adv. Chem. Series, **154** (1976) 326.
16. NOSHAY, A. and ROBESON, L. M., *Amer. Chem. Soc. Polym. Prepr.*, **15**(1) (1974) 613.
17. MCGARRY, F. J., ROWE, E. H. and RIEW, C. K., *32nd Ann. Tech. Conf., 1977*, Reinf. Plastics/ Composite Inst., SPE, USA, Section 16-C, p. 1.
18. CROSBIE, G. A. and PHILLIPS, M. G., *Proc. 5th Internat. Conf. Deformation Yield and Fracture of Polymers*, Plastics and Rubber Institute, London, 1977.
19. MEEKS, A. C., *Polymer*, **15** (1974) 675.
20. BASCOM, W. D., TING, R. Y., MOULTON, R. J., RIEW, C. K. and SIEBERT, A. R., *J. Mater. Sci.*, **16** (1981) 2657.
21. YEE, A. F. and PEARSON, R. A., *Toughening Mechanisms in Elastomer Modified Epoxy Resins*, NASA Contractor Report 3718, 1983.
22. BASCOM, W. D., COTTINGTON, R. L., JONES, R. L. and PEYSER, P., *J. Appl. Polymer Sci.*, **19** (1975) 2545.
23. ST. CLAIR, A. K. and ST. CLAIR, T. L., *Int. J. Adhesion Adhesives*, **2** (1981) 249.
24. KINLOCH, A. J., SHAW, S. J. and TOD, D. A., in *Rubber-Modified Thermoset Resins*, Ed. C. K. Riew and J. K. Gillham, Amer. Chem. Soc. Adv. Chem. Series, **208** (1984) 101.

25. SHAW, S. J. and KINLOCH, A. J., *Int. J. Adhesion Adhesives,* **5** (1985) 123.
26. BUCKNALL, C. B. and YOSHII, T., *Br. Polymer J.,* **10** (1978) 53.
27. SOLDATOS, A. C. and BURHANS, A. S., *Multicomponent polymer systems,* Ed. R. F. Gould, *Amer. Chem. Soc. Adv. Chem. Series,* **99** (1971) 53.
28. LEE, B. L., LIZAK, C. M., RIEW, C. K. and MOULTON, R. J., *12 Nat. SAMPE Conf.,* (Oct. 1966), 1116.
29. VISCONTI, S. and MARCHESSAULT, R. H., *Macromolecules,* **7** (1974) 913.
30. WANG, T. T. and ZUPKO, H. M., *J. Appl. Polym. Sci.,* **26** (1981) 2391.
31. KUNZ-DOUGLASS, S., BEAUMONT, P. W. R. and ASHBY, M. F., *J. Mater. Sci.,* **15** (1980) 1109.
32. KUNZ, S. and BEAUMONT, P. W. R., *J. Mater. Sci.,* **16** (1981) 3141.
33. KINLOCH, A. J., SHAW, S. J., TOD, D. A. and HUNSTON, D. L., *Polymer,* **24** (1983) 1341.
34. HUNSTON, D. L., BITNER, J. L., RUSHFORD, J. L., OROSHINK, J. and ROSE, W. S., *Elast. Plast.,* **12** (1980) 133.
35. SULTAN, J. N. and MCGARRY, F. J., *Polym. Eng. Sci.,* **13** (1973) 29.
36. KINLOCH, A. J. and SHAW, S. J., *J. Adhesion,* **12** (1981) 59.
37. BASCOM, W. D., TING, R. Y., MOULTON, R. J., RIEW, C. K. and SIEBERT, A. R., *J. Mater. Sci.* **16** (1981) 2657.
38. KINLOCH, A. J., SHAW, S. J. and HUNSTON, D. L., *Polymer,* **24** (1983) 1355.
39. PEARSON, R. A. and YEE, A. F., *Polym. Mater. Sci. Engng. Prepr.,* **49** (1983) 316.
40. GOODIER, J. N., *Trans. Amer. Soc. Mech. Eng.,* **55** (1933) 39.
41. BROUTMAN, L. J. and PANIZZA, G., *Int. J. Polym. Mater.,* **1** (1971) 95.
42. CHAN, L. C., GILLHAM, J. K., KINLOCH, A. J. and SHAW, S. J., in *Rubber-Modified Thermoset Resins,* Ed. C. K. Riew and J. K. Gillham, Amer. Chem. Soc. Adv. Chem. Series, **208** (1984) 261.
43. ARONHIME, M. T., PENG, X. and GILLHAM, J. K., *Polym. Mater. Sci. Engng.,* **51** (1984) 420.
44. KINLOCH, A. J. and WILLIAMS, J. G., *J. Mater. Sci.,* **15** (1980) 987.
45. BITNER, J. R., RUSHFORD, J. L., ROSE, W. S., HUNSTON, D. L. and RIEW, C. K., *J. Adhesion,* **13** (1982) 3.
46. HUNSTON, D. L., KINLOCH, A. J., SHAW, S. J. and WANG, S. S., in *Adhesive Joints,* Ed. K. L. Mittal, Plenum Press, New York, 1984, p. 789.
47. KINLOCH, A. J. and HUNSTON, D. L., *J. Mater. Sci. Lett.* (1986) in press.

6

Rigid-Particulate Reinforced Thermosetting Polymers

R. J. YOUNG*

Department of Materials, Queen Mary College, London, UK

NOTATION

A	constant
B	constant
D	interparticle separation
E_c	Young's modulus of composite
E_0	Young's modulus of matrix
E_p	Young's modulus of particles
G_{Ic}	fracture energy
K_{Ic}	fracture toughness
T	line energy
a_0	inherent flaw size
d	particle diameter
e_y	yield strain
m	$= E_p/E_0$
t_f	time to failure
w_m	weight of matrix
w_p	weight of resin
v_f	volume fraction
δ	crack opening displacement
ρ_m	density of matrix

* Present address: Department of Polymer Science and Technology, University of Manchester Institute of Science and Technology, Manchester, UK.

ρ_p density of particles
σ_f fracture strength of composite
σ_0 fracture strength of matrix

1. INTRODUCTION

Thermosetting polymers are finding increasing use in a wide range of applications such as adhesives, matrix materials in fibre-reinforced composites and potting compounds for electronic components. A large variety of different thermosetting polymers are employed, e.g. phenolics, unsaturated polyesters and polyimides, but by far the most widely studied materials are epoxy resins. Over recent years a great deal of effort has been expended in the investigation of methods of improving the mechanical properties of thermosetting polymers. One method is by rubber toughening, which has been described in detail in Chapter 5. Another technique that is widely employed is to incorporate a rigid second phase such as silica particles or glass spheres into the polymers. It is found, perhaps rather surprisingly, that the incorporation of these brittle materials into a brittle polymer can lead to significant degrees of toughening[1–12] by means of mechanisms such as crack front pinning. It is now recognised that the strength of the particle/matrix interface is of great importance in the action of the toughening effect and methods of modifying the interface are discussed in detail in the following text. The stiffness of the polymer is also increased by the presence of rigid particles, which has other advantages including the reduction of cost, exothermic temperature rise and coefficient of thermal expansion, and an increase in thermal conductivity.

An important recent development in this area is concerned with the preparation of hybrid-particulate composites[13–16] in which there are both rubbery and rigid particles. It is found that under certain circumstances synergism can be obtained, with several mechanisms of toughening operating and the composites having very impressive mechanical properties. The mechanical behaviour of the hybrid-particulate composites is discussed in detail in Sections 3–6 after the structural properties of the rigid-particle reinforced polymers have been described in Section 2. Most of the discussion will be concerned with epoxy resins reinforced with rigid particles such as silica or glass spheres although it is relevant to other systems with similar structures.

2. STRUCTURE

One of the problems with reviewing work upon structure/property relations in particle-filled polymers is that a wide range of thermosetting resins, types of particle of different size and particle surface treatments have been employed. This makes the determination of the effect of structure upon properties extremely difficult to evaluate unless the materials have been carefully characterised.

2.1. Fabrication

The moulding of particulate-reinforced polymers requires considerable care and attention as it is very easy to incorporate a third phase, air bubbles, into the material. These can act as a point of weakness in the structure, being initiation sites for cracks. The procedure for moulding a typical particulate-reinforced system such as an epoxy resin filled with glass beads has been described by several workers.[12,14] It involves taking the epoxy resin and carefully mixing in the curing agent and glass beads to minimise air entrapment. Since most particles have a wide distribution of sizes they are sometimes sieved into particular size ranges before moulding.[12] The glass particles should have been previously cleaned and dried in order to remove any contamination. The mixture is then heated to a higher temperature (typically >100°C) for a period of time depending upon the curing agent used. This enables any air bubbles to escape and allows the viscosity of the mixture to increase in order to prevent the glass from settling out during moulding. The mixture is again gently stirred and poured into preheated moulds and given the final heat treatment to cure the epoxy resin matrix fully. Typical moulds include ones consisting of parallel-sided plates of metal or glass which must be coated with release agent in order to remove the cast sheets. Variations on the procedure involve modification of the matrix polymer with the incorporation of third-phase rubber particles added as a liquid monomer along with the curing agent.[13–16] The surface of the glass beads is also often modified by the use of a silane-based bonding agent such as γ-glycidoxypropyltrimethoxysilane which is applied to the particles before moulding.[11,14] The use of bonding agents and primers is described in detail in Chapter 8.

2.2. Characterisation

The main factors that need to be established to characterise fully the structure of particulate-reinforced thermosetting polymers are the size and the volume fraction of the particles and their distribution throughout the specimens.

2.2.1. *Particle Size*

The particle size can be determined either before or after moulding. If it is done before moulding there are two relatively simple methods that can be employed. A representative sample of the particles can be sprinkled onto a scanning electron microscope (SEM) stub, which can then be examined in the microscope.[12] Measurements of the sizes of the particles can be performed either on the microscope screen or from micrographs, and as long as the magnifications are accurately calibrated a histogram of the particle sizes can be determined.[12] An alternative method which can be employed is to sieve the particles into fractions of different size ranges using calibrated sieves with differently sized meshes and to determine the weight of each fraction. It is often convenient to quote an average size, but the problem of translating the distribution into an average is the same as for polydisperse polymer samples where there is a range of molar mass.[17] It is clear that of the two methods outlined above the SEM method will produce a number-average particle size whereas sieving will produce a weight-average.

The particle size can be determined after moulding by sectioning, grinding and polishing the particulate-reinforced polymer, as shown in Fig. 1 for epoxy resins filled with either silica or spherical glass particles. It can be seen that there is a range of apparent particle sizes in the sections. However, care must be taken with the interpretation of such sections: even if all the particles were of exactly the same size, there would appear to be a range of diameters as they are sectioned at different positions relative to their centres. Various methods[18] can be used to convert the apparent range of diameters into an average value, but the interpretation is by no means trivial and assumptions often have to be made as to the form of the particle size distribution.

2.2.2. *Volume Fraction*

The volume fraction, v_f, of particles can be calculated from the weights of the ingredients making up the composites by means of the following equation:

$$v_f = w_p\rho_m/(w_p\rho_m + w_m\rho_p) \tag{1}$$

where w_p is the weight of particles, w_m the weight of thermosetting

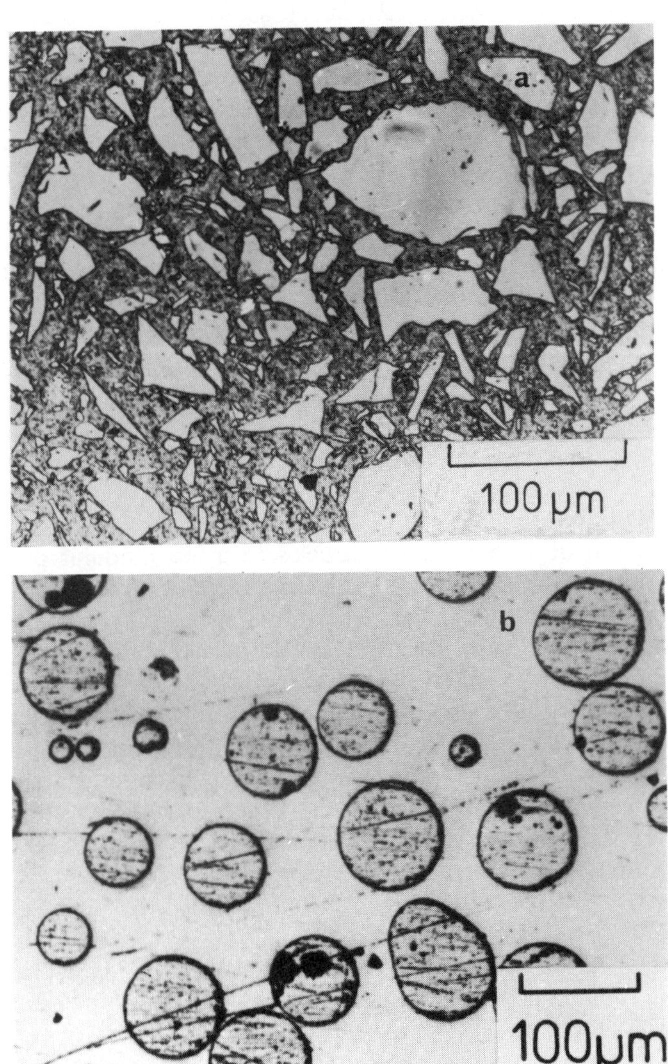

FIG. 1. Optical micrographs of polished sections of particulate-reinforced composites. (a) Silica particles in an epoxy resin ($v_f = 0{\cdot}47$, after ref. 8); (b) glass spheres in an epoxy resin ($v_f = 0{\cdot}30$, after ref. 10).

resin and hardener, ρ_p the density of the particles and ρ_m the density of the fully cured matrix polymer. It is wise also to measure the volume fraction of particles in the polymer after moulding by burning off the matrix polymer in a crucible and determining the weight loss. Along with the weighing data, determination of the density of the composite will also give an indication of the volume fraction of voids in the moulding; it should be kept to less than the order of 1% for accurate work.

As well as knowing the average volume fraction of particles it is essential to determine if there are any local variations in composition due to factors such as the settling of particles during moulding. This can be monitored by polishing random sections or from the burning and weighing of samples taken from different parts of the mouldings.

3. DEFORMATION

3.1. Elastic Deformation

The addition of high-modulus particles to a low-modulus polymeric matrix invariably leads to a considerable increase in modulus, E, of the composite over that of the matrix. There is a strong dependence of the modulus composites upon v_f, as shown in Fig. 2a for an epoxy resin reinforced with spherical glass particles. This behaviour has been well documented[8,14,19] and a variety of theories have been developed to explain the dependence of E upon v_f. Most theories[20–22] tend to produce two equations, one corresponding to uniform strain or displacement in the system and the other to uniform stress. For example, the simple 'rule of mixtures' equal-strain Voigt model predicts that the modulus of the composite, E_c, is given by

$$E_c = E_p v_f + E_0(1 - v_f) \tag{2}$$

where E_p is the modulus of the particles and E_0 is the modulus of the matrix. The uniform-stress Reuss model gives lower values of the composite modulus, as

$$1/E_c = v_f/E_p + (1 - v_f)/E_0 \tag{3}$$

However, these two models give rise to bounds which are too widely spaced to be of much use in rigid-particle reinforced systems. A more useful approach is to employ the equations of Ishai and Cohen,[20] who modelled the composite as cubic particles surrounded by a shell of

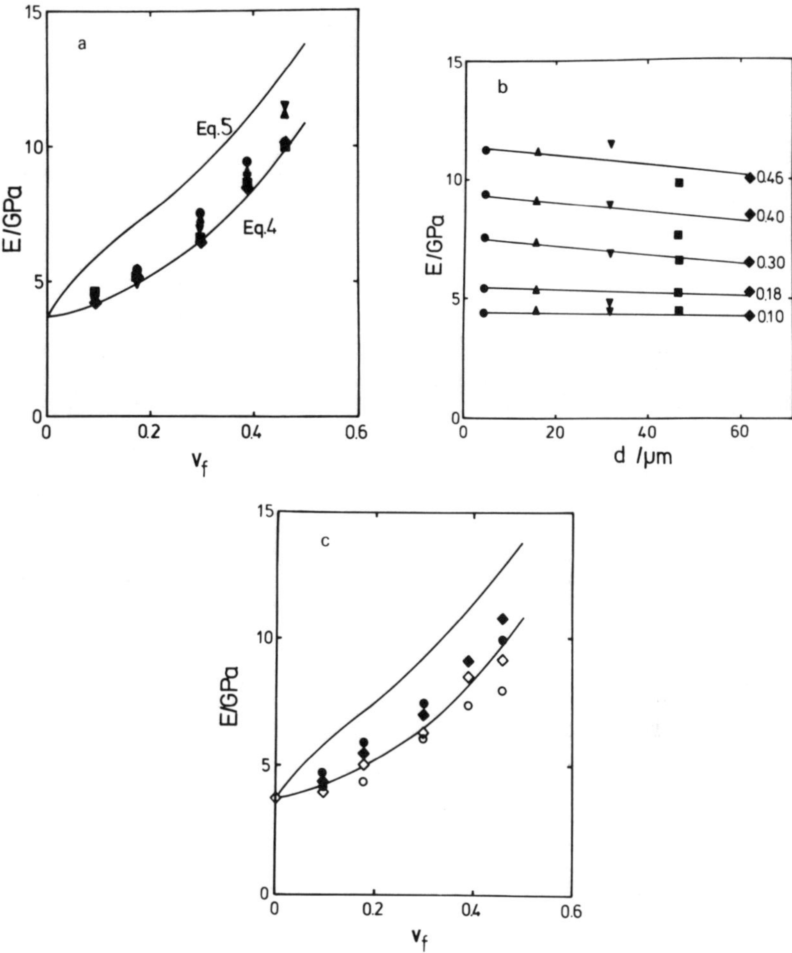

FIG. 2. Young's modulus (E) data for a glass-particle reinforced epoxy resin. (a) Dependence of E upon v_f. The different symbols refer to different average particle diameters (after ref. 10). (b) Dependence of E upon particle diameter, d, for the different volume fractions indicated. (after ref. 10). (c) Dependence of E upon v_f for different surface treatments. The open points are for particles treated with release agents and the closed ones for particles treated with bonding agents (after ref. 11).

matrix. If the cube boundary is subjected to a uniform displacement then the theory produces a lower bound given by

$$\frac{E_c}{E_0} = 1 + \frac{v_f}{[m/(m-1) - v_f^{1/3}]} \tag{4}$$

where E_c and E_0 are the composite and matrix moduli and $m = E_p/E_0$, the modular ratio. An alternative upper-bound equation is produced if the cube boundary is subjected to a uniform stress; this is

$$\frac{E_c}{E_0} = \frac{1 + (m-1)v_f^{2/3}}{1 + (m-1)(v_f^{2/3} - v_f)} \tag{5}$$

Equations (4) and (5) have been plotted in Fig. 2a assuming that $E_p = 73{\cdot}1$ GPa and taking E_0 as the value measured when $v_f = 0$. It can be seen that the experimental data fall closest to the lowest bound, corresponding to the case of uniform displacement at the particle/matrix boundary.

Most of the theories used to explain the dependence of the modulus of particulate composites upon the structure of the composite predict that for a given particle/matrix system the composite modulus depends only upon the volume fraction of particles and not upon the particle size. However, it can be seen from Fig. 2b that there is a clear decrease in the modulus of the composite with increasing particle diameter, *d*, for a given v_f. The relatively low value of E_c for large particle sizes has been attributed to a 'skin' effect. Lewis and Nielsen[23] suggested that when moduli are determined in flexure, as in the case of the data in Fig. 2b, then the properties of the surface are emphasised at the expense of the interior. This tends to be worse for the large particles, for which the surface 'skin' tends to be depleted of filler. This could explain why the modulus in Fig. 2b decreases with increasing particle size, and the error due to this effect can be removed by extrapolating to zero particle size.

It has also been found[11] that the value of modulus is also dependent upon the surface pretreatment of the particles, as can be seen in Fig. 2c where data are given for particles treated with either release agent or bonding agent. It is found that, for a given volume fraction, the composites containing particles having good adhesion with the epoxy matrix are significantly stiffer than composites containing particles treated with release agent.

3.2. Plastic Deformation

It is found that even such inherently brittle materials as epoxy resins are capable of undergoing a considerable amount of plastic deformation, especially when deformed in compression.[24,25] What is perhaps even more remarkable is that when such materials are toughened with brittle particles they can still exhibit considerable ductility in compression.[5,8,19] It can be seen from Fig. 3a that there is a steady increase in yield stress as the volume fraction of silica particles is increased in an anhydride-cured epoxy resin. Similar behaviour has been found for epoxy resins reinforced with spherical glass particles,[19] although in hybrid-particulate reinforced composites the increase in yield stress with volume fraction is only found when the particles are treated with a bonding agent.[19]

It is also found that the yield stress of the particulate-reinforced composites is strongly dependent upon temperature, as can be seen in Fig. 3b where a series of stress/strain curves are given for an amine-cured epoxy reinforced with spherical glass particles ($v_f = 0{\cdot}19$). The yield stress increases as the test temperature is reduced and it is found that such a system exhibits considerable ductility in compression even at −70°C.

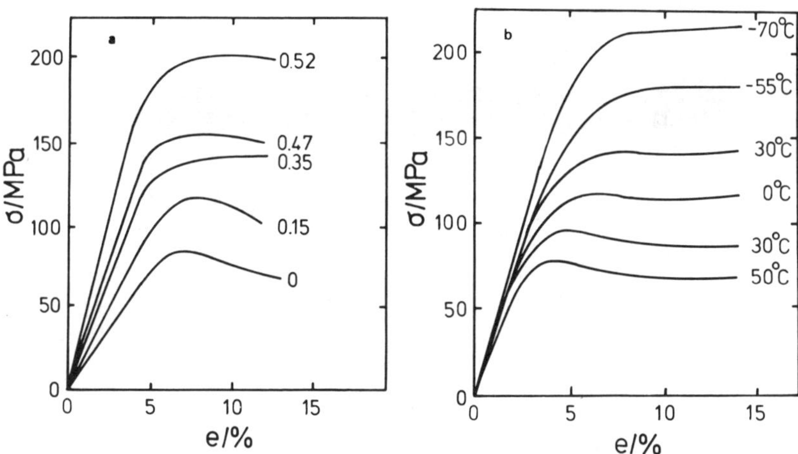

FIG. 3. Stress/strain curves for particle-reinforced epoxy resins determined in uniaxial compression. (a) Different volume fractions (indicated) of a silica-filled epoxy resin (after ref. 8). (b) Dependence of the behaviour of a glass-particle reinforced epoxy resin upon temperature (after ref. 19).

It is worthwhile at this stage to speculate upon the mechanisms of plastic deformation in particulate-reinforced thermosetting polymers, especially as it will be shown later that the fracture behaviour of these materials is controlled by their deformation characteristics. It is generally accepted that crazing[1] does not occur in such materials and it appears that the main deformation mechanism is shear yielding in the matrix polymer.[19] Investigations of polished sections taken from

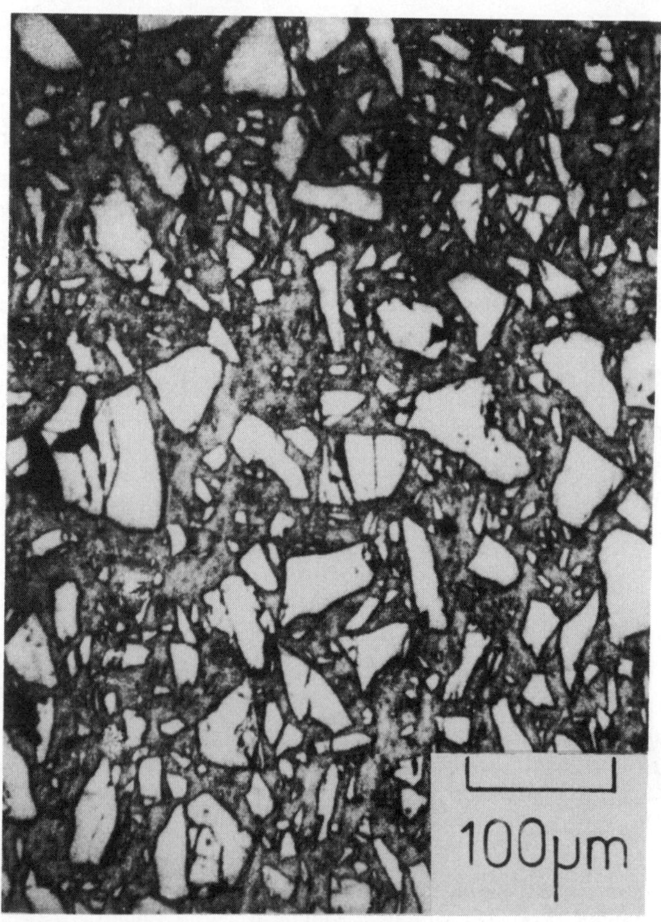

FIG. 4. Optical micrograph of a polished section of a silica-filled epoxy resin ($v_f = 0{\cdot}52$) after deformation in the vertical direction to a strain of 12% (after ref. 8).

specimens deformed in compression have shown that plastic deformation takes place by the matrix flowing around the rigid particles[8,19] (Fig. 4). The particles themselves are generally little affected by the flow of the matrix, which is not surprising since they have a considerably higher yield stress than the polymeric matrix. In the case of silica particles with irregular shapes, a small amount of particle fracture has been observed,[8] as can be seen from Fig. 4. This does not generally occur with spherical particles because of the absence of stress concentrations, but particle debonding has been found to take place[19] when pretreatments to improve surface bonding are not used.

The deformation mechanisms in particulate composites have been investigated recently by Young *et al.*[19] They determined the dependence of the Young's modulus and yield stress on temperature and volume fraction for several particulate-reinforced systems and showed that the data could be fitted to several theories[26–29] developed originally to explain shear yielding in amorphous polymers. Earlier, Yamini and Young[25] had shown that plastic deformation in thermosetting polymers such as epoxy resins took place by essentially the same mechanism as in amorphous thermoplastics such as polycarbonate and polystryene. The molecular parameters derived from such theories were found to be very similar in both thermoplastics and thermosets. The latest investigation[19] has shown that the approach can be extended to particulate-filled systems and parameters are produced which can be related to the molecular structure of the thermosetting matrix. The presence of the particles leads to an increase in the stiffness and yield stress but does not affect the underlying mechanism of plastic deformation which is shear yielding in the matrix polymer.

4. FRACTURE

The incorporation of brittle particles into thermosetting polymeric matrices to increase the toughness of the polymer will also have an effect upon the strength of the material. Unfortunately, although the toughness of the material can be significantly improved, this is often accompanied by a substantial reduction in strength.

4.1. Fracture Strength

Several workers[9,11,12,30–33] have been concerned with the effect of the incorporation of brittle particles upon the strength of thermosetting polymers. It is generally found that in the absence of any special particle surface pretreatments the strength of composite (measured for example in flexure) falls as the volume fraction of particles increases. If the particles are treated so that there is no adhesion with the matrix, then it is found that the fracture strength of the composite, σ_f, is given by an equation of the form

$$\sigma_f = \sigma_0(1 - 1{\cdot}21 v_f^{2/3}) \tag{6}$$

where σ_0 is the strength of the pure matrix. This equation can be derived using simple geometrical considerations[31] assuming that the particles sit unbonded in holes. It is plotted with the experimental data of Sahu and Broutman[30] in Fig. 5, in which data are also plotted for particles treated with a bonding agent to produce good adhesion where the strength of the composite is the same as that of the matrix and is independent of v_f (horizontal line). The two lines give the two bounds of the behaviour for a range of surface adhesion properties.

This approach is somewhat simplistic and takes no account of the particle size. For example, Hojo and co-workers[33] found that for a given volume fraction there was a decrease in the strength of silica-filled epoxy resins as the average size of the particles increased. They showed that the material obeyed a Petch-type[34] relationship of the form

$$\sigma_f = A + Bd^{-1/2} \tag{7}$$

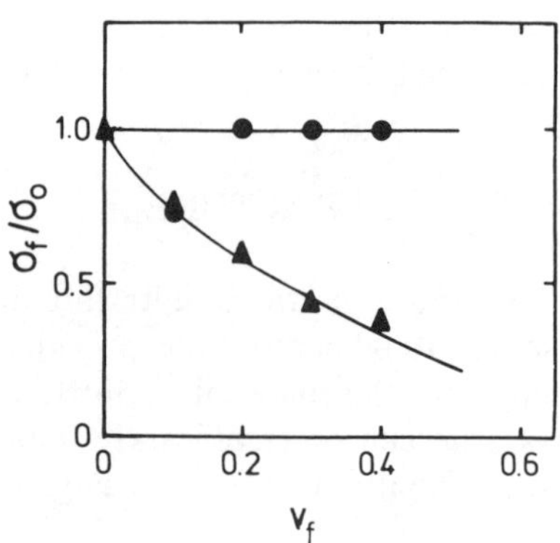

FIG. 5. Dependence of the tensile strength of a glass-filled epoxy resin upon v_f. The data are from the work of Sahu and Broutman.[30] The lower line corresponds to eqn (6) (after ref. 31).

where A and B are constants. This type of approach is consistent with the Griffith theory of fracture[1,35] where the particles or particle/matrix boundaries act as 'inherent flaws' in the structure giving rise to the $d^{-1/2}$ dependence.

Most practical systems have both a range of particle sizes and intermediate levels of adhesion and so their values of fracture strength fall between those of the two bounds in Fig. 5. Because of this, Leidner and Woodhams[32] developed a theory to account for both particle size and the level of particle/matrix adhesion. They went on successfully to apply the theory to their data on a polyester resin reinforced with different sizes of spherical glass particles with different levels of surface adhesion.

4.2. Impact, Fatigue and Time-Dependent Failure

It is found[7,9,33] that the dependence of the impact behaviour of particulate-filled thermosetting polymers upon structure is similar to that of the fracture strength, determined for example in flexure. The impact strengths of the composites are normally less than those of the pure matrix and it is generally found[9] that strong particle/matrix adhesion produces the highest impact strengths; this is demonstrated in Fig. 6a although the results of such experiments are naturally prone to high levels of scatter. These observations imply that the failure of these materials in impact depends upon the initiation of flaws in the structure rather than the propagation of cracks.

There have been few reports of the failure of thermosetting polymers subjected to cyclic fatigue deformation, and even fewer of fatigue in such materials reinforced with rigid particles. Brown[9] has presented some S–N data (maximum applied stress or strain versus the number of cycles to failure[1] for the fatigue of a polyester resin reinforced with various types of glass particles (Fig. 6b). The ranking of the materials in fatigue seems to be similar to that in normal fracture tests where the filled polymers are inferior to the pure resin and fatigue resistance is improved when bonding agents are used with the particles. This again implies that, as with impact, fatigue failure depends upon the initiation rather than the propagation of cracks.

One type of failure that has been reported[36] for particulate-filled thermosetting polymers is creep rupture in epoxy resins filled with silica particles subjected to constant stress for long periods of time. Subsequent investigations have been able to produce laboratory creep rupture data for such composites both at room temperature[36] and at

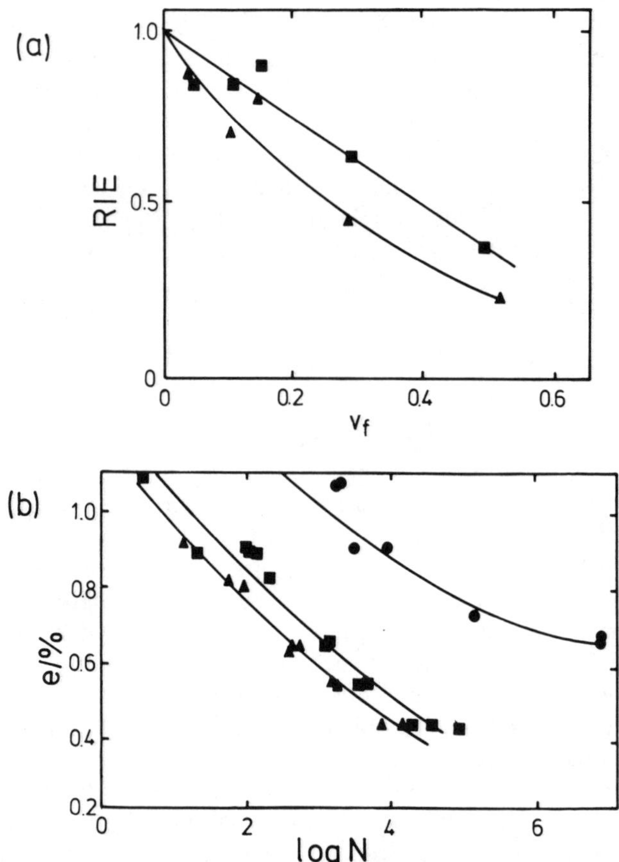

FIG. 6. Fracture data for a particulate-reinforced polyester resin (after ref. 9). (a) Dependence of relative impact energy (RIE) upon volume fraction for a calcite filler. ■, Good particle/matrix adhesion; ▲, poor particle/matrix adhesion. (b) Curves of applied strain versus number of cycles to failure for $v_f = 0{\cdot}25$ of glass particles. ●, Pure matrix ($v_f = 0$); ■, good particle/matrix adhesion; ▲, poor particle/matrix adhesion.

elevated temperatures[33] (Fig. 7), Young and Beaumont[36] interpreted such behaviour as a slow crack growth phenomenon and showed that it was possible to predict the creep rupture behaviour from crack propagation data. Hojo and co-workers[33] adopted a different approach and showed that Arrhenius plots of the rupture times for various filler sizes and contents converged to a characteristic point at the glass

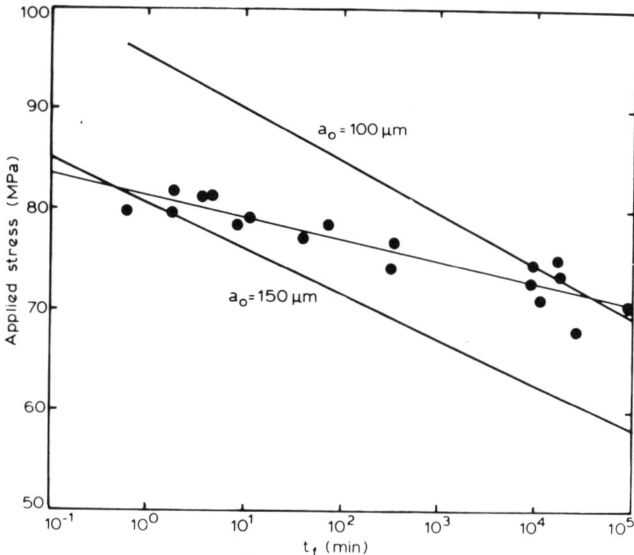

FIG. 7. Static fatigue data for an epoxy resin containing silica particles ($v_f = 0{\cdot}42$). The points are the experimental values and the two heavy lines are the theoretical predictions using an inherent flaw size of 100 μm and 150 μm (after ref. 36).

transition temperature of the matrix. They went on to produce a master curve which allows the prediction of the long-term strength of particulate-filled thermosetting composites as a function of rupture time, filler size and volume fraction.

5. CRACK PROPAGATION

Although it was shown in the preceding section that the addition of brittle particles to a thermosetting polymer tends to cause a reduction in the fracture strength of the material, it is well established[1–16] that crack propagation becomes more difficult in such materials. This leads to an increase in values of fracture parameters such as K_{Ic} and G_{Ic} on the addition of rigid particles which tend to impede a propagating crack. This is not inconsistent with the observations outlined earlier, since the presence of the particles leads to an increase in the inherent flaw size, a_0, for the material. Since the fracture strength depends on

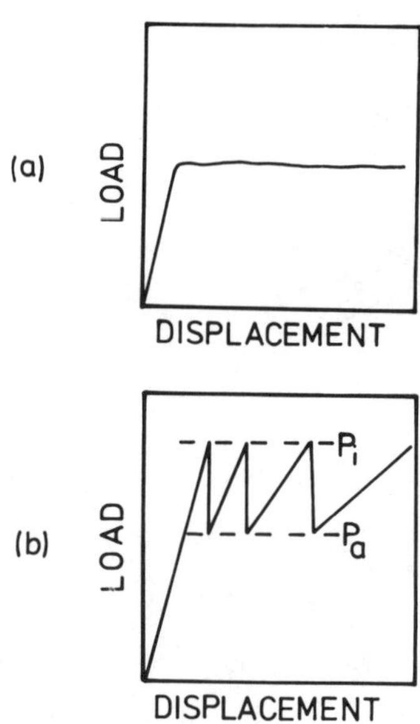

FIG. 8. Schematic load/displacement curves for thermosetting polymers fractured using a double torsion test-piece. (a) Stable continuous propagation. (b) Unstable stick/slip propagation. P_i and P_a are the initiation and arrest loads respectively.

both a_0 and K_{Ic},[1] it appears that any increase in strength due to a rise in K_{Ic} is offset by the corresponding increase in a_0.

Several types of fracture-mechanics specimens have been used to determine K_{Ic} and G_{Ic} for particulate-filled thermosets. They include both tapered double cantilever beam[6,7] and double torsion[10–15] specimens. The latter are now generally favoured since they are relatively easy to machine and do not use large quantities of material. For both types of specimen the value of K_{Ic} is independent of crack length.[1] It is found that for such specimens crack propagation can take place in two different ways, as shown in the schematic load/displacement curves shown in Fig. 8. The cracks can propagate in either a stable continuous or stick/slip discontinuous manner depending upon the material and the conditions of testing, and both types of behaviour are encountered in particulate-reinforced thermosets.

In Sections 5.1–5.4 the material and testing variables which control the nature of crack propagation and the values of K_{Ic} and G_{Ic} for these materials are discussed in detail.

5.1. Particle Volume Fraction and Size

Typical data[10] for the effect of v_f upon K_{Ic} for an epoxy resin reinforced with glass spheres of different average sizes are shown in Fig. 9a. Several important features can be seen. Firstly, although crack propagation is continuous and stable for the epoxy matrix, propagation in the composite is generally of the stick/slip variety, leading to initiation and arrest values of K_{Ic} being recorded for the composites. Secondly, it can be seen that K_{Ic} tends to rise with increasing particle volume fraction. Finally, it seems that the effect of particle size is rather complex, tending to have more effect upon the stability of propagation than on the value of K_{Ic}.

It can be seen from Fig. 9b that the dependence of G_{Ic} upon

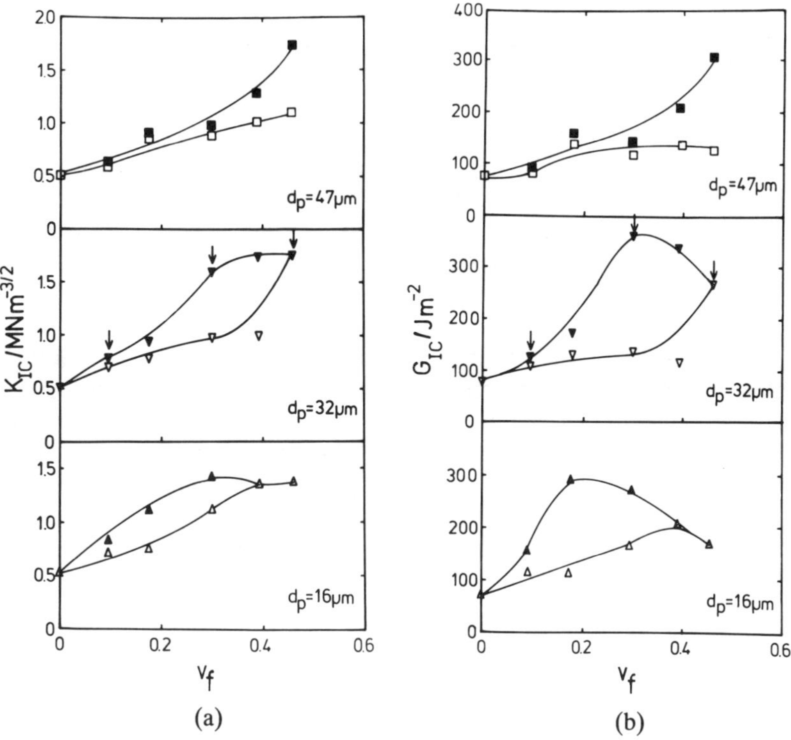

FIG. 9. Dependence of fracture-mechanics parameters upon particle size and volume fraction for an epoxy resin reinforced with spherical glass particles. The closed points are for crack initiation and the open ones are for arrest. The arrows indicate specimens which have their fracture surfaces shown in Fig. 10. (a) Variation of K_{Ic} with v_f; (b) variation of G_{Ic} with v_f.

particle volume fraction and size is rather different from that of K_{Ic}. The values of G_{Ic} have been determined using the relation:[1]

$$G_{Ic} \sim K_{Ic}^2/E \tag{8}$$

where E is the Young's modulus, the values of which have been taken from Fig. 2b. The initiation values of G_{Ic} peak for the smaller particles, with the values of v_f at the peak decreasing as the particle size is reduced. It appears therefore that particle size and volume fraction have a greater effect upon G_{Ic} than upon K_{Ic}.

At first sight this difference in behaviour for K_{Ic} and G_{Ic} might appear rather odd, as the two parameters normally show the same structural dependence for most materials. However, as E also varies with both particle size and volume fraction it is not surprising that K_{Ic} and G_{Ic} may vary in different ways. Many of the early investigations[3,6,7] into crack propagation tended to use G_{Ic} to characterise the behaviour. It appears that this may have led to a certain amount of confusion in determining the crack propagation mechanisms. For example, although K_{Ic} tends to rise with increasing volume fraction of filler regardless of the particle size, G_{Ic} peaks at a value of v_f which depends upon the particle size (Fig. 9). The underlying mechanisms controlling the dependence of K_{Ic} upon volume fraction will be discussed later (see Section 6).

The volume fraction of particles also has an important effect upon the appearance of the fracture surfaces of the specimens. Figure 10 shows scanning electron micrographs for the specimens in Fig. 9 marked by arrows. It can be seen that the fracture surfaces show both glass spheres and hemispherical holes. For low volume fractions there are 'tails'[10–12] behind the particles pointing in the direction of crack propagation, whereas for the highest volume fractions there are a large number of particles on the surface and no tails can be seen. The significance of the tails in suggesting the mechanism of propagation will be discussed later.

5.2. Particle/Matrix Adhesion

It has been recognised for several years that the degree of adhesion between the particles and the matrix has a significant effect upon the crack propagation behaviour of particulate-reinforced composites. However, there has been some confusion as to whether or not the best properties are obtained with good or poor adhesion. It was clearly demonstrated in Section 4 that the best tensile, impact and fatigue

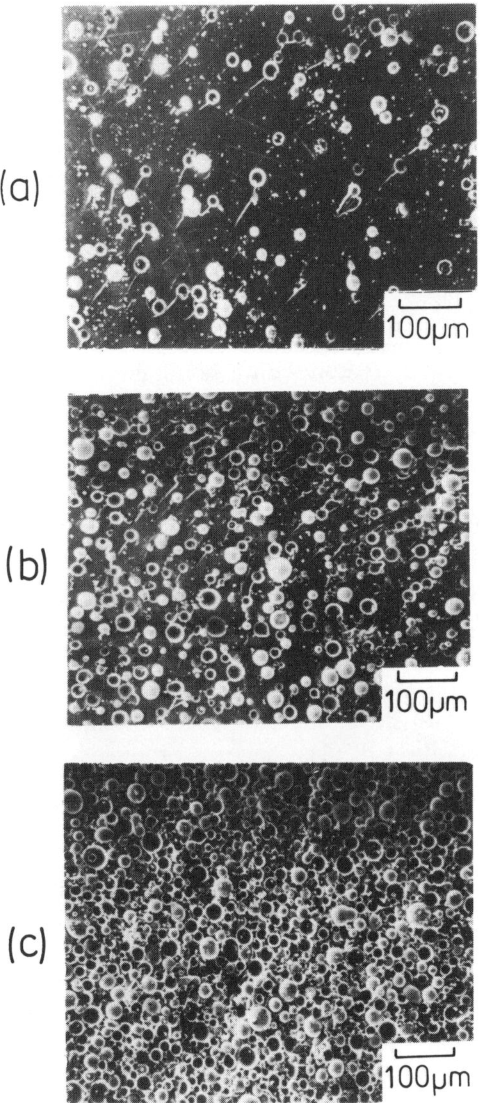

FIG. 10. Scanning electron micrographs of the fracture surfaces of the specimens indicated in Fig. 9 with a particle diameter of 32 μm: (a) $v_f = 0{\cdot}10$; (b) $v_f = 0{\cdot}30$; (c) $v_f = 0{\cdot}46$.

properties are obtained with the use of bonding agents to provide good adhesion. However, the indications are that with brittle matrix materials the best resistance to crack propagation, in terms of the highest values of K_{Ic}[6] or G_{Ic},[11] is obtained when there is poor adhesion between the particles and matrix. Although this would appear to be something of a contradiction, it can be rationalised when it is realised that although crack propagation is made more difficult, the weak interface between the particles and matrix leads to a larger inherent flaw size and a lower fracture strength.

The influence of the adhesion at the particle/matrix interface upon the fracture behaviour can be seen from the fracture surfaces in Fig. 11. The untreated particles appear to have debonded from the epoxy matrix, whereas those treated with a bonding agent are covered with resin and are clearly well bonded to the matrix. It appears therefore that the debonding may make crack propagation more difficult but crack initiation easier.

5.3. Matrix Toughness

It is of considerable interest to know how the toughness of a particle-reinforced thermoset is affected by matrix toughness, so that some idea can be gained of the potential for increasing toughness in such systems. The problem has been examined by Spanoudakis,[12] who determined K_{Ic} and G_{Ic} for an epoxy resin cured with two different types of hardener to give different values of matrix toughness. He found that the addition of glass spheres produced an increase in K_{Ic} and G_{Ic} in both cases but that the greatest percentage increase in these parameters was for the more brittle matrix (i.e. that with the lower value of K_{Ic}).

Very recently, Kinloch *et al.*[13–15] have followed the effect of matrix toughness for a particle-reinforced system by modifying the matrix using rubber-toughening techniques (Chapter 5). Typical results for such investigations (for $v_f(\text{glass}) = 0{\cdot}1$) are given in Table 1. It can be seen that in such 'hybrid-particulate' composites the toughness of the matrix increases with the addition of either glass or rubber and that the hybrid system shows the greatest values of G_{Ic}, indicating a degree of synergism. It can also be seen that at 50°C the toughness is further increased by the use of a silane bonding agent. In addition, it should be noted that the use of the glass particles in the hybrid system tends to keep the value of Young's modulus similar to that of the matrix epoxy and offsets the reduction in modulus caused by the addition of rubber particles.

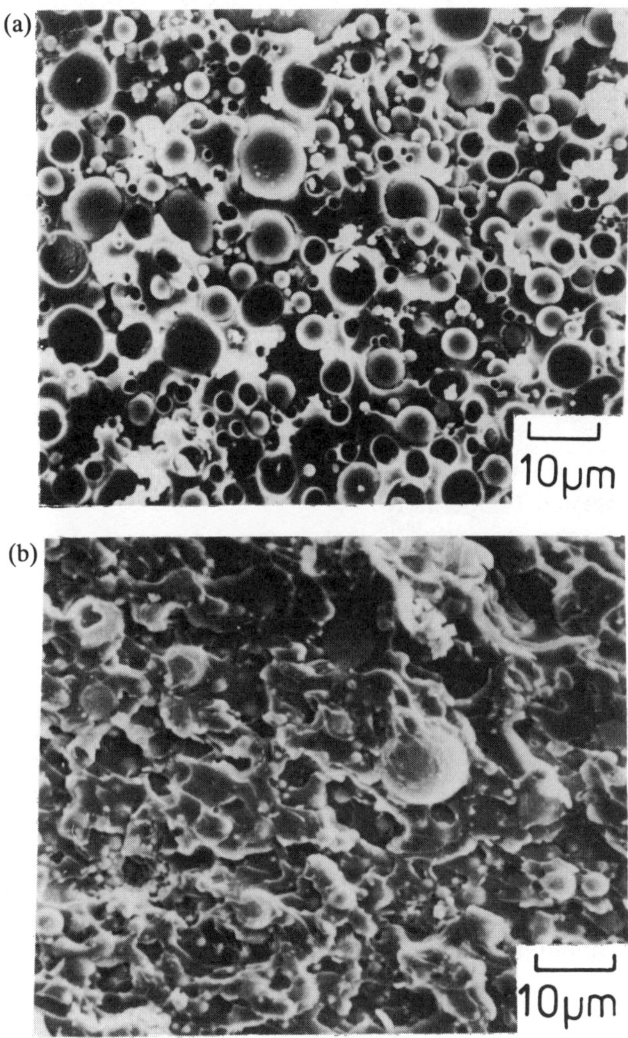

FIG. 11. Scanning electron micrographs of the fracture surfaces of a glass-particle reinforced epoxy resin (after ref. 11). (a) Untreated particles; (b) particles treated with bonding agent.

TABLE 1
COMPARISON OF MODULUS AND FRACTURE ENERGY VALUES FOR DIFFERENT EPOXY PARTICULATE COMPOSITES (AFTER REF. 14)

Composite	*−70°C*		*50°C*	
	E (GPa)	G_{Ic} *(kJ/m²)*	*E (GPa)*	G_{Ic} *(kJ/m²)*
Epoxy	4·1	0·15	2·8	0·50
Epoxy/glass[a]	4·3	0·26	3·1	1·0
Epoxy/rubber[b]	3·3	0·60	2·2	3·3
Hybrid[c]	4·1	0·80	2·6	4·3
Hybrid[c] (using silane-treated glass particles)	4·3	0·75	2·5	5·8

[a] v_f(glass) = 0·1
[b] Rubber, 15 phr.
[c] v_f(glass) = 0·1; rubber, 15 phr.

5.4. Rate and Temperature

It is only very recently[10–15] that systematic measurements have been made of the effect of testing variables such as rate and temperature upon crack propagation in particle-reinforced thermosets. It is found that in general the behaviour reflects that of the matrix polymer. For example, Spanoudakis and Young[10] examined the effect of rate upon a glass-particle reinforced epoxy which underwent either stick/slip or stable brittle propagation depending upon the particle size and volume fraction. They demonstrated that K_{Ic} increased with rate of testing for stable propagation but decreased with an increasing rate for stick/slip propagation. This is similar to the behaviour of the matrix epoxies without the addition of glass particles where it is found[37–39] that the criterion for propagation is that it occurs when a critical stress is reached at a critical distance ahead of the crack.

It is found that temperature has a profound effect upon crack propagation in particulate-reinforced thermosets. This is shown in Fig. 12a, where K_{Ic} is plotted as a function of test temperature for a hybrid-particulate composite. It can be seen that K_{Ic} increases with test temperature and also that the mode of crack propagation changes from being unstable and brittle to stable and ductile. The effects of both temperature and volume fraction upon G_{Ic} are demonstrated in Fig. 12b. In this case the value of G_{Ic} increases as temperature increases for all values of v_f whereas G_{Ic} peaks at a volume fraction of the order of 0·1–0·2 for all temperatures. The detailed crack propagation behaviour for particulate composites will depend upon the exact type of particle, matrix and surface treatment. However, it is thought that the behaviour shown in Fig. 12 is typical of that of most particulate-reinforced thermosetting polymers.

6. TOUGHENING MECHANISMS

The observations described in Section 5 show that both the fracture toughness, K_{Ic}, and the fracture energy, G_{Ic}, of thermosetting polymers can be significantly increased by the inclusion of rigid particulate fillers. In this section the mechanisms that have been proposed as being responsible will be described in detail.

6.1. Crack Pinning

In 1971 Lange and Radford[3] suggested that the increase in fracture energy that they found on the inclusion of a rigid-particulate filler (alumina trihydrate) in an epoxy resin was due to a crack pinning process. The mechanism itself had been proposed earlier by Lange[40] and was subsequently modified by Evans[4] and Green and coworkers.[41,42] The pinning process is illustrated schematically in Fig. 13; basically it assumes that cracks can be impeded by rigid, impenetrable, well bonded particles. This arises since when a crack meets an array of such obstacles it becomes pinned and tends to bow out between the particles forming secondary cracks. Thus a new fracture surface is formed and the length of the crack front is increased due to its change of shape between the pinning positions. Now, energy is required not only to create the new fracture surface but, by analogy with the theory of dislocations, it must also be supplied to the newly formed non-linear crack front, which is assumed to possess a line energy. This latter factor is particularly suggested as leading to the enhanced crack resistance often observed when impenetrable particles are well bonded into a brittle matrix.

Fractographic studies have provided strong evidence that pinning is an operative mechanism in particle-reinforced thermosets.[10–14] This can be seen from optical micrographs of fracture surface such as Fig. 14a, which clearly shows crack bowing between glass particles and suggests that the pinned crack assumes an elliptical shape at breakaway. Further, when such secondary cracks do eventually break away

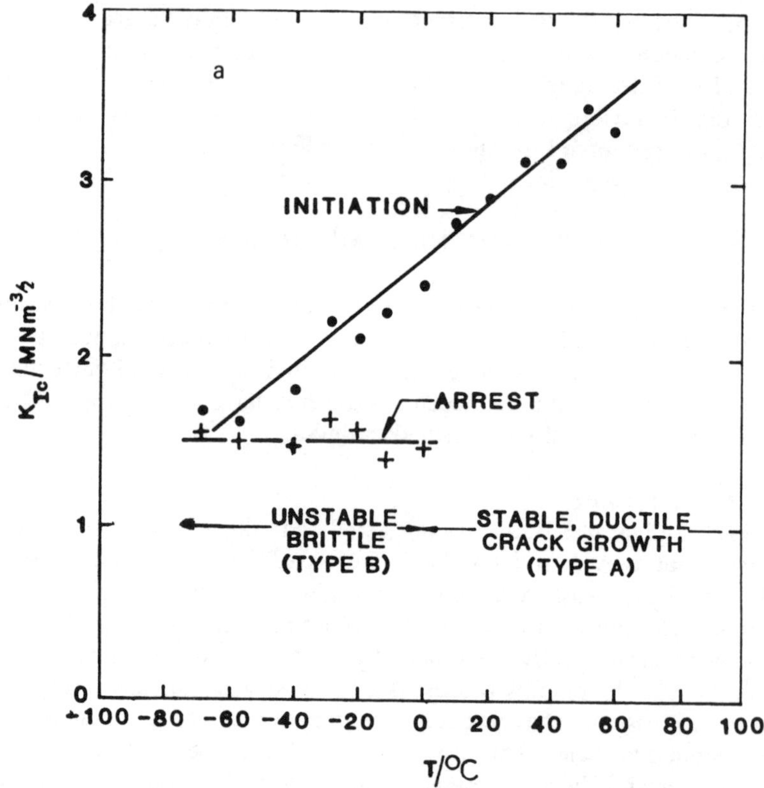

FIG. 12. Fracture behaviour of a hybrid-particulate epoxy composite with 15 phr of rubber (after ref. 14). (a) Dependence of K_{Ic} upon testing temperature; v_f(glass) = 0·085. The different types of crack propagation behaviour are indicated.

from the pinning positions, this frequently leads to the formation of characteristic 'tails' (mentioned earlier and shown in Fig. 10) which are caused by the meeting of two arms of the crack front from different fracture planes. For high volume fractions (e.g. Fig. 10c) the tails or steps are rarely observed. However, it is thought[10,14] that this does not necessarily mean that no pinning occurs, as such features will be obscured at high volume fractions due to considerable overlap of secondary cracks. It is worth also considering the difference between the fracture surfaces in Fig. 11 for samples with different degrees of interfacial adhesion. In the case of poor adhesion (Fig. 11a) the cracks

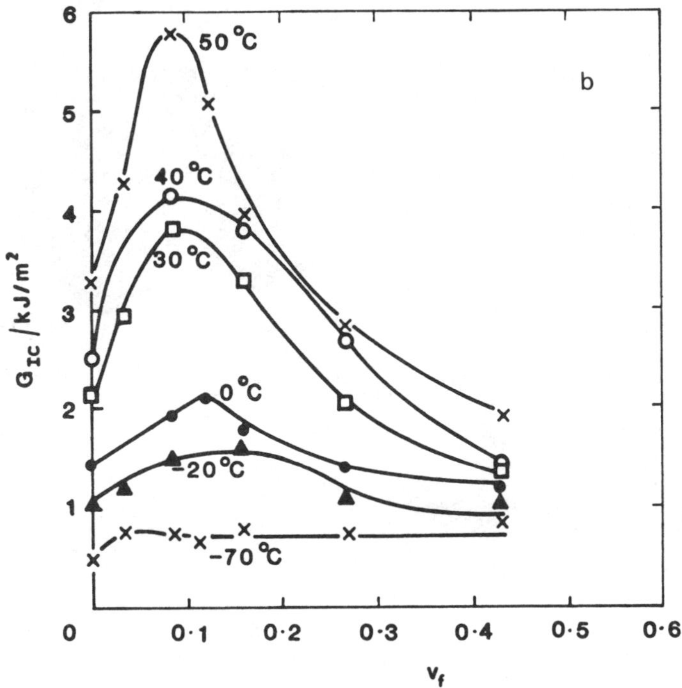

FIG. 12—*contd.* (b) Dependence of fracture energy, G_{Ic}, upon glass particle volume fraction and temperature.

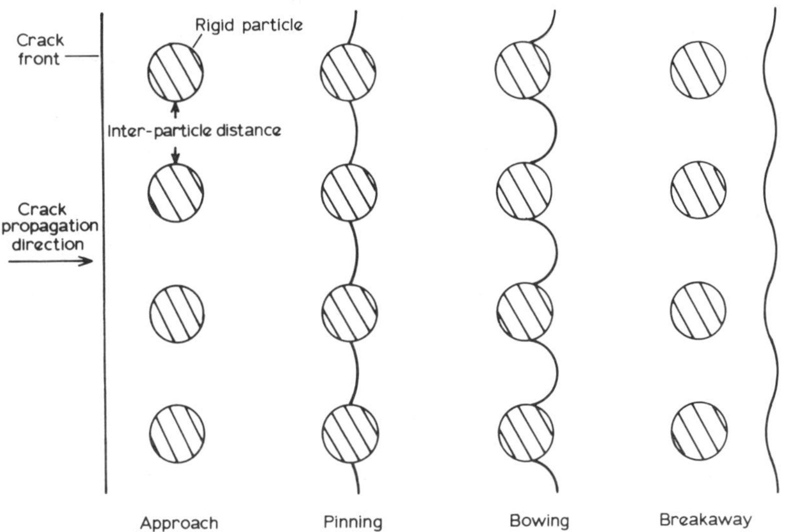

FIG. 13. Schematic representation of the crack pinning process.

FIG. 14. Crack pinning in a glass-particle reinforced epoxy resin (after ref. 14). (a) Optical micrograph showing pinned crack front bowing between particles. The arrow indicates the direction of crack propagation.

are attracted towards the equators of the particles, whereas they are attracted towards the poles in the composites containing the silane-coated glass particles (Fig. 11b). This can be explained by considering the local stresses around the rigid glass particles in an epoxy resin under an applied stress. As the interfacial adhesion between the particle and matrix is increased, then the maximum tensile stress moves from the equatorial regions to the poles of the particles.[10,14,43] Thus, propagating cracks will be attracted to the equatorial regions of the particles when the interfacial adhesion is relatively low, but towards the poles when it is higher, due for example to the use of a reactive silane bonding agent.

6.2. Analysis of Pinning

The theoretical approach to the analysis of pinning was developed originally by Lange,[40] who derived an expression relating the fracture energy, G_{Ic}, of the composite to that of the matrix, G_{Ic0}, through the line energy, T, per unit length of bowed crack front. By assuming that the crack front breaks away from the pinning positions when it attains a radius of $D/2$, where D is the interparticle spacing, Lange obtained

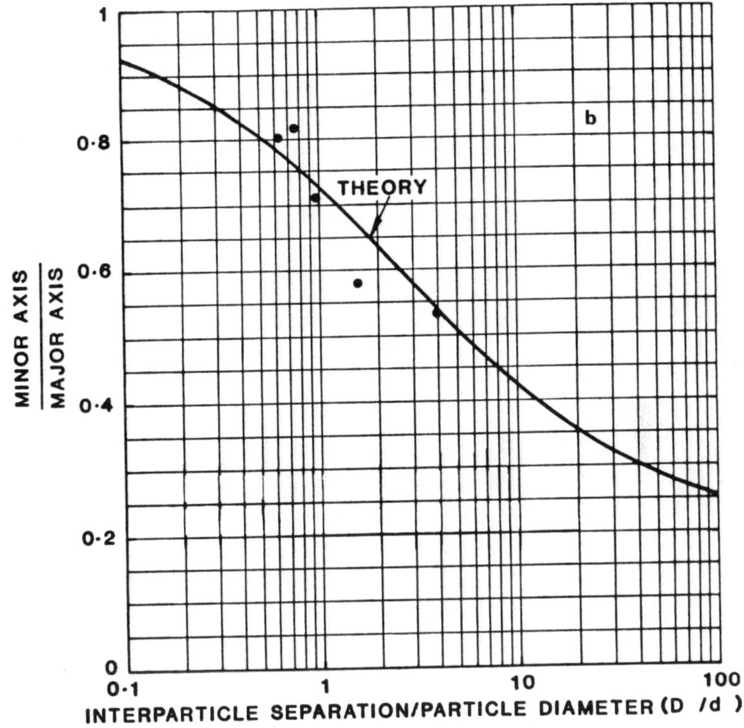

FIG. 14—*contd.* (b) Shape of the pinned crack front at breakaway as a function of D/d. The solid curve is the theoretical line (after ref. 42) and the points are experimental measurements (after ref. 14).

the relation

$$G_{Ic} = G_{Ic0} + 2T/D \tag{9}$$

However, this equation has been found[1,3,5,10] to be inadequate to describe the crack front pinning mechanism. For example, the initial slopes of plots of G_{Ic} versus D^{-1} are often a function of the volume fraction of particles and thus T is not invariant with particle size as suggested by eqn (9).

More satisfactory models have been derived by several groups of workers[4,42] who have calculated from first principles the stresses required to propagate a crack through a matrix reinforced by an array of rigid particles. The first attempt was made by Evans,[4] who demonstrated that the line energy contribution is important for rigid,

TABLE 2
PREDICTED VALUES OF σ_c/σ_0 AS A FUNCTION OF d/D FOR PARTICLE-FILLED COMPOSITES

d/D	σ_c/σ_0			
	Non-interacting elliptical cracks		*Interacting elliptical cracks*	
	Evans[4]	*Green et al.*[42]	*Evans*[4]	*Green et al.*[42]
0	1	1	1	1
0·25	1·85	2·02	1·19	1·19
0·50	2·18	2·52	1·65	1·80
1·00	2·55	3·05	2·23	2·52
1·25	2·70	3·25	2·40	2·86
2·00	2·90	3·75	2·75	3·52

impenetrable particles but that the value of T is indeed a function of particle size and shape. Evans' analysis also enables the stresses required to propagate cracks to be predicted. The results are given in Table 2, where the ratio of the stress required to propagate a crack in the composite to the stress required for a crack in pure matrix, σ_c/σ_0, is given in terms of the ratio of the particle diameter to the interparticle separation, d/D. Values are given for elliptical secondary crack fronts and for both the assumptions of no interactions between elliptical secondary cracks, and of crack interactions. Green *et al.*[42] have modified the analysis of Evans and, in particular, have taken into account the exact position of the secondary crack front when it breaks away from the pinning positions. They obtained a similar dependence of σ_c/σ_0 with d/D as Evans, as can be seen from Table 2, but their analysis leads to rather higher values of σ_c/σ_0.

This theoretical work of Evans[4] and Green *et al.*[42] has also enabled the shape of the pinned crack front at breakaway to be calculated as a function of the ratio d/D. The results of such calculations are given in Fig. 14b and show that for high volume fractions of particles the crack shape at breakaway (defined by the ratio of length of the minor axis of the elliptical crack to that of the major axis) tends to be circular. In contrast, for low volume fractions the crack front tends to remain straight. The shape of the crack front can be determined from micrographs such as Fig. 14a, and experimental measurements[14] are also plotted in Fig. 14b. The ratio d/D can be calculated from

$$d/D = 3v_f/2(1 - v_f) \tag{10}$$

It can be seen that they are in good agreement with the theoretical calculations and this gives further confirmation of the role of crack pinning and confidence in the theoretical analyses.[4,42]

It is also possible to fit the experimental measurements of K_{Ic} for the composites to the theoretical predictions from Table 2. It is found[14] that this can be done successfully for composites with brittle matrix materials where pinning is the major toughening mechanism operating. For a particulate composite containing primary cracks of the same length as those in the unmodified matrix, then the ratio of σ_c/σ_0 is equivalent to the ratio of K_{Ic}(composite)/K_{Ic}(matrix). Measured values[10] of this ratio (using data from Fig. 9a) are plotted versus d/D in Fig. 15, along with the theoretical predictions of Green *et al.* taken from Table 2. It may be seen that the agreement between theory and experiment is good. In contrast, when there is more plastic deformation at the crack tip due to a more ductile type of matrix (e.g. rubber-toughened[14]) or higher test temperatures being employed, then it is found that the agreement is not so good. However, this does not necessarily mean that pinning is not taking place in this case. Increases in toughness due to pinning may be swamped by those caused by other mechanisms taking place simultaneously.

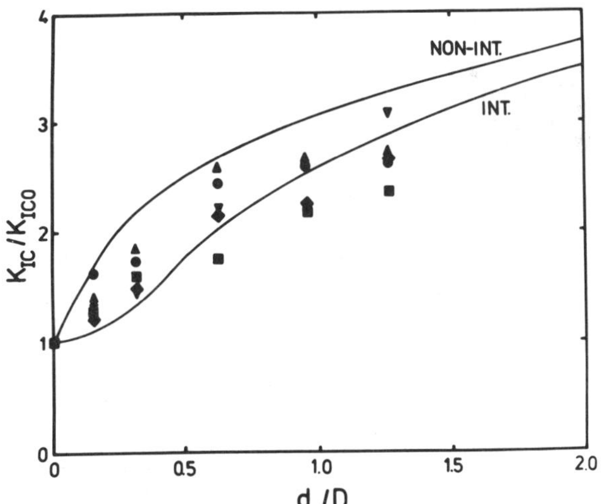

FIG. 15. Variation of the ratio of K_{Ic} of the composite to K_{Ic} for the matrix (K_{Ic0}) for a glass-particle reinforced epoxy resin as a function of d/D. The points correspond to different particle diameters and the lines represent the theoretical predictions of Green *et al.*[42] (after ref. 10).

6.3. Crack Opening Displacement

It is well established[1,44] that crack propagation in many glassy polymers is controlled by a critical crack opening displacement criterion. It has also been shown[37,45] that stable brittle crack propagation in thermosetting polymers such as epoxy resins is governed, to a first approximation, by such a criterion until crack tip blunting takes place leading to unstable propagation. The crack opening displacement, δ, is given by:[1]

$$\delta = K_{Ic}^2/\sigma_y E \tag{11}$$

or alternatively, for a material which shows linear elastic behaviour, by

$$\delta = K_{Ic}^2/E^2 e_y \tag{12}$$

where σ_y is the yield stress and e_y is the yield strain.

It has been found[10,14] that particulate reinforced thermosetting polymers tend to obey a constant critical δ_c criterion when crack propagation is stable and brittle. An example is shown in Fig. 16(a), where K_{Ic} is plotted against E for the data for a glass-reinforced epoxy resin system taken from Figs 2a and 9a. The straight line relationship shows that the criterion is obeyed since the e_y does not vary with the volume fraction of particles (cf. Fig. 3). It is generally found[10,14] that for particulate composite systems undergoing brittle stable crack propagation the value of δ_c is constant and similar to that of the matrix. It should be remembered that in such cases the principle mechanism of toughening is pinning, which must therefore take place at a critical crack opening displacement. The result of this is that K_{Ic} must increase with particle volume fraction similarly to the way in which E increases with v_f (compare Fig. 2a with Fig. 9a).

When more ductile thermosetting matrix materials are used and propagation is unstable or stable and ductile (e.g. Fig. 12a), then the constant crack opening displacement criterion is found to no longer apply.[14] This is shown in Fig. 16b, where the value of δ_c is plotted against temperature for a hybrid-particulate composite and the types of propagation observed are also indicated. The increase in δ_c with increasing temperature clearly demonstrates that localised plastic deformation is taking place at the crack tip, leading to a considerable amount of crack tip blunting. The consequences of this are discussed in Section 6.4.

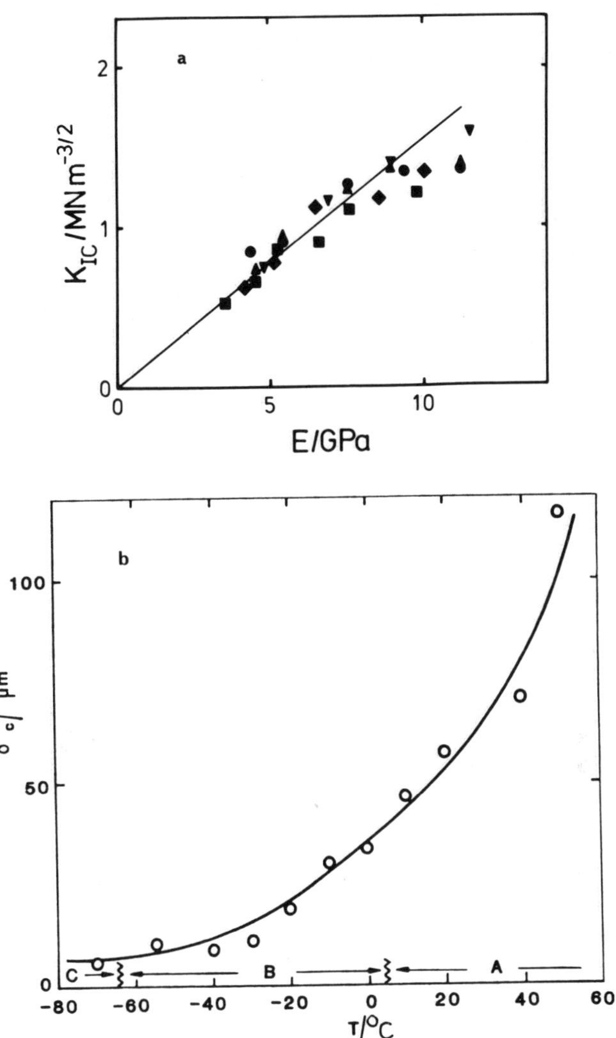

FIG. 16. Crack opening displacement criterion for particulate-reinforced epoxy resins. (a) Variation of K_{Ic} with E for a glass-particle reinforced epoxy resin undergoing brittle stable propagation. The straight line implies that δ_c is constant (after ref. 10). (b) Variation of δ_c with temperature for a hybrid-particulate epoxy composite undergoing different types of crack propagation: A, stable brittle; B, unstable brittle; and C, stable ductile (after ref. 14).

6.4. Localised Plastic Deformation

The highest levels of toughness are obtained in particulate-reinforced thermosetting polymers when both pinning and localised plastic deformation take place simultaneously. This is demonstrated in Fig. 17, where G_{Ic} is plotted against temperature for several different systems. It can be seen that the use of rubber particles in the hybrid system leads to the highest fracture energies. It is known (Chapter 5) that the rubber particles act as initiation sites for shear yielding in the epoxy resin matrix. However, there is also evidence that the glass particles may also induce some extra shear deformations.[14,15] This can be seen from Fig. 18a, which is an optical micrograph of a crack propagating through a hybrid-particulate composite. The large dark holes are where the glass particles have been removed during polishing of the sample. The light circular regions are glass particles with the dark arcs at the poles showing where debonding has occurred ahead of the crack tip. The rubbery particles appear as very small dark

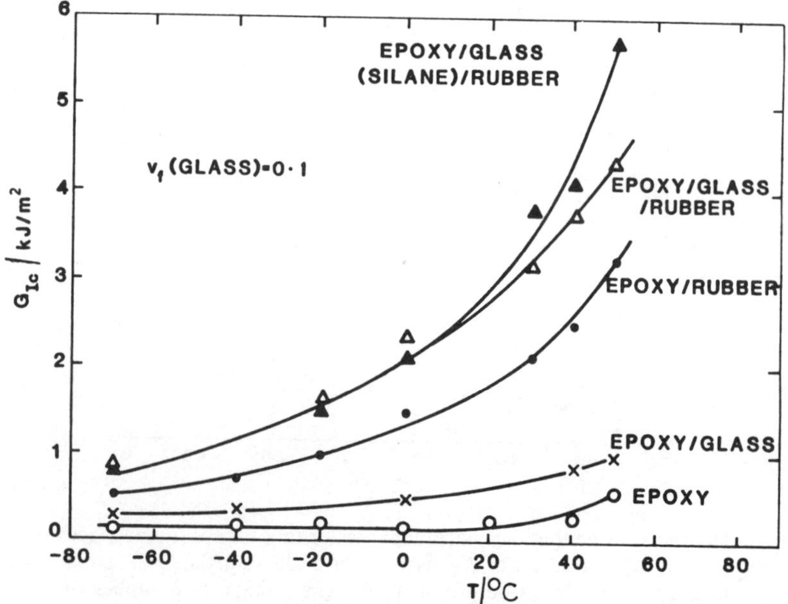

FIG. 17. Dependence of fracture energy upon temperature for different types of matrix materials and particulate composites; $v_f = 0{\cdot}1$ (after ref. 14).

particles. It may be seen that many fine black lines connect the rubbery particles in the vicinity of the crack and tend to lie at an angle of about 45° to the maximum principal tensile stress, indicating that they are plastic shear bands. This has recently been confirmed[15] by observing similar specimens in the scanning electron microscope after deformation as shown in Fig. 18b. It can be seen that the rubber particles are joined by 'furrows' where the surface material has been sucked in during the formation of shear bands. These furrows scatter light in the optical microscope and therefore appear as black lines in Fig. 18a. Thus, the local stress concentrations introduced by the glass particles appear to assist in initiating shear deformation in the matrix. Hence it is possible that the glass particles may increase the toughness through this mechanism as well as by the crack pinning mechanism. It would be expected that the initiation of shear deformation would be more significant at higher test temperatures when the yield stress of the matrix is relatively low.

The relationship between the plastic deformation of the composite, the fracture toughness and the type of crack growth can be expressed quantitatively by relating the values of the stress intensity factor, K_{Ic}, with the yield stress of the composite (Section 3.2). This is done in Fig. 19, where values of K_{Ic} normalised by the values of K_{Ic} at the onset of crack growth (i.e. the sharp crack value) are plotted against the yield stress for a wide variety of compositions of a hybrid composite tested at a variety of temperatures. It can be seen that there is a good correlation between the toughness, the type of crack growth and the yield stress. This emphasises the importance of localised plastic deformation in increasing the toughness, especially at higher temperatures and with tough matrix materials.

It was demonstrated in Section 6.3 that when there is considerable plastic deformation a critical crack opening displacement cannot be applied (Fig. 16b). It has been shown recently by Kinloch *et al.*[14] that in this case the onset of crack growth is governed by a failure criterion based upon the attainment of a critical stress acting over a certain distance ahead of the crack tip. This expression has been found to represent a good description of the crack blunting mechanism in unfilled epoxy[37–39] and polyester[46] resins undergoing unstable crack propagation. Its application to rigid-particulate reinforced systems has served to unify the analysis of crack propagation in a wide variety of thermosetting polymeric systems.

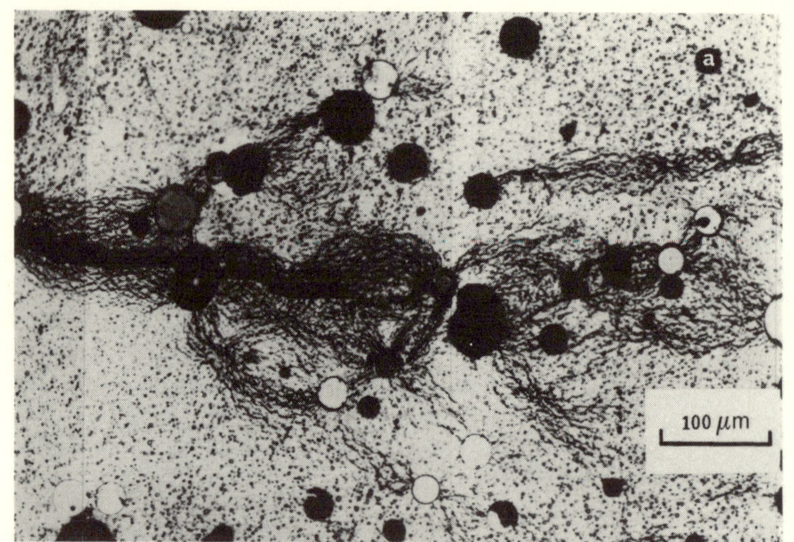
a
100 μm

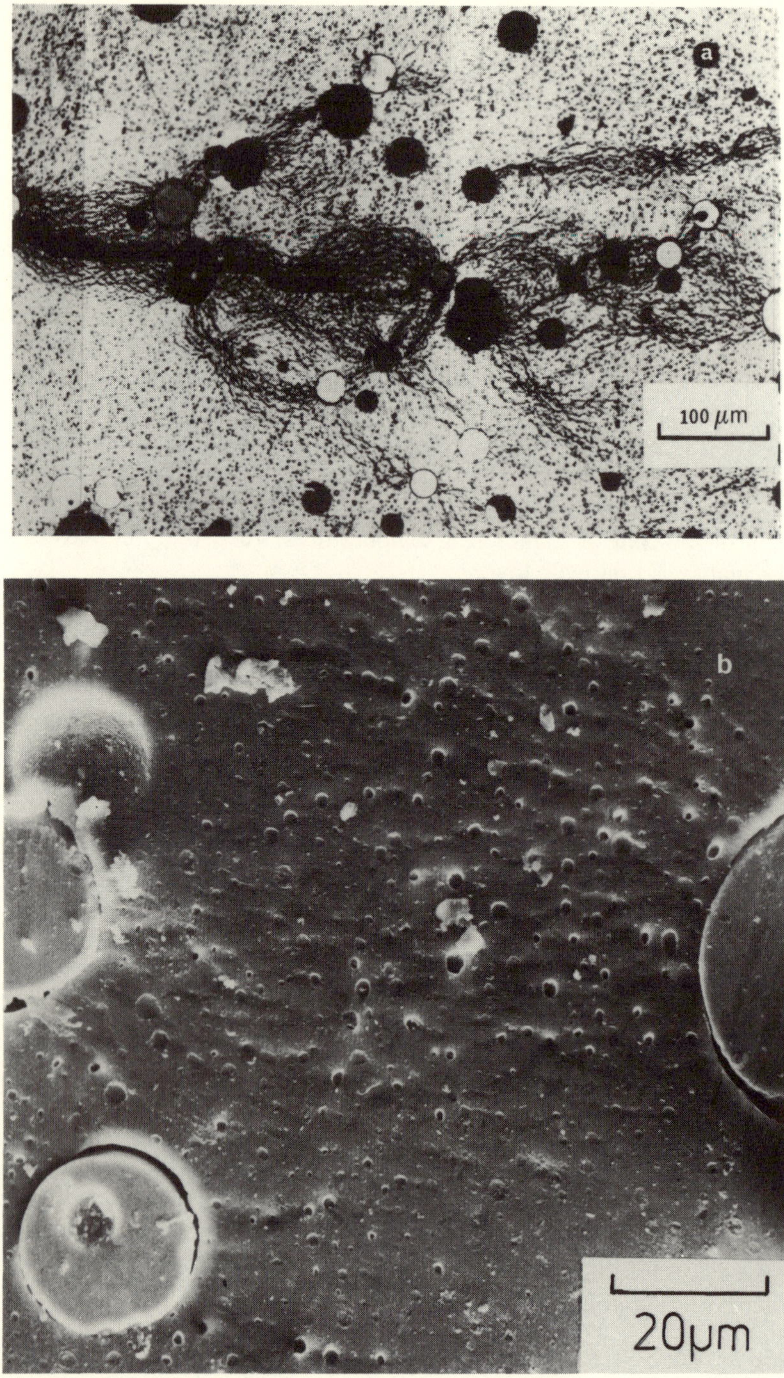

b
20μm

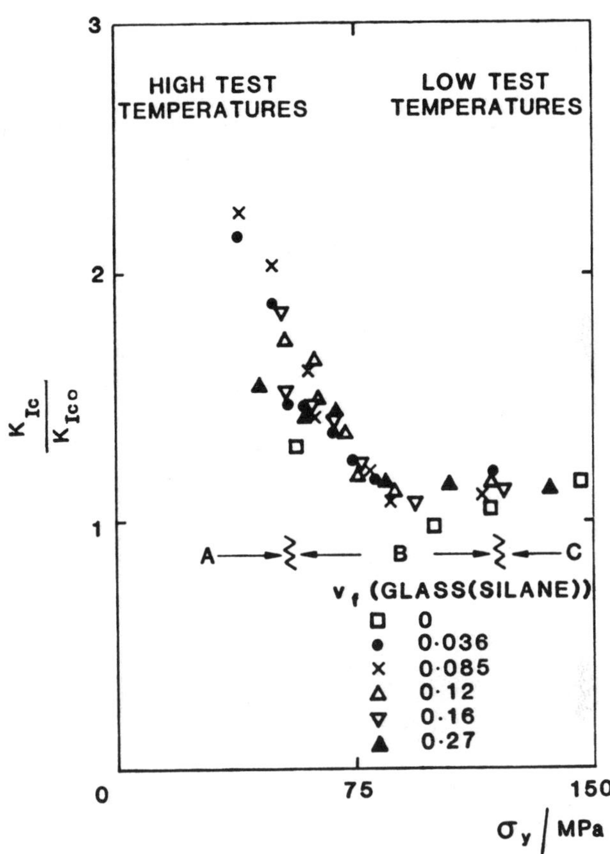

FIG. 19. Dependence of the ratio K_{Ic}/K_{Ic0} upon the yield stress of the material for hybrid-particulate epoxy composites (after ref. 14).

FIG. 18. Micrographs obtained from the free surface of a hybrid-particulate composite after crack growth. (a) Optical micrograph of the region in the vicinity of the crack tip (after ref. 14). (b) Scanning electron micrograph of the same region as in (a). The specimen was coated after deformation (after ref. 15).

7. CONCLUDING REMARKS

One of the main reasons for the addition of rigid filler particles to thermosetting polymers is the resultant reduction in cost. However, it has been shown above that this does not necessarily lead to a deterioration in properties but can often lead to an improvement. Although most studies have been concerned with spherical glass particles, the indications are that such observations are of relevance to other types of particulate fillers. It appears that the best property improvements are generally obtained with the use of bonding agents to provide good particle/matrix adhesion. This leads to the composites having higher values of Young's modulus, compressive yield stress, fracture, impact and fatigue strengths and generally improved resistance to crack propagation than without the use of bonding agents. Recent developments in the area of rigid-particulate reinforced thermosetting polymers have included the incorporation of a rubbery phase in the matrix. This has been found to lead to a considerable improvement in toughness due to the optimisation of crack pinning and plastic deformation, which are the main toughening mechanisms in such materials. It is clear that thermosetting polymers reinforced with particulate fillers are now poised to be utilised as moulding compounds or engineering adhesives for their inherently better properties.

ACKNOWLEDGEMENTS

The author has collaborated over the years with several people in the area of particulate reinforcement of thermosetting polymers. They include P. W. R. Beaumont, J. Spanoudakis, D. L. Maxwell and A. J. Kinloch. He is grateful to them for their stimulating discussions and permission to use their results and ideas in this chapter.

REFERENCES

1. Kinloch, A. J. and Young, R. J., *Fracture Behaviour of Polymers*, Applied Science Publishers, London, 1983.
2. Bucknall, C. B., *Adv. Polym. Sci.*, **27** (1978) 121.
3. Lange, F. F. and Radford, K. C., *J. Mater. Sci.*, **6** (1971) 1197.
4. Evans, A. G., *Phil. Mag.*, **26** (1972) 1327.

5. Moloney, A. C., Kausch, H. H. and Steiger, H. R., *J. Mater. Sci.*, **18** (1983) 208.
6. Broutman, L. J. and Sahu, S., *Mater. Sci. Eng.*, **8** (1971) 98.
7. Mallick, P. K. and Broutman, L. J., *Mater. Sci. Eng.*, **18** (1975) 63.
8. Young, R. J. and Beaumont, P. W. R., *J. Mater. Sci.*, **12** (1977) 684.
9. Brown, S. K., *Br. Polym. J.*, **14** (1982) 1.
10. Spanoudakis, J. and Young, R. J., *J. Mater. Sci.*, **19** (1984) 473.
11. Spanoudakis, J. and Young, R. J., *J. Mater. Sci.*, **19** (1984) 487.
12. Spanoudakis, J., *Fracture in Particulate-Filled Epoxy Resins*, Ph.D. Thesis, University of London, 1981.
13. Maxwell, D. L., Young, R. J. and Kinloch, A. J., *J. Mater. Sci. Lett.*, **3** (1984) 9.
14. Kinloch, A. J., Maxwell, D. L. and Young, R. J., *J. Mater. Sci.*, **20** (1985) 4169.
15. Kinloch, A. J., Maxwell, D. L. and Young, R. J., *J. Mater. Sci. Lett.*, **4** (1985) 1276.
16. Moloney, A. C., Kausch, H. H., Kaiser, T. and Beer, H. R., *J. Mater. Sci.*, in press.
17. Young, R. J., *Introduction to Polymers*, Chapman and Hall, London, 1981.
18. Brandon, D. G., *Modern Techniques in Metallography*, Butterworths, London, 1966, p. 249.
19. Young, R. J., Maxwell, D. L. and Kinloch, A. J., *J. Mater. Sci.*, **21** (1986) 380.
20. Ishai, O. and Cohen, L. J., *Int. J. Mech. Sci.*, **9** (1967) 539.
21. van der Poel, C., *Rheol. Acta*, **1** (1958) 198.
22. Smith, J. C., *J. Res. NBS*, **80A** (1976) 45.
23. Lewis, A. G. and Nielsen, L. E., *J. Appl. Polym. Sci.*, **14** (1970) 1449.
24. Bowden, P. B. and Jukes, J., *J. Mater. Sci.*, **7** (1972) 52.
25. Yamini, S. and Young, R. J., *J. Mater. Sci.*, **15** (1980) 1814.
26. Bowden, P. B. and Raha, S., *Phil. Mag.*, **29** (1974) 149.
27. Thierry, A., Oxborough, R. J. and Bowden, P. B., *Phil. Mag.*, **30** (1974) 527.
28. Argon, A. S., *Phil. Mag.*, **28** (1973) 839.
29. Kitagawa, M., *J. Polym. Sci., Polym. Phys. Ed.*, **15,** (1977) 1601.
30. Sahu, S. and Broutman, L. J., *Polym. Eng. Sci.*, **12** (1972) 91.
31. Nicolais, L. and Nicodemo, L., *Polym. Eng. Sci.*, **13** (1973) 469.
32. Leidner, J. and Woodhams, R. T., *J. Appl. Polym. Sci.*, **18** (1974) 1639.
33. Hojo, H., Toyoshima, W., Tamura, M. and Kawamura, N., *Polym. Eng. Sci.*, **14** (1974) 604.
34. Petch, N. J., *Phil. Mag.*, **1** (1956) 186.
35. Griffith, A. A., *Phil. Trans. Roy. Soc.*, **A221** (1920) 163.
36. Young, R. J. and Beaumont, P. W. R., *J. Mater. Sci.*, **10** (1975) 1343.
37. Kinloch, A. J. and Williams, J. G., *J. Mater. Sci.*, **15** (1980) 987.
38. Yamini, S. and Young, R. J., *J. Mater. Sci.*, **15** (1980) 1823.
39. Young, R. J., in *Developments in Reinforced Plastics—1,* Ed. G. Pritchard, Applied Science Publishers, London, 1980.

40. Lange, F. F., *Phil. Mag.*, **22** (1970) 983.
41. Green, D. G., Nicholson, P. S. and Embury, J. D., *J. Mater. Sci.*, **14** (1979) 1413.
42. Green, D. G., Nicholson, P. S. and Embury, J. D., *J. Mater. Sci.*, **14** (1979) 1657.
43. Agarwal, B. D. and Broutman, L. J., *Fibre Sci. Tech.*, **7** (1974) 63.
44. Williams, J. G., *Fracture Mechanics of Polymers,* Ellis Horwood, Chichester, 1984.
45. Gledhill, R. A., Kinloch, A. J., Yamini, S. and Young, R. J., *Polymer,* **19** (1978) 574.
46. Gatward, C. H., Hogg, P. J. and Hull, D., paper presented at *6th International Conference on Deformation, Yield and Fracture of Polymers,* PRI, London, 1985.

7

Microstructure in Crosslinked Epoxy Polymers

G. C. STEVENS

Central Electricity Research Laboratories, Leatherhead, Surrey, UK

NOTATION

E	epoxide group concentration
G_{Ici}	critical strain energy release rate for crack initiation
H	hydroxyl group concentration
I_B	Brillouin peak intensity
I_{Hu}, I_{Vu}	90° unpolarized (u) scattered light intensity obtained using horizontally (H) or vertically (V) polarized incident light
I_R	Rayleigh peak intensity
I'	X-ray scattered intensity
$I(\theta)$	light scattered intensity at an angle θ to the forward direction
K_{Ic}	fracture toughness
M'	high-frequency real part of the longitudinal acoustic modulus
M'_0	low-frequency real part of the longitudinal acoustic modulus
P	extent of reaction
$P(90°)$	particle scattering factor for 90° scattering
R^{90}	Rayleigh ratio at 90° scattering angle
R_{an}	anisotropic Rayleigh ratio
R_{is}	isotropic Rayleigh ratio
T	temperature
T_g	glass transition temperature
V	volume
V_f	volume fraction of aggregates

f_E	epoxy resin epoxide group functionality
f_A	crosslinking agent functionality
k	chemical reaction rate constant
l_e	polymer chain length between entanglements
m	solvent/aggregate refractive index ratio
n_0	solvent-phase refractive index
p_A	fraction of reactive crosslinking agent groups at the gel point
p_E	fraction of reactive epoxide groups at the gel point
t	time
v	aggregate volume
z	light-scattering envelope dissymmetry
γ	ratio of specific heats C_p/C_v
θ	scattering angle
λ	incident optical wavelength
λ_{cr}	craze fibril extension ratio
λ^e_{max}	maximum extension ratio of the entanglement network
ρ	electron density
ρ_u	depolarization ratio

1. INTRODUCTION

Epoxy resins form the most varied and versatile group of thermosetting crosslinked polymers and are widely employed in adhesive compositions and particulate and fibre composites, as discussed in Chapters 3, 5 and 6. In many mechanical and electrical applications there is an increasing need to understand the chemical and physical factors influencing the properties and performance of these materials, particularly the long-term performance and defect mechanisms of failure. However, although the chemistry and physical properties of epoxy resins have been studied extensively and a qualitative prediction of some properties is possible,[1–4] a quantitative understanding of the dependence of properties on the crosslinked network structure is still largely not available.

This is attributed in part to the difficulties of fully characterizing the network structure, particularly of the more common and highly crosslinked materials.[5] Strictly, this characterization should not only include the chemical composition of the network and average network characteristics such as crosslink density, molecular weight distribution between crosslinks and number of elastically active network chains; it should also include their spatial arrangement and local organization in the matrix.[5–7] Intuitively, the involvement of 'morphology' seems obvious but is countered by our experience of these materials as polymeric glasses which 'appear' isotropic and homogeneous. Such a view is coupled with a dearth of detailed studies; those which exist neither clearly support nor refute the universal presence of morphology in epoxy resins. Consequently this has led to considerable debate over whether or not morphology and network inhomogeneity does or can exist at all in epoxy resin systems in general.

This contribution will examine this debate by considering some of our current understanding of the unreacted prepolymers and monomers used in epoxy resin systems, the factors influencing network formation during reaction and the evidence for bulk microstructure or network inhomogeneity in fully cured systems. Finally, proposed correlations between network microstructure and physical properties will be examined and the possible role of microstructure in influencing behaviour and properties will be briefly discussed.

2. NETWORK FORMATION

Epoxy resin networks may be formed from a vast number of polyfunctional monomeric and polymeric compounds containing epoxide groups (usually 1,2-epoxy, CH_2—CH—R, with O bridging CH_2 and CH) reacted with an equally vast range of base- and acid-type curing agents. Excellent reviews exist[1,2] covering the chemical variety and reaction mechanisms within the different systems and these will not be repeated here. Rather, two common classes of epoxy resin system based on the same type of epoxy prepolymer will be considered. These provide very different examples of reaction mechanisms and network formation and they reflect the characteristics of many of the systems examined in microstructure studies.

2.1. Typical Materials and Reaction Mechanisms

The two classes of system considered include the organic acids based on compounds containing carboxyl groups (R—C(=O)—OH) and the Lewis bases based on primary (R—NH_2) and secondary (R—NH) amines. The most common epoxy resins used with these systems are diepoxide prepolymers based on diglycidyl ether of bisphenol A (DGEBA) having the chemical structure I. Oligomers with n-values extending to

$$\overset{O}{\overbrace{CH_2-CH}}-CH_2-\left[O-C_6H_4-\underset{CH_3}{\overset{CH_3}{\underset{|}{\overset{|}{C}}}}-C_6H_4-O-\overset{OH}{\overset{|}{CH}}-CH_2-\right]_n O-C_6H_4-\underset{CH_3}{\overset{CH_3}{\underset{|}{\overset{|}{C}}}}-C_6H_4-O-CH_2-\overset{O}{\overbrace{CH-CH_2}} \qquad \text{(I)}$$

12,[8] and above,[9] can be present and their distribution controls the number of epoxide and hydroxyl groups available for reaction. In general, the oligomers have a very low degree of side-chain branching[10] and they are hindered and rodlike.[11,12] This prevents them from behaving as classic Gaussian chains, particularly at molecular weights less than 2500 (about $n = 8$) as revealed by neutron scattering.[12] In amine crosslinked systems it is common to use $n = 0$ prepolymer and this fraction will crystallize and exhibit a melting point around 315 K.[11] Crystallization may be useful in purifying an $n = 0$ prepolymer[13,14] but this can present an unwanted inhomogeneity in low-n systems which are initially cured at room temperature.[15]

2.1.1. Organic Acid Reactions

Simple end-linking of diepoxide DGEBA oligomers would occur if the addition esterification reaction only were present:

$$-\overset{O}{\overset{\|}{C}}-OH + \overset{O}{\overbrace{>C-C<}} \longrightarrow -\overset{O}{\overset{\|}{C}}-O-\underset{|}{\overset{|}{C}}-\overset{OH}{\overset{|}{C}}- \qquad \text{(i)}$$

However, the resulting alcoholic hydroxyl, or the DGEBA hydroxyl group, may also preferentially react with epoxide groups in addition etherification,

$$-\underset{|}{\overset{|}{C}}-OH + \overset{O}{\overbrace{>C-C<}} \longrightarrow -\underset{|}{\overset{|}{C}}-O-\underset{|}{\overset{|}{C}}-\underset{|}{\overset{OH}{\overset{|}{C}}}- \qquad \text{(ii)}$$

or with organic acid groups in condensation esterification[1,2]

$$-\overset{O}{\overset{\|}{C}}-OH + -\underset{|}{\overset{|}{C}}-OH \longrightarrow -\overset{O}{\overset{\|}{C}}-O-\underset{|}{\overset{|}{C}}- + H_2O \qquad \text{(iii)}$$

So, in general, DGEBA oligomers with $n > 0$ will act as polyfunctional units of functionality $f = 2 + n$ and these will contain at least two reaction centres of differing reactivity, i.e. epoxide and hydroxyl groups. Thus, for diepoxide prepolymers reacted with bifunctional organic acids, the products resulting from reactions of type (i) will produce linear polymers whose molecular weight will increase before slower reactions of type (ii) and (iii) lead to branching and eventually crosslinking. Crosslinked networks are established more effectively with resins of higher epoxide group functionality and polyfunctional organic acids.

Clearly, with such competitive reactions the intermediate and final network structures can be complex and varied. Additional complexity is introduced when an initiating step is required to form the organic acid and this is most commonly met in anhydride-crosslinked DGEBA systems; these reactions have been widely studied.[1,2,8,16–20] In this case DGEBA free-hydroxyl groups are required to ring-open the anhydride to form a monoester linkage and a reactive carboxyl group before a diester crosslink can be formed. The following reactions and structures result when phthalic anhydride is used.

Addition esterification:

monoester formation

$$R_1\text{—}CH(OH)\text{—}R_2 + C_6H_4(CO)_2O \xrightarrow{k_1} R_1R_2HC\text{—}O\text{—}C(=O)\text{—}C_6H_4\text{—}C(=O)\text{—}OH \qquad \text{(iv)}$$

diester formation

$$R_1R_2HC\text{—}O\text{—}C(=O)\text{—}C_6H_4\text{—}C(=O)\text{—}OH + \overset{O}{CH_2\text{—}CH}\text{—}R_3 \xrightarrow{k_2}$$

$$R_1R_2HC\text{—}O\text{—}C(=O)\text{—}C_6H_4\text{—}C(=O)\text{—}O\text{—}CH_2\text{—}CH(OH)\text{—}R_3 \qquad \text{(v)}$$

TABLE 1

EPOXY RESIN PREPOLYMERS AND ANHYDRIDE AND AMINE CURING AGENTS DISCUSSED IN THIS REVIEW

Name	Abbreviation	Chemical structure
Diglycidyl ether of bisphenol A	DGEBA	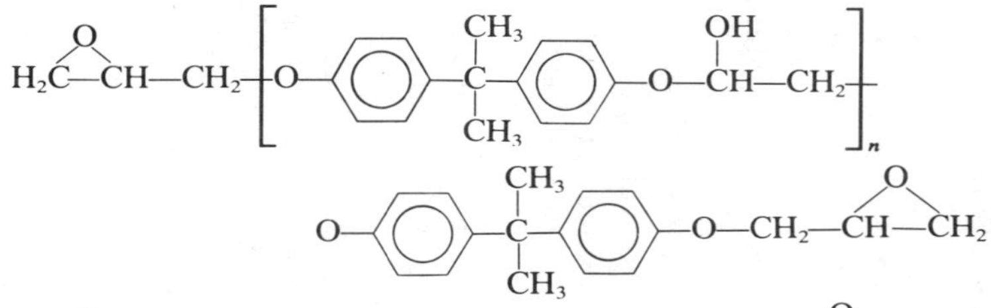 $H_2C(O)CH{-}CH_2{-}[O{-}C_6H_4{-}C(CH_3)_2{-}C_6H_4{-}O{-}CH(OH){-}CH_2]_n{-}O{-}C_6H_4{-}C(CH_3)_2{-}C_6H_4{-}O{-}CH_2{-}CH(O)CH_2$
Diethylene glycol diglycidyl ether	DDGE	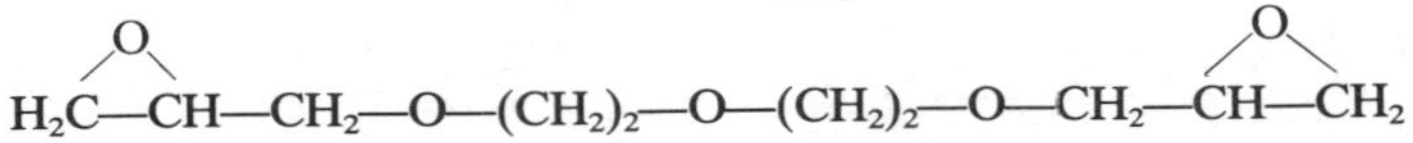 $H_2C(O)CH{-}CH_2{-}O{-}(CH_2)_2{-}O{-}(CH_2)_2{-}O{-}CH_2{-}CH(O)CH_2$
Resorcinol diglycidyl ether	RDGE	 $H_2C(O)CH{-}CH_2{-}O{-}C_6H_4{-}O{-}CH_2{-}CH(O)CH_2$
Triglycidyl ether of glycerol	TGEG	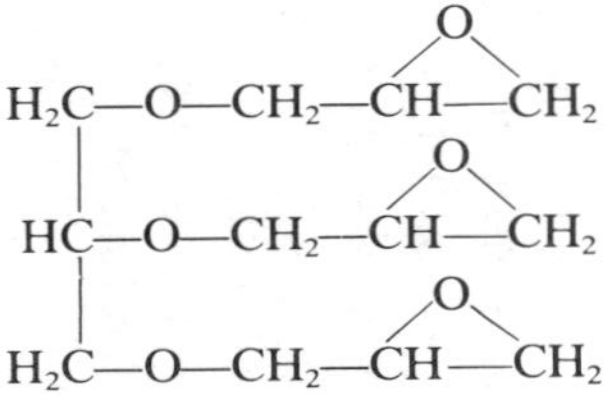 $H_2C{-}O{-}CH_2{-}CH(O)CH_2$ / $HC{-}O{-}CH_2{-}CH(O)CH_2$ / $H_2C{-}O{-}CH_2{-}CH(O)CH_2$
Dodecenylsuccinic anhydride	DDSA	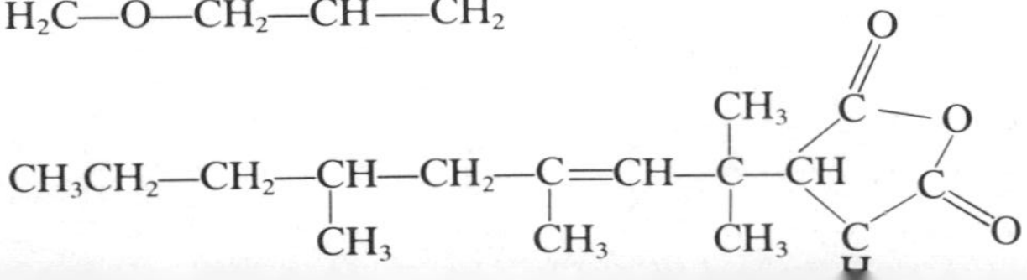 $CH_3CH_2{-}CH_2{-}CH(CH_3){-}CH_2{-}C(CH_3){=}CH{-}C(CH_3)_2{-}CH$ (succinic anhydride ring: $C(=O){-}O{-}C(=O){-}C$)

Hexahydrophthalic anhydride	HHPA	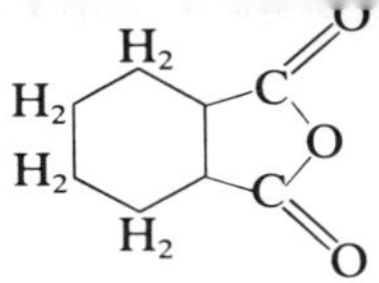
Methyltetrahydrophthalic anhydride	MTHPA	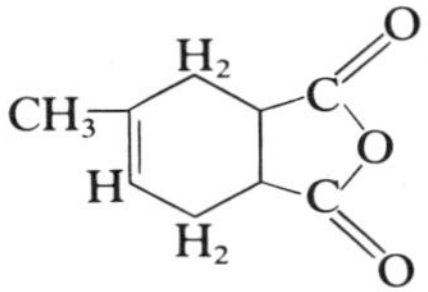
Methylnadic anhydride	MNA	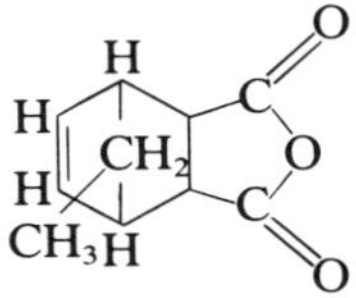
Phthalic anhydride	PA	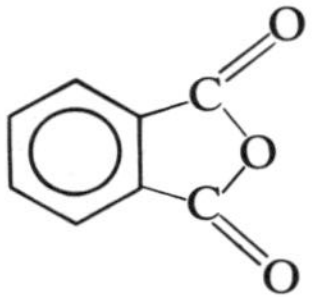
Tetrahydrophthalic anhydride	THPA	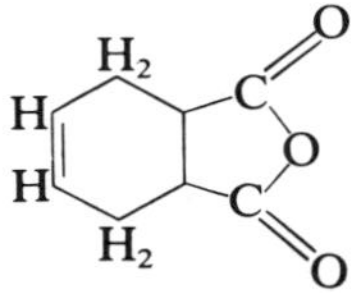

TABLE 1 (*contd.*)

Name	*Abbreviation*	*Chemical structure*
n-Dodecylamin	DDA	$C_{12}H_{25}NH_2$
Ethylenediamine	EDA	$H_2N—(CH_2)_2—NH_2$
Hexamethylenediamine	HMDA	$H_2N—(CH_2)_6—NH_2$
Diethylenetriamine	DETA	$H_2N—(CH_2)_2—NH—(CH_2)_2—NH_2$
Triethylenetetramine	TETA	$H_2N[(CH_2)_2NH]_2(CH_2)_2—NH_2$
Tetraethylenepentamine	TEPA	$HN(CH_2—CH_2—NH—CH_2—CH_2—NH_2)_2$
4′,4′-Diaminodiphenyl sulfone	DDS	H_2N O S O NH_2
4′,4′-Methylenedianiline (also 4,4′-diaminodiphenylmethane)	MDA (DDM)	H_2N CH_2 NH_2
Benzyldimethylamine	BDMA	$CH_2N(CH_3)_2$
m-Phenylenediamine	mPDA	NH_2 NH_2

Addition etherification:

$$\underset{\substack{|\\R_2}}{\overset{\substack{R_1\\|}}{HC}}\text{—OH} + \overset{O}{\overbrace{CH_2\text{—}CH}}\text{—}R_3 \xrightarrow{k_3} \underset{\substack{|\\R_2}}{\overset{\substack{R_1\\|}}{HC}}\text{—O—}CH_2\text{—}\overset{\substack{OH\\|}}{CH}\text{—}R_3 \qquad \text{(vi)}$$

Under the conditions normally employed for these reactions, addition etherification occurs[8,19,20] and condensation esterification is minimal. Thus on completion of reactions (v) and (vi) the free-hydroxyl group concentration is conserved and these groups are able to participate in further reactions.

Some of the components of organic-acid cured epoxy resins mentioned here are shown in Table 1.

2.1.2. Polyamine Reactions

Reviews of polyamine reactions[1,2] indicate that primary and secondary amines react with epoxide groups in similar ways although the reactivity of the former is higher. The following reactions for a primary amine illustrate the general reaction scheme.

Primary to secondary amine:

$$R_1NH_2 + \overset{O}{\overbrace{CH_2\text{—}CH}}\text{—}R_2 \xrightarrow{k_1} R_1NH\text{—}CH_2\text{—}\overset{\substack{OH\\|}}{CH}\text{—}R_2 \equiv R_1'NH \qquad \text{(vii)}$$

Secondary to tertiary amine:

$$R_1'NH + \overset{O}{\overbrace{CH_2\text{—}CH}}\text{—}R_2' \xrightarrow{k_2} R_1'N\text{—}CH_2\text{—}\overset{\substack{OH\\|}}{CH}\text{—}R_2' \qquad \text{(viii)}$$

Reaction (viii) will proceed if the secondary amine is not too sterically hindered. In the presence of proton donors these reactions are accelerated. So the initial presence and development of free-hydroxyl groups will influence the rate of reaction. Although some authors ignore etherification reactions (ii) and (vi) in polyamine systems,[13,21] others suggest that they may occur under conditions of excess of epoxide over amine, i.e. at epoxide concentrations above that required for stoichiometric reaction.[22–24]

Hence, the course of a polyamine–DGEBA reaction can be kept relatively simple by reducing the epoxy prepolymer free-hydroxyl group concentration through choice of an $n \approx 0$ resin. In this case diepoxide prepolymers will form linear structures with bifunctional amines (e.g. a single primary or bifunctional secondary amine). Network formation requires a polyfunctional amine and in such cases a functionality $f_A \geqslant 4$ is usually employed. If etherification reactions are excluded, simple network structures will result. Table 1 summarizes some of the systems that will be mentioned later.

2.2. Ideal and Non-Ideal Networks

2.2.1. Ideal Networks

Flory[25] identified the concept of an infinite network, which in an ideal case would extend to the macroscopic dimensions of the specimen. Using a statistical treatment of a homogeneous reaction between two groups of equal reactivity in the presence of intermolecular reactions only, Flory derived a condition for gelation (i.e. the formation of an infinite network). For an epoxy system with an epoxide group functionality f_E and a crosslinking agent functionality f_A the following equation should hold in the absence of differing group reactivities and intramolecular reactions:

$$(p_E p_A)\text{gel} = 1/(f_E - 1)(f_A - 1) \tag{1}$$

where p_A and p_E are the respective fractions of reactive groups at the gel point.

However, more information is required to identify whether a network is ideal. In their review of structure in polymer networks, Dusek and Prins[26] identified a structural requirement by defining an ideal network as 'a collection of Gaussian chains between f-functional junction points (crosslinks) under the condition that all functionalities of the junction points have reacted with the ends of all and different chains'. This is in complete accord with the general process of random intermolecular crosslinking reactions where, as explained by Flory,[25] with increasing degree of branching the molecular weight distribution becomes broader and at the gel point the number-average degree of polymerization remains finite whereas the weight-average becomes infinite. If the reaction is able to continue beyond the gel point the unreacted sol fraction should coexist with the infinite network and eventually become attached to it as the reaction proceeds to completion.

2.2.2. Non-Ideal Networks

Real networks are rarely ideal. They suffer from a number of structural imperfections which may be an intrinsic feature of the system prior to crosslinking or may form as a result of crosslinking. Dusek and Prins[26] identified four basic types of defects which have a bearing on structure and morphology in epoxy resins.

(i) *Pre-existing order:* This could be natural, involving partial ordering of molecules due to aggregation or better defined with the presence of crystallites. Alternatively it could be imposed by external means, e.g. mechanical shear or liquid convection currents.

(ii) *Network defects*: The most common are illustrated in Fig. 1 and include unreacted functional groups, intramolecular reactions leading to closed loops and permanent chain entanglements.

(iii) *Inhomogeneous network formation:* In the absence of an etherification reaction, an $n = 0$ DGEBA diepoxide prepolymer reacted with a simple tetra-functional crosslinking agent would form a homogeneous network with two discrete crosslink chain lengths (if the conditions for an ideal network

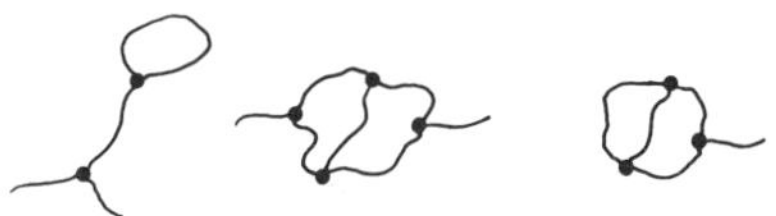

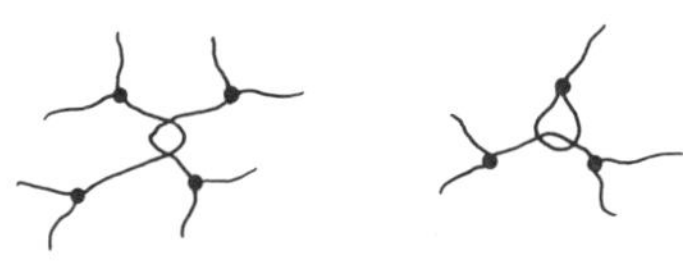

FIG. 1. Common types of network defects[26] including unreacted functional groups, intramolecular reactions leading to loops and cyclization, and physical entanglements which may be permanently trapped by chemical crosslinks.

were achieved). More generally the prepolymer chains will be polydisperse and there will always be a distribution of crosslink sizes. In principle this is not important if the crosslink chain segments exhibit Gaussian behaviour (and affine deformation holds). Although this is unlikely in most highly crosslinked epoxy networks it has been demonstrated that some weakly crosslinked epoxy–amine networks (reacted with an excess of diamine) appear Gaussian.[27,28] A broad distribution of crosslink chain lengths could be homogeneously distributed but a number of factors may produce a spatial distribution of chain lengths and an inhomogeneous network. These could include pre-existing order, localization of the heat of reaction producing more highly reactive centres,[29] local gelation with resultant increases in local reaction rates leading to the formation of microgel[30,31] and reactions which are intrinsically inhomogeneous because intramolecular reactions are favoured.[32,33]

(iv) *Phase separation:* In solution, reactions must proceed at localized points and true macroscopic gelation requires that the individual regions become connected before local separation of the developing gel from the unreacted solvent phase occurs. Of course the latter may be suppressed because the activity of the unreacted solvent phase in the gel may remain high throughout the reaction. If it does not, or if some critical value of crosslinking density is achieved locally, phase separation will occur. In this case the gel phase will shrink as the solvent phase is rejected and a two-phase or multiphase structure could result, i.e. crosslinks may be introduced at different volumes of the network phase or reactions may proceed from the surfaces of microgel particles. Dusek and co-workers have called this incompatibility process macro- and micro-syneresis.[26,34,35]

Intrinsic formation of an inhomogeneous network during reaction is different from that due to phase separation, as Fig. 2 illustrates. The two appear similar when extreme inhomogeneous reaction occurs. This produces a discrete microgel particulate phase which may continue to be reactive and become the dominant phase or may become unreactive but connected by a more loosely crosslinked intergel network.

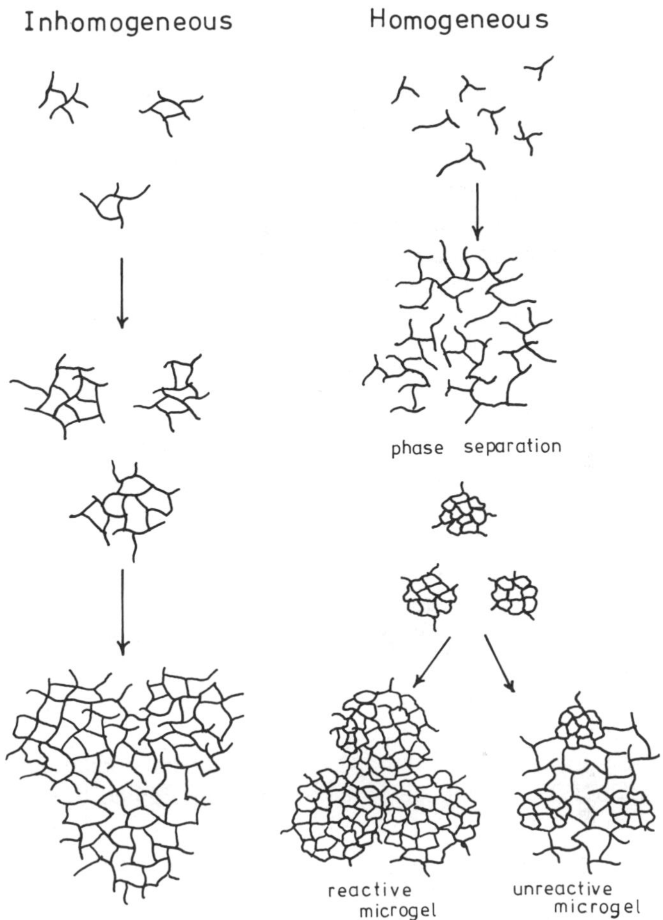

FIG. 2. Essential difference between an intrinsically inhomogeneously reacting network and one that reacts homogeneously but experiences phase separation to form a microgel. Phase separation will in general produce better defined boundaries.

2.3. Observed Reaction Behaviour

As discussed above, the crosslinking reaction mechanism has often been held responsible for the development of inhomogeneous networks and morphology. This can be tested by theoretically treating the reactions involved and comparing predictions with observations. A number of statistical treatments exist to predict reaction characteristics.[23–26,36,37] However, the observed behaviour may depart from that predicted for several reasons. The most obvious is over-simplification of the reaction mechanism due to neglect of certain reactions or the inability of the model to handle complex reactions. Similarly, most treatments are simplified by setting the reactivities of functional groups or their corresponding reactivity ratios as constants during the course of reaction; this cannot always be verified. Most treatments assume homogeneous chemical control of the reaction, but in highly viscous media (or beyond the gel point) diffusion control may be important. In the event that the reactive groups are inhomogeneously distributed, it is likely that the reactions and the resulting network will be spatially inhomogeneous, particularly if competitive simultaneous reactions (e.g. esterification and etherification) are possible in the system.

2.3.1. Homogeneous Reactions

In spite of theoretical advances it is still difficult to treat crosslinking systems which have a variety of reactive groups and those where the reactivity of primary or secondary groups of the same type differ or change during the course of reaction. Consequently, few epoxy resin studies have been undertaken which can clearly demonstrate that homogeneous reaction has occurred.

The best documented system where homogeneous reaction has been demonstrated is the diepoxide/diamine system studied by Dusek and co-workers.[13,38] Pure $n = 0$ DGEBA prepolymer was cured with aliphatic and aromatic amines including hexamethylenediamine (HMDA), ethylenediamine (EDA) and n-dodecylamine (DDA). Most attention was directed at the HMDA system at a variety of epoxide/amine group ratios. Applying the vector probability-generating-function theory of Gordon[36] and ignoring the possibility of etherification reactions[38] it was possible to show that the course of the reactions up to the gel point followed a homogeneous course. For a range of reactive-group stoichiometries the conversion at the gel point (cf. eqn (1)), the concentration of tertiary amine groups (cf. reaction (viii)) and the number-average molecular weight over the course of the reaction agreed reasonably well with the predictions.

However, above the gel point the results were less conclusive because many of the systems considered were not fully reacted. Charlesworth[14,39] corrected this by studying essentially an $n \approx 0$

DGEBA prepolymer with stoichiometric and non-stoichiometric systems containing EDA and diaminodiphenylmethane (DDM) to nearly complete reaction. In this case he applied the simpler expectation theory of Macosko and Miller[37] and compared its predictions with the weight fraction of soluble material, the unreacted functional-group concentration and the number-average molecular weight of the soluble fraction.[14] Significantly, he also compared the theoretical and experimental distribution of individual chemical species present in the mixture before gelation and in the soluble fraction following gelation.[39] Both studies confirmed that the reactions were closely homogeneous over the complete course of reaction. This indicated that diffusion control of the reaction following gelation was absent or not the rate-determining step.

Conversion at the gel point is sensitive to intramolecular reactions. Lunak and Dusek found little evidence for their presence even when the system was diluted by up to 60% in chloroform.[13] However, Charlesworth found evidence that in stoichiometric systems, one in 16 of the bonds formed at the theoretical gel point were loop producers and in systems containing diluent the fraction of intramolecular bonds increases as the polymer concentration decreases.[39] If epoxy groups are in excess the critical conversion at the gel point is fractionally higher in these systems. This point was noted by Lunak and Dusek[13] and explained by Bokare and Gandhi using an expectation theory treatment which included the etherification reaction.[24] Bokare and Gandhi's results agreed closely with the original work, but showed that free-hydroxyl group reactivity only becomes significant in diepoxide/diamine systems when the amine group concentration becomes low, i.e. at the later stages of a stoichiometric reaction and earlier for a non-stoichiometric (epoxide excess) reaction.

In summary, then, some diepoxide/diamine systems appear to react randomly and homogeneously, and etherification reactions, when present, do not produce any significant departures from this. However, a homogeneous reaction mechanism does not necessarily imply a fully homogeneous and fully interconnected matrix. Some observations by Charlesworth bear this out. In stoichiometric and excess diamine (non-stoichiometric) systems he noted that just above the theoretical gel points the systems were totally soluble but that upon agitation the chloroform solutions appeared cloudy.[39] He thought this was due to fragmentation of a low crosslink-density macrogel. It could also result from a microgel phase which had not been interconnected.

2.3.2. *Inhomogeneous Reactions*

In line with the experience of demonstrating homogeneous reaction behaviour, few studies have been able to demonstrate that inhomogeneous epoxy resin crosslinking reactions occur. One exception has been the study of anhydride-crosslinked DGEBA prepolymers based on commercial systems. Such materials and more model systems have been widely studied[1,2,16–19] but conflicting reaction mechanisms and kinetics have been reported.

A more recent study[8,20] considered that inhomogeneous reactions could be favoured in systems containing $n > 0$ DGEBA prepolymers where the free-hydroxyl groups were required to ring-open anhydride molecules before complete crosslinking could occur [reactions (iv) and (v)]. Two systems were studied having very different epoxy/hydroxyl (E/H) group ratios; the average DGEBA n-values were 1·7 (low E/H) and 0·36 (high E/H). As shown in Table 2, the prepolymer molecular weight distributions and resulting epoxide- and hydroxyl-group concentrations are significantly different. These commercial systems were crosslinked with phthalic anhydride (PA) in the case of the $\langle n \rangle \sim 1{\cdot}7$ resin (where $\langle n \rangle$ represents the average DGEBA oligomer chain repeat length) (commercial system CT200-HT901) and a eutectic mixture of PA and tetrahydrophthalic anhydride (THPA) in the $\langle n \rangle \sim 0{\cdot}36$ resin (commercial system CY207–HT903); the respective epoxide : anhydride reaction stoichiometries of 1 : 0·85 and 1 : 0·91 were those recommended for this type of system.[1]

The reaction mechanism was found to follow generally the consecutive-step addition esterification and simultaneous addition etherification scheme represented by reactions (iv) to (vi) in agreement with that proposed earlier by Fisch and Hofmann.[16] These reactions may be rewritten as

$$\mathrm{A} + \mathrm{B} \xrightarrow{k_1} \mathrm{C} \tag{ix}$$

$$\mathrm{C} + \mathrm{D} \xrightarrow{k_2} \mathrm{E(A')} \tag{x}$$

$$\mathrm{A} + \mathrm{D} \xrightarrow{k_3} \mathrm{F(A')} \tag{xi}$$

where A' represents regenerated or secondary free-hydroxyl groups which may go on to participate in further reactions. In the $\langle n \rangle \sim 0{\cdot}36$ system, equations (x) and (xi) are repeated with B′ and C′ species because PA and THPA are present and have different reactivities.

TABLE 2
CHARACTERIZATION OF THE DGEBA PREPOLYMERS WITH $\langle n \rangle \sim 1{\cdot}7$ (COMMERCIAL SYSTEM CT200) AND $\langle n \rangle \sim 0{\cdot}36$ (COMMERCIAL SYSTEM CY207)

Variable	*CT200* $\langle n \rangle \sim 1{\cdot}7$	*CY207* $\langle n \rangle \sim 0{\cdot}36$
Epoxide content (mol/kg)[a]	2·4 ± 0·1	4·5 ± 0·1
Hydroxyl content (mol/kg)[b]	2·1–2·2	0·7–1·2
Epoxide/hydroxyl molar ratio	1·12	4·74
DGEBA % mass fractions:[c]		
$n=0$	15·4	61·2
$n=1$	18·4	10·2
$n=2$	14·0	5·8
$n=3$	13·7	2·6
$n=4$	10·6	2·7
$n>4$	19·3	8·4
$n=0$ end-groups:		
α-Glycol	3·5	8·8
or		
Chlorohydrin	3·7	9·2

[a] Determined by the potassium iodide method.
[b] Inferred from GPC data; end-group impurities account for the range.
[c] Calculated from GPC data; α-glycol groups assigned to end-group impurities.

A simplifying element is introduced into the kinetic treatment of these complex reactions because anhydride consumption is the key reaction step. In the $\langle n \rangle \sim 1{\cdot}7$ system the initial concentrations of hydroxyl, anhydride and epoxide groups are nearly equal, i.e. $[A] \sim [B] \sim [D]$, so during the initial phase of reaction anhydride consumption should be second-order

$$\frac{dB}{dt} = k_1[A][B] \tag{2}$$

if homogeneous reaction occurs. If hydroxyl- or anhydride-group

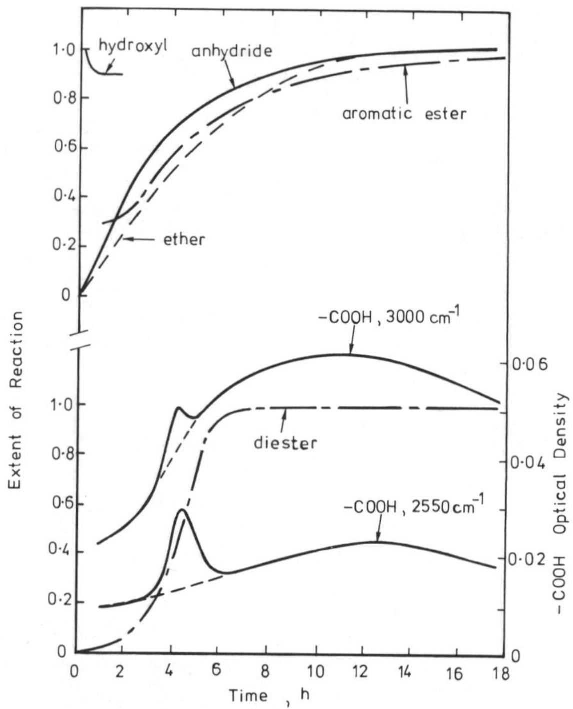

FIG. 3. Summary of the extent-of-change behaviour of key chemical groups in the reaction of DGEBA ($\langle n \rangle \sim 1{\cdot}7$)/PA (commercial system CT200-HT901) at 125°C. These were determined by infrared spectroscopy[8] and the carboxylic-acid group optical density is shown directly.

availability is restricted, first-order kinetics would be expected:

$$\frac{\mathrm{d}B}{\mathrm{d}t} = k_1[\mathrm{B}] \tag{3}$$

In the $\langle n \rangle \sim 0{\cdot}36$ system $[\mathrm{A}] \ll [\mathrm{B}] \sim [\mathrm{D}]$ and in the absence of competition between PA and THPA first-order anhydride consumption kinetics would be expected for a homogeneous reaction.

The extent of reaction

$$P(t) = 1 - \frac{x(t) - x(\infty)}{x(0) - x(\infty)}$$

for various key groups x as a function of reaction time for the $\langle n \rangle \sim 1{\cdot}7$ system is shown in Fig. 3. Analysis of the PA consumption curve demonstrated that first-order kinetics were followed up to 94%

conversion. However, this followed an initial rapid reaction in which 8% of the free hydroxyl groups were consumed and provided a carboxylic acid group reservoir. If eqns (ix) and (x) are represented as consecutive first-order reactions and the regenerated free hydroxyl groups (A′) are ignored, then the sequence

$$M_1 \xrightarrow{k_{12}} M_2 \xrightarrow{k_{23}} M_3 \qquad \text{(xii)}$$

applies and this has been shown to have an analytic solution.[40] In this case M_1 (anhydride consumption) follows first-order kinetics, M_2 (carboxylic acid) undergoes a peaking behaviour and M_3 (diester formation) should follow S-shaped kinetics with an induction period that depends on k_{12} and k_{23}. This behaviour is observed in Fig. 3. Analysis of the kinetics of the total aromatic ester content also gave first-order behaviour and slower ether formation followed half-order kinetics. Retrospective kinetic analysis of a closely similar system studied by other authors[16] confirmed that the same kinetic behaviour occurred.[8]

The reaction procedure and behaviour of the $\langle n \rangle \sim 0{\cdot}36$ system was more complex.[20] In summary, S-shaped kinetic behaviour occurred for most groups during partial reaction in a 'gelation' phase of cure and this followed an initial rapid reaction of anhydride which consumed about 14% of the available hydroxyl groups. This phase of the reaction was unusual in that an apparent equilibrium or induction period of several hours was observed before the reactions 'switched on' again and progressed to a new equilibrium stage. Postcuring at a higher temperature led to gelation after several hours and PA and THPA consumption was found to be competitive with each following half-order kinetics. However, total anhydride consumption and aromatic ester formation followed first-order kinetics and slower etherification reactions occurred following half-order kinetics.

Comparing the extent of reaction at the gel point for the two systems it was found, contrary to the expectation based on functionality [eqn (1)], that the extent of anhydride reaction was higher in the $\langle n \rangle \sim 1{\cdot}7$ system. All of the results were consistent with both systems exhibiting inhomogeneous reaction behaviour of an hydroxyl-group limited colloid type, i.e. $X + Y \rightarrow Y$. Such reaction behaviour is in line with Waite's consideration of bimolecular reaction rates in viscous liquids or condensed phases.[41,42] He also showed that in condensed phases where the bimolecular reaction $X + Y \rightarrow XY$ is diffusion-limited, the behaviour should remain second-order but the rate constant would become time- and viscosity-dependent; this type of behaviour is not observed. In the colloidal model the free-hydroxyl groups would be inhomogeneously distributed, possibly being retained in or around the colloid particle Y.

These studies confirm that anhydride-cured DGEBA prepolymers can exhibit inhomogeneous reaction behaviour and, from the discussion above, an inhomogeneous but connected matrix could be expected. However, reaction studies do not provide insight into network spatial structure; other bulk techniques are required and these will be discussed below.

2.4. Uneven Curing

Inhomogeneous structures and reaction behaviour could result from incomplete mixing of multicomponent epoxy resins or excessive exothermic heating during reaction. Inadequate mixing has been investigated for triethylenetetramine (TETA)-cured DGEBA resins using dinitrofluorobenzene-stained sections and ultraviolet light microscopy.[43] This work confirmed that hand mixing led to a non-uniform distribution of amine but vigorous high-shear mixing would produce uniform materials.

An extensive study was carried out by Bell on diethylenetriamine (DETA) and polyamide-cured DGEBA resins.[44] In this case he consolidated earlier work by Tateosian and Royer (see cross-reference in ref. 44) which had shown a 58% improvement in impact strength following dynamic mixing. Bell compared hand mixing, hand plus concentric-cylinder (high-shear) mixing and also the latter with the addition of a high voltage across the concentric cylinders. He considered their effect on tensile strength, glass transition temperature T_g and microscopic fracture-surface morphology. Improved mixing in the polyamide system produced no significant improvement in properties but in the DETA system T_g was found to increase by 7°C and the tensile strength (in the degassed case) increased from 44 to 67 MPa. Hand mixing produced finely structured (10–100 nm) fracture surfaces but the electromechanical mixing produced a more homogeneous product with little evidence of microstructure.

These studies confirm that thorough mixing is essential if a homogeneous product with optimum properties is to be obtained. The same is true if a correct view of the reaction behaviour is sought. Many studies on epoxy resin systems do not specify their purification and mixing methods and in such cases microstructural claims may be called into question. Interestingly, the electromechanical system used by Bell reportedly provides shear on an Ångström scale.[45–48] This is below the size of possible network microstructure in epoxy systems and raises the question of what effective mixing scale is required to achieve a homogeneous material, assuming homogeneous reactions are favoured? Further work on this topic could be informative and lead to improved materials properties.

3. BULK MICROSTRUCTURE

In crosslinked polymers many levels of order may potentially exist, ranging from local chain or network configurational order at a molecular level through a variety of intermediate structures up to that of an ideal three-dimensional network occupying the macroscopic dimensions of the sample. In the absence of a distinct crystalline phase, partial ordering of chains or local differences in packing or network density will affect homogeneity. Thus the region associated with a particular microstructural entity (above the molecular level) should exhibit a density different from its surroundings. In this case the lower limit of discrimination is imposed by thermal density fluctuations[48–51] or non-equilibrium structures which are dissipative in time.[52] Such microstructural regions may be fully interconnected in the matrix and unless phase separation has occurred distinct structure boundaries would not necessarily be expected. Whatever the nature of the boundary, microstructure based on density difference will be amenable to study using scattering techniques such as light scattering and small-angle X-ray scattering where density differences below 1% can be detected.

It is possible that the density of a network is uniform and homogeneous in comparison with thermal fluctuations but that it is not homogeneously connected. In this case a microstructural entity would be differentiated in the bulk by its type and concentration of crosslinks. This could be discerned by microscopic observations of swelling behaviour or by indirect means such as fracture surface topography or rubber-phase mechanical properties. However, in general, this type of non-ideal network microstructure is difficult to detect, particularly when the differences between the entity and its surroundings are small.

3.1. Microstructure in Linear Amorphous Polymers

Local order and microstructure in linear glassy polymers has been more hotly debated than that in crosslinked polymer networks, and some of the experience gained on linear systems is helpful. Many investigations based largely on electron microscope observations asserted that a purely random chain model of amorphous polymers was not appropriate and that microstructure in the form of locally ordered bundles, nodules, micelles or paracrystals exists in some polymers with a scale around 5–10 nm; evidence supporting this has been widely reviewed.[53,54] However, many of the properties of these materials are explained by the random coil model[55] and evidence from electron microscopy, X-ray and neutron scattering[55–57] supports this view.

More recently a Faraday Discussion, which included a session on amorphous polymers, brought the supporters of each side of the debate together. New evidence supporting the absence of extensive microstructure in many non-interacting linear polymers was presented and discussed vigorously.[58] However, weak orientational correlation was found in liquid *n*-alkanes (model Gaussian chains) with a correlation volume $<(10\,\text{Å})^3$ by light-scattering methods[59] and the same work used small-angle neutron scattering and Raman spectroscopy to show that, in spite of the small segmental correlation, the average conformational behaviour of single chains was random. Similarly, using small-angle X-ray scattering and high-resolution electron microscopy, Uhlmann[60] found little evidence to support the significant presence of bulk microstructure in polycarbonate, polymethyl methacrylate, polyethylene terephthalate, polyvinyl chloride and polystyrene.

Thus, whilst not suggesting that the debate on microstructure in linear amorphous polymers is closed, this work has shown that great care is required in interpreting electron-microscopy observations of microstructure and bulk scattering techniques should be applied to 'clean' materials to elucidate unambiguously bulk inhomogeneity or microstructure. In the following sections the chronological order of experimental observations will be reversed and the evidence for microstructure in epoxy resins from scattering experiments will be discussed first.

3.2. Scattering Observations of Inhomogeneity

Few detailed light-scattering or X-ray scattering experiments have been reported for crosslinked epoxy resins. However, local order and differential density in the bulk will give rise to valuable scattering information provided the results are not confused or masked by the presence of contaminants or extrinsic defects.

In recent work,[61–64] Rayleigh-type elastic light scattering has been used to study pre-existing order and reaction behaviour. In this technique regions in the bulk that exhibit a different refractive index from their surroundings scatter light at the same wavelength as the incident beam. So fluctuations in density and concentration due to permanent structures or transient processes in the bulk will give rise to light scattering. The angular dependence of scattered light intensity (usually recorded in the plane of the incident beam) forms a 'scattering envelope' which contains information on the scatterer size and concentration. The polarization of the scattered light also gives information on scatterer shape and optical anisotropy. A homogeneous medium such as a simple liquid would exhibit a small scattering intensity due to thermal-density fluctuations and, provided these were very small, the scattering envelope would be symmetrical. Pure water is an example of a weakly scattering homogeneous medium; it has a Rayleigh ratio (ratio of scattered to incident light intensity) of about $1 \times 10^{-6}\,cm^{-1}$.

Elastic light scattering has been complemented by Brillouin spectroscopy to elucidate pre-existing order and reaction behaviour.[61,62,64] This technique looks at both the elastically scattered light and that scattered inelastically (at a different frequency from that of the incident light) by propagating density fluctuations associated with GHz-frequency acoustic phonons. Thus, it provides complementary information on static scatterers and additional information on phonon velocity and attenuation which are related to the bulk properties of the medium and its molecular relaxation characteristics at GHz frequencies. In this technique the scattered intensity is recorded over a limited frequency range (typically 100 GHz centred about the incident optical frequency) at a fixed scattering angle. In its simplest form a spectrum is obtained which consists of an elastically scattered 'Rayleigh' peak surrounded by two weaker, frequency-shifted 'Brillouin' peaks. The relative intensity, peak position and peak width give information on static and phonon scattering.

Small-angle X-ray scattering has also been applied to fully reacted materials[60,65,66] and provides information on microstructural size and concentration. All of these techniques also give information on the level of thermal-density fluctuations.

3.2.1. Pre-existing Order

One study[61,62] has used light scattering to investigate the presence of local order in DGEBA prepolymers in an effort to gauge its influence on the formation of inhomogeneity during anhydride crosslinking (as discussed above). The $\langle n \rangle \sim 0{\cdot}36$ and 1·7 prepolymers characterized in Table 2 were studied following exhaustive filtration through 0·2 μm filters. The prepolymers were found to contain pre-existing order which was temperature-dependent and also reduced in size and concentration on dissolution; some key observations are repeated here.

At room temperature the lower molecular weight prepolymer, $\langle n \rangle \sim 0{\cdot}36$, with the smaller number of free-hydroxyl groups, exhibited a symmetric scattering envelope whose dissymmetry $Z = I(45°)/I(135°)$ was unity, a value characteristic of small scatterers with dimensions less than about 17 nm. In comparison, the higher molecular weight (higher number of free-hydroxyl groups) $\langle n \rangle \sim 1{\cdot}7$ prepolymer exhibited a temperature-independent dissymmetry of 1·28, a value characteristic of scatterers greater than 20 nm. The fully corrected 90° scattering intensity allows calculation of the Rayleigh ratio R^{90},[67] and determination of its isotropic (R^{90}_{is}) and anisotropic (R^{90}_{an}) components,

$$R^{90} = R^{90}_{\mathrm{is}} + R^{90}_{\mathrm{an}} \tag{4}$$

where R^{90}_{is} may be calculated from[68]

$$R^{90}_{\mathrm{is}} = R^{90}(6 - 7\rho_{\mathrm{u}})/(6 + 6\rho_{\mathrm{u}}) \tag{5}$$

and the depolarization ratio is obtained from

$$\rho_{\mathrm{u}} = I_{\mathrm{Hu}}/I_{\mathrm{Vu}} \tag{6}$$

where I_{Hu} and I_{Vu} are the 90° unpolarized scattered intensities obtained using horizontally and vertically polarized incident light.

R_{is} and R_{an} for the $\langle n \rangle \sim 1{\cdot}7$ resin were twice as large as the values for the $\langle n \rangle \sim 0{\cdot}36$ resin and both decreased with increasing temperature in the range 300–420 K; the isotropic scattering behaviours are shown in Fig. 4. The $\langle n \rangle \sim 0{\cdot}36$ resin exhibits an R_{is} transition from a T^{-1} to a T dependence above 363 K and R_{an} approaches a constant value. The behaviour of both resins is consistent with a structural liquid transforming to a more unstructured state at higher temperatures. In the case of the $\langle n \rangle \sim 0{\cdot}36$ resin, it becomes nearly fully unstructured and thermal-density fluctuations begin to play a sig-

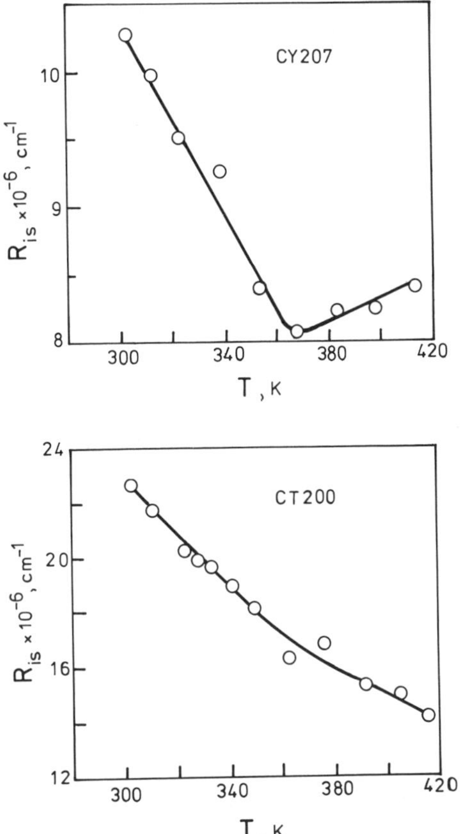

FIG. 4. The temperature dependence of the isotropic Rayleigh ratio (R_{is}) for the $\langle n \rangle \sim 1{\cdot}7$ resin (CT200) and the $\langle n \rangle \sim 0{\cdot}36$ resin (CY207).

nificant role in resin inhomogeneity above 363 K. This results in an isotropic scattering intensity which is directly proportional to temperature, a result expected for a non-interacting liquid.

In Brillouin spectroscopy a central (frequency-unshifted) Rayleigh peak (I_R) is observed along with, usually, two phonon frequency-shifted Brillouin peaks (I_B) symmetrically positioned about I_R; the resulting Landau–Placzek ratio $I_R/2I_B$ is a sensitive probe of local structure. The fully corrected temperature dependence of $I_R/2I_B$ for the two prepolymers is shown in Fig. 5: these resins were found to be undergoing hypersonic (GHz) relaxation in the temperature range

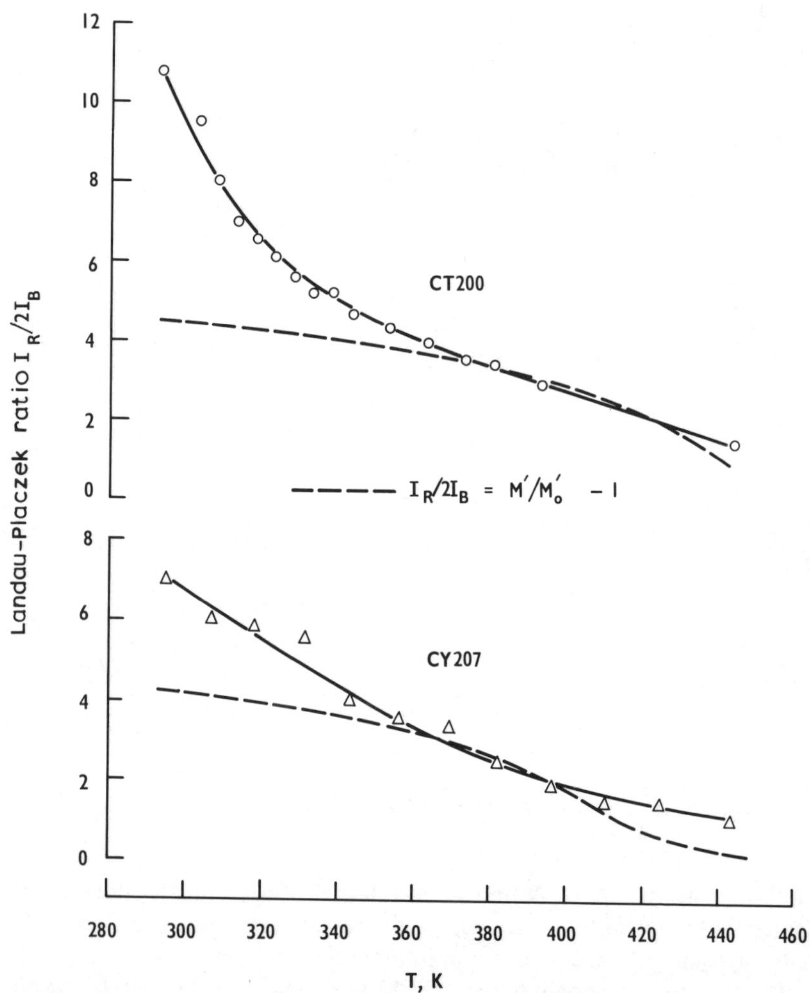

FIG. 5. The temperature dependence of the experimental Landau–Placzek ratios (full lines, ○, △) for the $\langle n \rangle \sim 1{\cdot}7$ resin (CT200) and the $\langle n \rangle \sim 0{\cdot}36$ resin (CY207), compared with that expected for hypersonic relaxation (———). (Reproduced with permission from ref. 62)

studied and this will influence the contributions of the unstructured liquid to $I_R/2I_B$. This contribution to the Landau–Placzek ratio was calculated using the model of Pinnow *et al.*,[69] who have shown that

$$I_R/2I_B = \gamma \frac{M'}{M'_0} - 1 \tag{7}$$

where γ is the specific heat ratio, C_p/C_v.

The temperature-dependent contribution was calculated from the high-frequency real part of the longitudinal modulus $M' = M'(T)$ and the low-frequency part $M'_0 = M'_0(T)$. These liquid-phase contributions are shown as dashed curves in Fig. 5. The structured-phase contributions to the Landau–Placzek ratio at lower temperatures are now clearly seen. The resulting excess scattering ratio of the two resins $\langle n \rangle \sim 1{\cdot}7/\langle n \rangle \sim 0{\cdot}36$ is 2·1 from the Landau–Placzek ratio and 2·2 from the isotropic Rayleigh ratio (R_{is}). This agreement confirms the higher structured level in the higher molecular weight resin.

Dissolution of the resins in chloroform gave a complex and significant change in R_{is} and a continuously decreasing change in R_{an} with increasing solvent content. By comparing the excess depolarization ratios of Sicotte and Rinfret[70] it was shown that both resins move progressively to an unstructured phase above a solvent content of about 65% and the $\langle n \rangle \sim 1{\cdot}7$ resin experiences a further change in liquid-phase correlation above 90%.[62]

These observations of thermally unstable and soluble scattering entities parallels the infrared spectroscopy behaviour of epoxide–hydroxyl group association in these resins.[8,20] This suggests that the scatterers are DGEBA molecular aggregates exhibiting hydrogen-bonding and that the extent of this bonding and the size and concentration of the aggregates is related directly to the free-hydroxyl group content of the prepolymer. If this is the case, the $\langle n \rangle \sim 1{\cdot}7$ resin should contain larger aggregates and/or a higher concentration of aggregates.

Inhomogeneity appraisal of these resins included a discrete sphere case and a non-discrete sphere case (i.e. an inhomogeneous Gaussian radial distribution of density[71]). These gave a dissymmetry inferred upper size limit for the aggregates of 20 nm for the $\langle n \rangle \sim 0{\cdot}36$ resin and 60–70 nm for the $\langle n \rangle \sim 1{\cdot}7$ resin. Molecular models combined with $n = 0$ DGEBA crystal data[11] gave a minimum aggregate size of about 5 nm. These sizes were combined with a realistic [aggregate]/

[$n = 0$, DGEBA solvent] two-phase model to obtain the possible range of aggregate volume fractions from the isotropic Rayleigh ratio data using

$$R_{is} = \frac{9}{2}\pi^2\left(\frac{n_0^4}{\lambda^4}\right)[(m^2-1)/m^2+2)]^2 V_f v P(90°) \tag{8}$$

where n_0 is the solvent-phase refractive index, m is the ratio of the solvent/aggregate refractive index, V_f is the volume fraction of aggregates of volume v and $P(90°)$ is the particle scattering function for non-discrete spheres.[71]

From refractive index measurements on the two prepolymers an estimate was made of the maximum density difference between an $n = 0$ phase and a phase containing all other oligomers with $n > 0$. This gave a maximum density difference of 4·3% for distinctly phase-separated prepolymers. However, in practice this is unlikely because the aggregates must be diffuse and exist as dynamic-equilibrium structures. So, for convenience and in line with similar considerations in linear amorphous polymers, a 1% density difference was chosen as a second case. A simple calculation[62] showed that a 5 nm thermal-density fluctuation in these prepolymers at 293 K would have a mean density variance of 0·76%. Using these two cases and eqn (8), the possible ranges of aggregate volume fractions were calculated and the room-temperature results are shown in Fig. 6. Assuming a simple case of monodisperse 10 nm aggregates exhibiting a mean density difference of 1%, the resulting volume fractions appropriate to the $\langle n \rangle \sim 0{\cdot}36$ and $\langle n \rangle \sim 1{\cdot}7$ resins would be 7% and 13% respectively. These would be higher if the density differences were smaller.

Although the size and concentration of aggregates will decrease with increasing temperature, they will be present in both resins at normal reaction temperatures; their persistence should be greater for larger-n oligomers. These pre-existing ordered regions should be rich in free-hydroxyl groups if hydrogen-bonding accounts for their formation, and as such they may act as preferential sites for reaction. This is certainly the case for anhydride crosslinking where the free-hydroxyl group initiates the reaction through anhydride ring-opening [reaction (iv)]. Primary and secondary reactions will continue to be favoured at these sites [cf. reactions (v) and (vi)] so an inhomogeneous reaction mechanism would be expected which could lead to an inhomogeneous network.

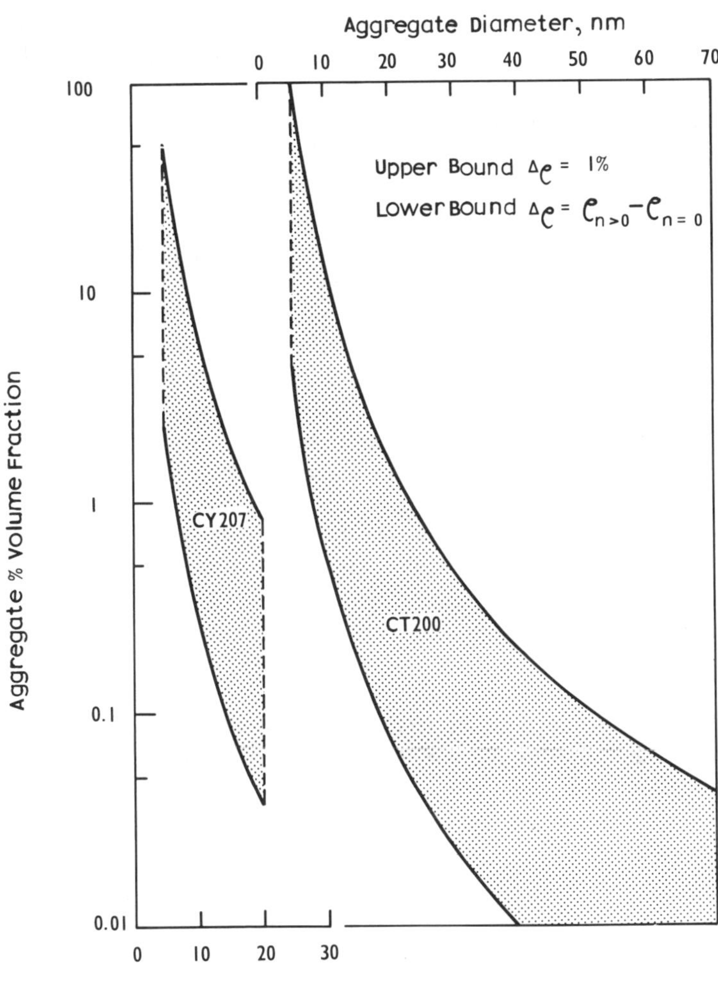

FIG. 6. Volume fractions of molecular aggregates for particular aggregate sizes to account for the observed isotropic Rayleigh ratios in the $\langle n \rangle \sim 1{\cdot}7$ resin (CT200) and the $\langle n \rangle \sim 0{\cdot}36$ resin (CY207). Each upper bound curve is for an aggregate–solvent phase density difference of 1% and each lower bound is for a density difference obtained for DGEBA $n = 0$ and $n > 0$ phase separation. (Reproduced from ref. 62.)

3.2.2. *Inhomogeneity Development During Reaction*

Two studies have used light scattering to follow the development of inhomogeneity during reaction in four different epoxy systems. The first[64] included the two DGEBA resins discussed above, which were anhydride-crosslinked (cf. Section 2.3.2). The second[63] involved two diamine-cured diepoxide systems whose oligomers contained no free-hydroxyl groups.

A full account of the Rayleigh and Brillouin scattering results obtained during the reaction of the anhydride-cured DGEBA resins will be given elsewhere. However, the Rayleigh scattering envelope dissymmetry ($Z = I(45°)/I(135°)$) and the isotropic Rayleigh ratio (R_{is}) do provide a useful insight into the development and change of inhomogeneity during reaction and these are shown in Figs 7 and 8. In both cases the $\langle n \rangle \sim 0{\cdot}36$ system (CY207–HT903) reacts more slowly with its 80°C 'gelation' phase and 120°C postcure, whereas the $\langle n \rangle \sim 1{\cdot}7$ system (CT200–HT901) reacts more rapidly at 125°C; the physical gelation points are correspondingly different.

The most striking initial observation is that on addition of the anhydride at the reaction temperature a significant change in dissymmetry occurs: from a prepolymer value of 1·28 to a reaction mixture value of 2·2 for the $\langle n \rangle \sim 1{\cdot}7$ system and from 1 to 1·8 for the $\langle n \rangle \sim 0{\cdot}36$ system. This behaviour coincides with the initial rapid hydroxyl–anhydride group reaction. This reaction then proceeds more slowly and the dissymmetry changes in line with it, decreasing to a constant value. During the initial rapid reaction the scattering intensity of the original prepolymer does not change dramatically; R_{is} increases from 8·5 to 10·2 in the $\langle n \rangle \sim 0{\cdot}36$ system and decreases from 15 to 12·5 in the $\langle n \rangle \sim 1{\cdot}7$ system. As reaction proceeds R_{is} increases steadily. However, in the $\langle n \rangle \sim 0{\cdot}36$ case R_{is} and z exhibit periods of constancy in the gelation phase which coincide with periods of little chemical change. On moving into the postcure phase, the higher temperature reduces R_{is} before it continues to rise but z changes more slowly, in line with a reaction-induced change, before increasing to a high final value. The time scales of change of these scattering parameters parallel closely the more rapid anhydride consumption reactions, whereas the more slowly varying refractive index (and density) changes parallel the slower diester and ether reactions. Finally, the more rapidly reacting and initially more inhomogeneous $\langle n \rangle \sim 1{\cdot}7$ system retains a relatively high isotropic scattering and dissymmetry, whereas the more homogeneous $\langle n \rangle \sim 0{\cdot}36$ system retains a high dissymmetry but a reduced isotropic scattering.

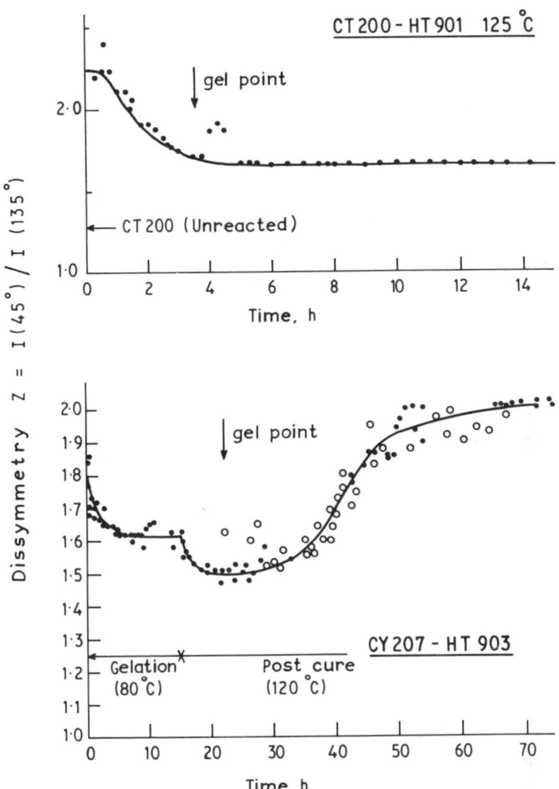

FIG. 7. Time dependence of the light-scattering envelope dissymmetry during the reaction of the DGEBA/anhydride systems CT200–HT901 and CY207–HT903. The gelation points and the unreacted resin dissymmetry values are indicated; for unreacted CY207, $z = 1$.

Interpreting these observations is not simple because above the gelation point internal stresses may develop in these materials which could influence the behaviour, particularly during the final stages of reaction. However, up to and just beyond the gel point the behaviour can be reconciled with inhomogeneity developing from pre-existing order. The simplest scheme is outlined in Fig. 9. Addition of the anhydride results in immediate reaction at DGEBA aggregate sites which contain a relatively high concentration of free-hydroxyl groups. This could be viewed as an outer-shell reaction in the first instance which reveals the 'true' extent of the aggregates. These rapid reactions account for 14% of the available free-hydroxyl groups in the $\langle n \rangle \sim$

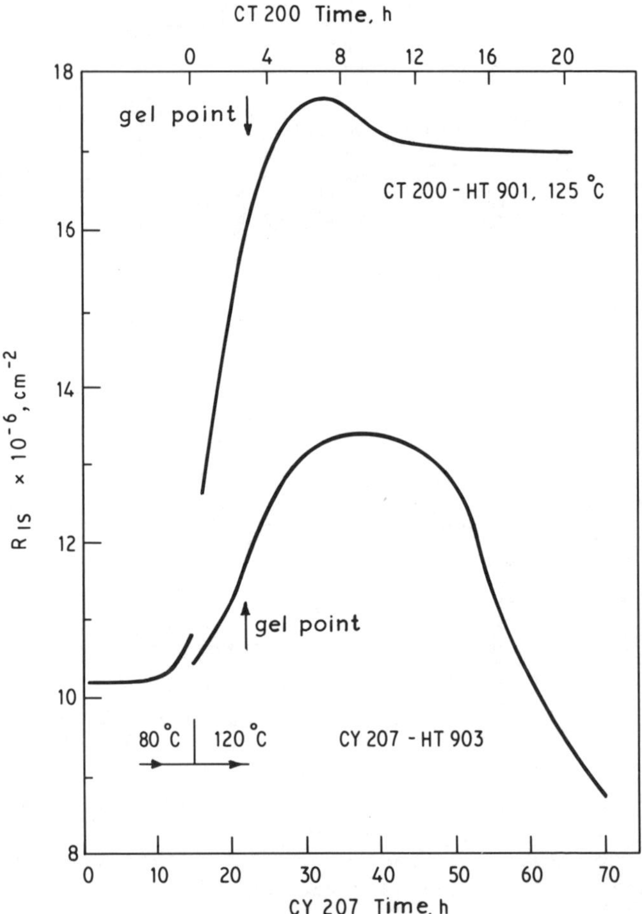

FIG. 8. Time dependence of the isotropic Rayleigh ratio during the reaction of the DGEBA/anhydride systems CT200–HT901 and CY207–HT903.

0·36 resin compared with 8% in the $\langle n \rangle \sim 1{\cdot}7$ resin, a result consistent with the presence of smaller aggregates in the lower-$\langle n \rangle$ resin. Applying both a non-discrete and a discrete aggregate model (cf. Section 3.2.1), it is found that the initial reacting aggregate size range for the $\langle n \rangle \sim 1{\cdot}7$ system extends up to 80–130 nm; for the $\langle n \rangle \sim 0{\cdot}36$ resin it is up to 70–110 nm. Here, the difference in size is smaller than

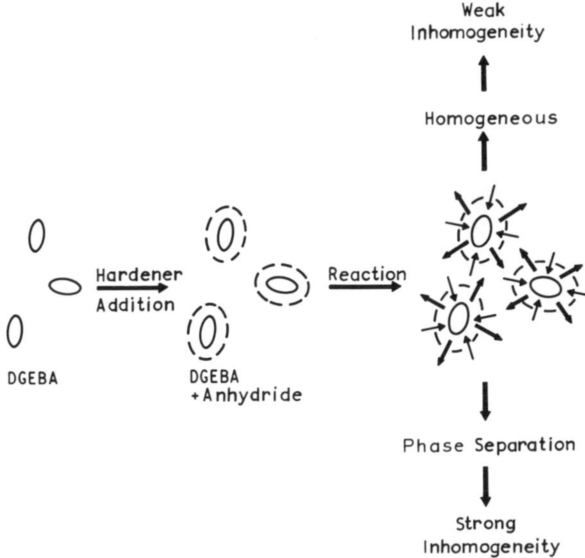

FIG. 9. Inhomogeneous reaction at pre-existing DGEBA molecular aggregate sites. Anhydride addition reveals the 'true' extent of the aggregates and reaction proceeds within and away from these sites. The eventual development of weak or strong inhomogeneity will depend on local conditions.

that in the unreacted prepolymers, suggesting that the 'true' aggregates are highly diffuse and do not contain well defined boundaries. If this is the case, anhydride reactions would be expected to proceed within and at the 'surface' of these structures. The reaction would then appear inhomogeneous and colloid-like (cf. Section 2.3.2). Such structures would remain thermally unstable until gelation had occurred locally. However, as reaction proceeds the inhomogeneities would begin to merge and appear similar in density (and refractive index) to their surroundings. This would cause both the dissymmetry and the isotropic scattering to decrease. In this respect the increase in dissymmetry that occurs after the macroscopic gel point in the low-$\langle n \rangle$ system is not understood. Taking the minimum dissymmetry values, the high- and low-$\langle n \rangle$ resin systems apparently contain inhomogeneities up to 70–100 nm (CT200–HT901) and 60–90 nm (CY207–HT901) respectively. Also, the isotropic scattering intensities suggest that the $\langle n \rangle \sim 1{\cdot}7$ system retains a higher density variance than the $\langle n \rangle \sim 0{\cdot}36$ system. As discussed later, this probably results from very slow reactions proceeding in the low-$\langle n \rangle$ system between inhomogeneity sites. These may be chemically dissimilar to earlier reactions but produce a similar final density to that of the inhomogeneities, so the material appears more homogeneous.

The question of density difference and chemical difference between an inhomogeneity and its host matrix in epoxy resins is important because the density changes which result during cure can be very small even though a dramatic change in the chemical and physical state of the initial components has occurred. This is illustrated for the systems considered above in Fig. 10 for the density during cure (inferred from refractive-index measurements[64]) and Fig. 11 for the resulting room-temperature density as a function of cure time. During cure, both systems exhibit density changes of about 4%, which in the case of the high-$\langle n \rangle$ system (CT200–HT901) produce a net change of 0·8% in the room-temperature density and 1·6% for the low-$\langle n \rangle$ system (CY207–HT903). These small room-temperature density changes and the unusual time dependence of the high-$\langle n \rangle$ system can be explained by the change in the thermal expansion coefficient that occurs in passing through the glass transition and the dependence of the coefficients on

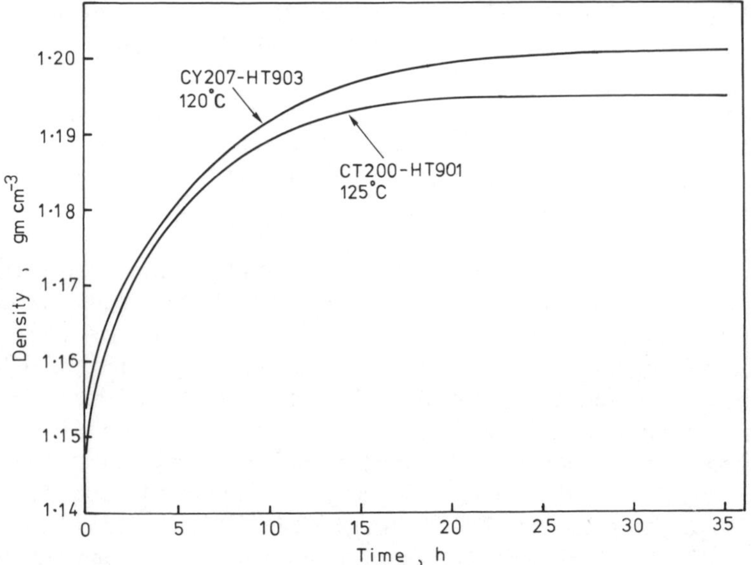

FIG. 10. Density change during the reaction of CT200–HT901 at 125°C and CY207–HT903 during postcure at 120°C.

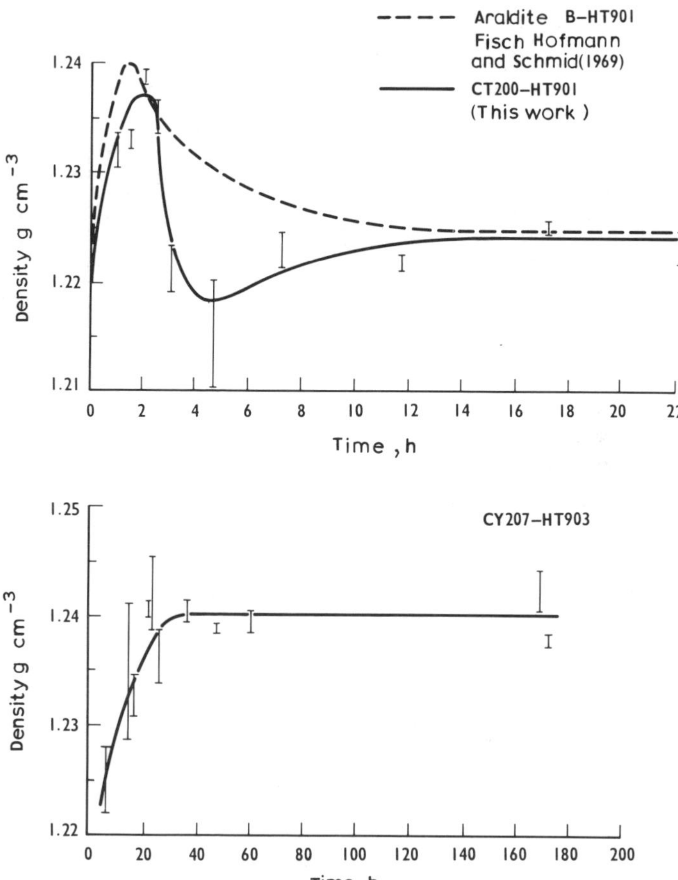

FIG. 11. Room-temperature density change of CT200–HT901 during cure at 125°C and CY207–HT903 during postcure at 120°C. — — —, Behaviour of a closely similar system studied by Fisch *et al.*[18]

the reaction temperature and extent of reaction.[18] However, at room temperature the difference in density between the two fully reacted systems is only about 1·2% in spite of the respective T_g values, which differ by 13 K,[72] and the molecular weights between crosslinks, which differ by typically 30%.[72] The final chemical differences between these two systems are also substantial.[8,20] These comparisons indicate that subtle density changes related to significant chemical and physical differences can exist in epoxy resins and similar (or more subtle) differences could be expected between an inhomogeneity and its surroundings.

The second light-scattering study[63] involved the stoichiometric reaction of resorcinol diglycidyl ether (RDGE) with *m*-phenylenediamine (mPDA) and diethylene glycol diglycidyl ether (DDGE) with 4,4′-diaminodiphenyl sulfone (DDS); see Table 1. The prepolymers nominally contain no free-hydroxyl groups and amine–epoxide reactions (vii) and (viii) followed by possible etherification reactions (ii) could occur. Both resin systems cured with DDS experienced an increase in scattering intensity during reaction as shown in Fig. 12. Instead of dissymmetry, the authors used the approach of Debye and Bueche to calculate the inhomogeneity size from the scattered light envelope.[73] The fully cured sizes were comparable with those determined by electron microscopy of argon-ion plasma etched surfaces, as shown in Table 3.

Using the work of Arutunyan *et al.*[74] and Blyachman *et al.*[75] the authors considered that the scattering arose from inhomogeneities formed as a result of aggregation of reaction products due to hydroxyl-group formation and resultant hydrogen-bonding. This was tested by imposing step temperature changes to the strongly scattering RDGE/mPDA system and also reacting it in the presence of polar solvents (butanol and dimethyl formamide). The scattered intensity responded to the step changes in temperature in a manner consistent

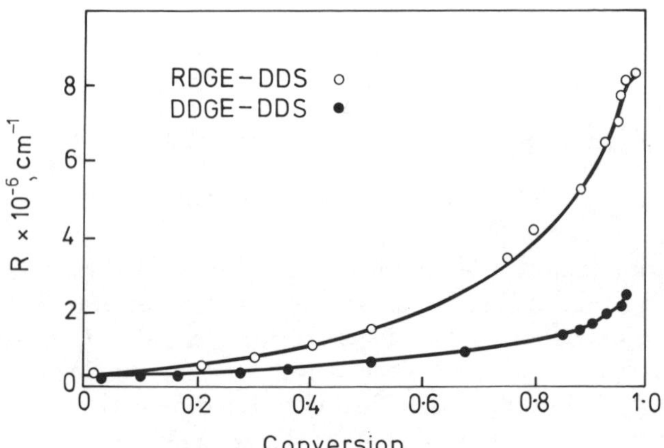

FIG. 12. Variation of the Rayleigh ratio with the extent of conversion during the reaction of RDGE/DDS at 140°C and DDGE/DDS at 138°C.

TABLE 3

COMPARISON OF THE NODULE SIZES DETERMINED BY ELECTRON MICROSCOPY (a_{em}) AND LIGHT SCATTERING (a_{ls}) AND THE MAGNITUDE OF THE RAYLEIGH RATIO (R) FOR THREE AMINE-CURED EPOXY RESIN SYSTEMS[a]

Epoxy resin system	*Cure temperature (°C)*	$R \times 10^5$ (cm^{-1})	*Nodule size (nm)*	
			a_{em}	a_{ls}
RDGE/DDS	140	82	140	100
DDGE/DDS	140	16	80	50
RDGE/mPDA	48	41	70–80	95
	70	65	50–60	60
	90	18	40	40

[a] Taken from the work of Bogdanova *et al.*[63]

with thermally unstable hydrogen-bonded aggregates. Addition of dimethyl formamide to the reaction mixture reduced the light-scattering intensity and at 54% solvent concentration it was eliminated completely. This notion was also tested by reacting the RDGE/mPDA system at significantly different temperatures and looking at the reaction behaviour of the light-scattering intensity; the results are shown in Fig. 13. Such reaction behaviour is probably an overall response to at least three factors: (a) the number and rate of formation of reacting centres, both of which increase with increasing temperature; (b) the reaction rate locally which increases with increasing temperature and increasing local concentration of proton-donating free-hydroxyl groups; and (c) lower reaction temperatures which will allow larger aggregates to be stabilized and an opportunity for etherification reactions to occur. Thus the size of hydrogen-bonded molecular aggregates would always decrease with increasing temperature, but their concentration could vary depending on the reaction conditions. The results of Fig. 13 and Table 3 support this model and suggest that for the RDGE/mPDA system, higher-temperature reaction leads to a more homogeneous material; the same would be expected for a DGEBA/anhydride system.

In summary, both studies indicate the important role of hydrogen-bonding in the formation of structural inhomogeneities. These may be formed by the presence of molecular aggregates in the resin prior to reaction. As such, they may initiate inhomogeneous reaction which

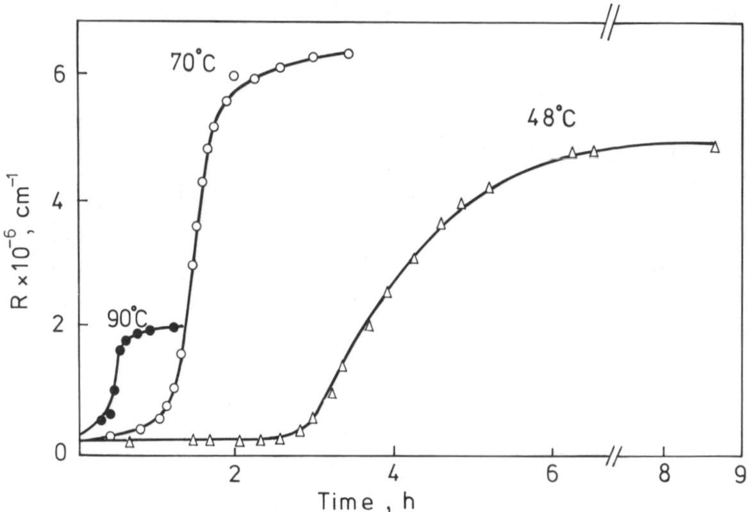

FIG. 13. Dependence of the Rayleigh ratio on cure temperature and time for the RDGE-mPDA system at 48°C, 70°C and 90°C.

could lead to the formation of microgel and an inhomogeneous network. Alternatively, inhomogeneity may occur as a consequence of reaction. In this case progressive molecular aggregation (arising from hydrogen-bonding) occurs as the reaction proceeds and results in the formation of microgel and an inhomogeneous network.

Such behaviour is not restricted to epoxy resins, and interesting parallels exist with the polyesterification reactions of alkyd resins where reaction homogeneity and the constancy of hydroxyl-group reactivity have been questioned. Indeed, Solomon and co-workers[76–78] invoke inhomogeneous reaction and the formation of microgel particles to explain their observations on systems with functionality greater than two. During the early stage of polymerization, micelles are formed; polyesterification reactions proceed at their surfaces and eventually microgel particles are formed. These particles may then exhibit reduced activity if phase separation or phase inversion occurs.

3.2.3. Small-Angle X-Ray Scattering (SAXS)

SAXS has been used to appraise inhomogeneity in fully cured epoxy resins and to the author's knowledge no observations during reaction have been attempted. Dusek and co-workers[65] examined DGEBA systems cured with the tetrafunctional amines hexamethylenediamine (HMDA) and 4,4′-diaminodiphenylmethane (DDM) and bifunctional hexahydrophthalic anhydride (HHPA). Previous work had suggested that these systems react homogeneously and the intention was to compare the structures of apparently homogeneous networks with those of common amorphous polymers using electron microscopy and SAXS.

Transmission electron microscopy (TEM) of surface replicas obtained from plasma etching in air revealed nodular-type structure in the range 20–40 nm, and the nature and size of this microstructure appeared to be a characteristic of the material and not an etch artifact. However, similar structures were also found in polymethyl methacrylate (PMMA) and polystyrene (PS). The corrected SAXS results are reproduced in Fig. 14. In the angular range $2\theta = 35–90'$ the SAXS intensity of dry networks varied from 0·2 to 0·4 $el^2/Å^3$ (where el represents electrons) in line with that expected from thermal-density fluctuations with domain sizes smaller than 1–2 nm. However, below $2\theta = 35'$ the scattering intensity increased and the authors indicated that this was consistent with the presence of inhomogeneities of colloid dimensions (~100 nm). To investigate the possible presence of regions of different crosslink density, HMDA-crosslinked materials were swollen in tetrahydrofuran, which has an electron density much lower

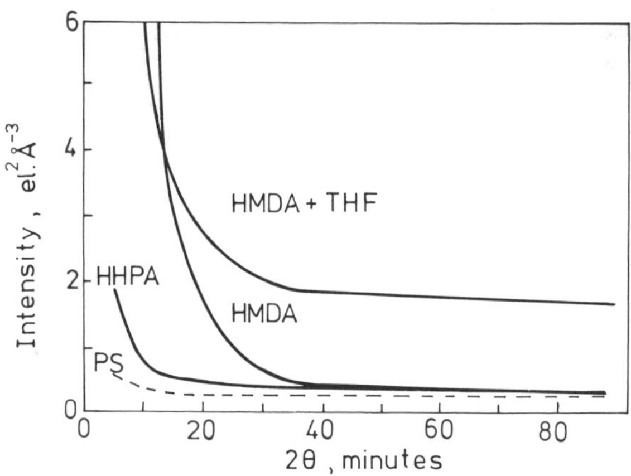

FIG. 14. Small-angle X-ray scattering curves of Dusek *et al.*[65] obtained from polystyrene (PS) and cured DGEBA/HHPA and DGEBA/HMDA. Addition of tetrahydrofuran to the latter system increases the scattering intensity at larger angles.

than that of the resin. This resulted in an increase in the thermal-density fluctuation contribution to the scattering as shown in Fig. 14, but at lower angles the scattering did not increase. However, such a change would only be expected if a significant difference in crosslink density existed between different regions; this appears unlikely in systems that were chosen because they apparently reacted homogeneously. These authors applied a semicrystalline two-phase model to the scattering results to obtain the possible volume fraction of inhomogeneities. In this case the mean square electron density fluctuation $\langle(\Delta\rho)^2\rangle$ from a system with volume fractions V_1 and $V_2 = 1 - V_1$ and a difference of electron densities $\Delta\rho = \rho_1 - \rho_2$ is given by

$$\langle(\Delta\rho)^2\rangle = (\Delta\rho)^2 V_1(1 - V_1) \tag{9}$$

A semicrystalline polymer $\Delta\rho$-value of 0·05 mol/cm^3 (~0·1 g/cm^3) produced an inhomogeneity volume fraction of 10^{-3} and the authors concluded that epoxy resins do not exhibit higher inhomogeneity than PS and PMMA. However, as the discussion on density differences above demonstrates, lower values could be expected. If $\Delta\rho$ values of 0·005 mol/cm^3 occurred, inhomogeneity volume fractions greater than 1% could result.

Uhlmann and co-workers[60,66] conducted similar SAXS studies on some systems previously examined by Koutsky in the electron microscope (see Section 3.3). These included a DGEBA resin cured with triethylenetetramine (TETA) and an anhydride system methylnadic anhydride (MNA) with 1% benzyldimethylamine (BDMA) as an accelerator; the third system was based on a TGEG resin cured with dodecenylsuccinic anhydride (DDSA), MNA and BDMA. Interestingly, the same hand mixing method was used as that in the electron microscopy work.

The corrected SAXS results for the three systems are compared in Fig. 15. As in the previous study a constant scattering intensity was observed at large angles ($2\theta > 40'$) and this was accounted for by the liquid-like thermal-density fluctuations frozen in at the glass transition temperature. The contribution at smaller angles is comparable with that observed by Dusek *et al.* above, but in this case the scattered intensity was calculated using an inhomogeneity model based on spherical scatterers where the excess scattering above that from thermal-density fluctuations is given by:

$$I'(h) = V_1(1 - V_1)V \,.\, (\Delta\rho)^2 \,.\, 3 \,.\, \left\{\frac{\sin(hR) - hR\cos(hR)}{h^3R^3}\right\} \tag{10}$$

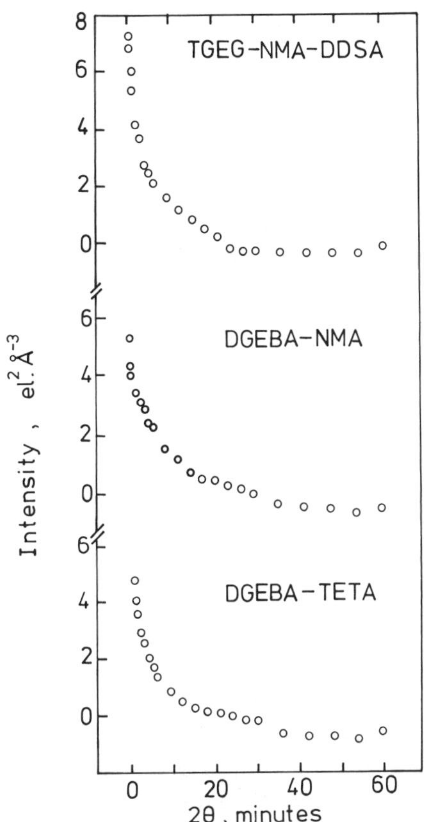

FIG. 15. Small-angle X-ray scattering curves of Matyi *et al.*[66] obtained from cured DGEBA/TETA, DGEBA/MNA and TGEG/MNA/DDSA.

where R is the inhomogeneity radius, $h = 4\pi \sin \theta/\lambda$ and λ is the X-ray wavelength. The presence of inhomogeneities greater than 25 nm is required to account for the small-angle results and volume concentrations between 1 and 30% of a range of sizes ($2R$) from 30 nm to 800 nm could explain the results if a 1% inhomogeneity density difference is applied. However, SAXS is very sensitive to microvoids and a volume concentration of 10^{-3}% could account for the excess scattering.

In summary, whilst impurities or microvoids could account for the observed SAXS it is possible that structural inhomogeneities could be present in reasonable volume concentrations. This would require their

Epon 825 (DGEBA) + DETA at various stoichiometries (6–11 phr)	20–40 nm	TEM of replicas from fracture surfaces with acetone etching in some cases. Size decreases with increasing amine content.	Mijovic and Koutsky[87,88] Mijovic and Tsay[89]
Epon 828 (DGEBA) + mPDA at various stoichiometries (7·5–25 phr)	5–100 nm	SEM and TEM of replicas from plasma-etched fracture surfaces, also TEM and STEM of stained microtomed sections.	Gupta *et al.*[132]
Epon 828 (DGEBA) + DETA: hand mixing	10–100 nm	TEM of replicas from fracture surfaces. Structure absent on vigorous mixing.	Bell[44]
RDGE + mPDA (amine) RDGE + DDS (amine)	40–80 nm 80–140 nm	Electron microscopy of argon-ion-plasma-etched surfaces. Similar sizes found by light scattering.	Bogdanova *et al.*[63]
DGEBA + anhydride (unspecified)	~100 nm	SEM of fracture surfaces with and without ion etching.	Schmid[90]
Araldite F (DGEBA) + anhydride: 403 K cure 435 K cure	 ≤300 nm ~50 nm	SEM of fracture surfaces with and without DMF swelling.	Luttgert and Bonart[29]
Epicote 828 (DGEBA) + anhydride (MCDA) with an accelerator	~50 nm	TEM of replicas from fracture surfaces with and without ion etching. Catalyst-only cured resins were less structured.	Takahama and Geil[91]
Epicote 828 (DGEBA) + catalyst only (2-ethyl-4-methylimidazole; 3 phr)	27–70 nm	TEM of replicas from fracture surfaces with ion etching; size increases with increasing reaction time. Little structure seen in unetched surfaces.	Takahama and Geil[92]
CT200 (high-$\langle n \rangle$ DGEBA) + anhydride (PA) components vigorously mixed in liquid phase at the reaction temperature, 398 K.	Median range 140–210 nm	SEM of gold–palladium-shadowed fracture surfaces (liquid N_2). Median of size distributions varied during cure.	Stevens (unpublished)
CY207 (low-$\langle n \rangle$ DGEBA) + anhydride (THPA) mixed as above, gelation phase 353 K, postcure 393 K.	Median range 120–140 nm		

TABLE 4

SUMMARY OF THE ELECTRON MICROSCOPY OBSERVATIONS OF NODULE FEATURES IN AMINE- AND ANHYDRIDE-CURED EPOXY RESIN SYSTEMS

Epoxy resin type and conditions	*Structural size*	*Structure technique and comments*	*Reference*
Unspecified	82–85 nm	TEM of ultramicrotome sections and fracture surface replicas.	Erath and Robinson[79]
Polyol-modified DGEBA + HHPA: Low-temperature cure High-temperature cure	 90 μm 20 μm	'Floccules' revealed by optical microscopy of chromic-acid-etched and THF-swollen surfaces.	Cuthrell[80]
SU205 + amines and anhydride (MTHPA): fully cured and thermally aged	15–20 nm	'Globules' by TEM of replicas from fracture surfaces or oxygen-etched surfaces.	Basin *et al.*[81]
ED-20 + polyamine with and without dibutyl phthalate; 100–120 μm films	10–30 nm	TEM of replicas from oxygen-etched surfaces. Larger structure observed in plasticized resins.	Stratsev *et al.*[82]
Epon 828 (DGEBA) + DETA: manual mixing Epon 812 (TGEG) + anhydrides (DDSA, MNA)	'Nodules' 'Aggregates' 10–60 nm <0·5 μm	TEM of replicas from room-temperature and liquid-N_2 fracture surfaces etched in acetone or sulfuric acid.	Racich and Koutsky[83]
MY750 (DGEBA) + TETA (7 phr): Cured Postcured MY750 (DGEBA) + anhydride (PA): 60°C undercure	'Nodules' 5 nm 'Aggregates' 17 nm 'Aggregates' 40 nm 'Aggregates' 40 nm	TEM of low-angle shadowed ultramicrotome sections; TEPA structure similar to that in TETA-cured resins.	Aspbury and Wake[84]
DER332 (DGEBA) + DETA (9–13 phr) Fracture surfaces Films	20–35 nm 'Nodules' 6–9 nm 'Aggregates' 20–35 nm	SEM of gold-coated fracture surfaces. TEM of films stressed *in situ* revealing 'nodules' in craze fibrils.	Morgan and O'Neal[85]
Range of Epon DGEBA resins + MDA amine	Microgel 25–200 nm	SEM and TEM of replicas, from chromic-acid-etched surfaces; microgel particulates found in the etchant.	Misra *et al.*[86]

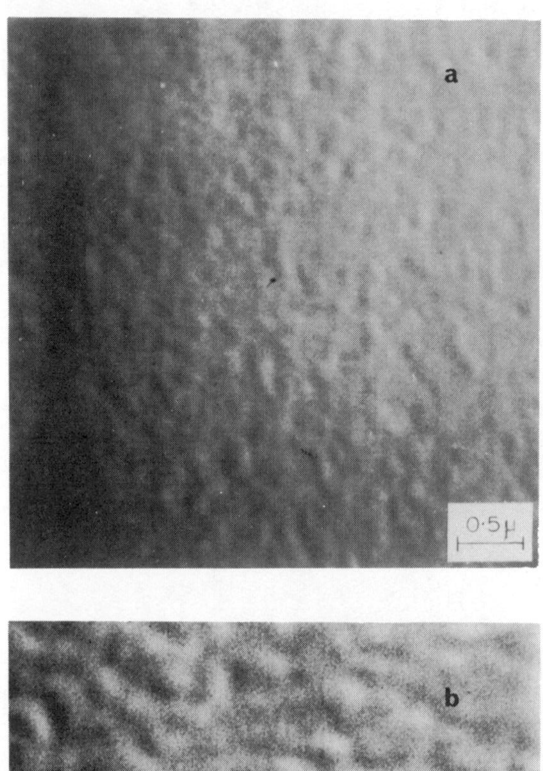

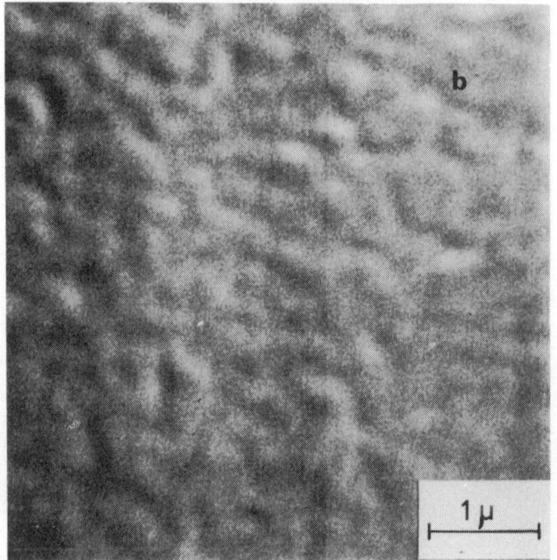

FIG. 16. Nodular structure observed by SEM of the slow crack growth mirror-zone region of DGEBA/anhydride epoxy resins fractured at liquid nitrogen temperatures: (a) CT200–HT901 after 2 h cure at 125°C; (b) as in (a) but after 4 h.

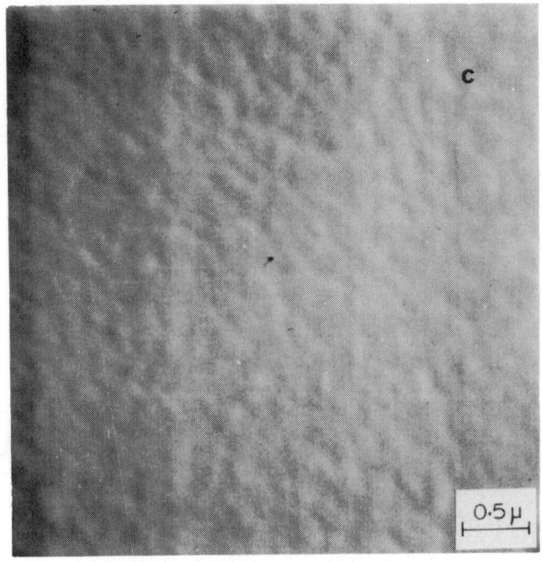

FIG. 16—*contd.* (c) CY207–HT903 after gelation + 24 h postcure at 120°C. ($1\mu = 10^{-6}$ m.)

density to be around 1% different from that of the bulk and as we have seen this is not inconsistent with the bulk density differences observed over the course of cure or between generically similar resin systems.

3.3. Microscopy of Bulk and Surface Structure

The initial interest in microstructure was born from several electron microscope observations of free and fracture surfaces. These exhibited a granular appearance typically in the size range 10–100 nm and the structure could be improved by various forms of acid and ion etching or swelling. Some of these studies are summarized in Table 4 and some typical features for anhydride-cured epoxy resins fractured at liquid nitrogen temperatures are shown in Figs 16 and 17,

These features have been interpreted as being related to underlying 'nodular' microstructures or their grouping into 'aggregates'. However, most of the features which result from surface etching and replication could be caused by the same factors that produce artifact structures in linear amorphous polymers.[53,54,58,65] Similarly, it has been suggested that fracture surface microstructure could result from factors

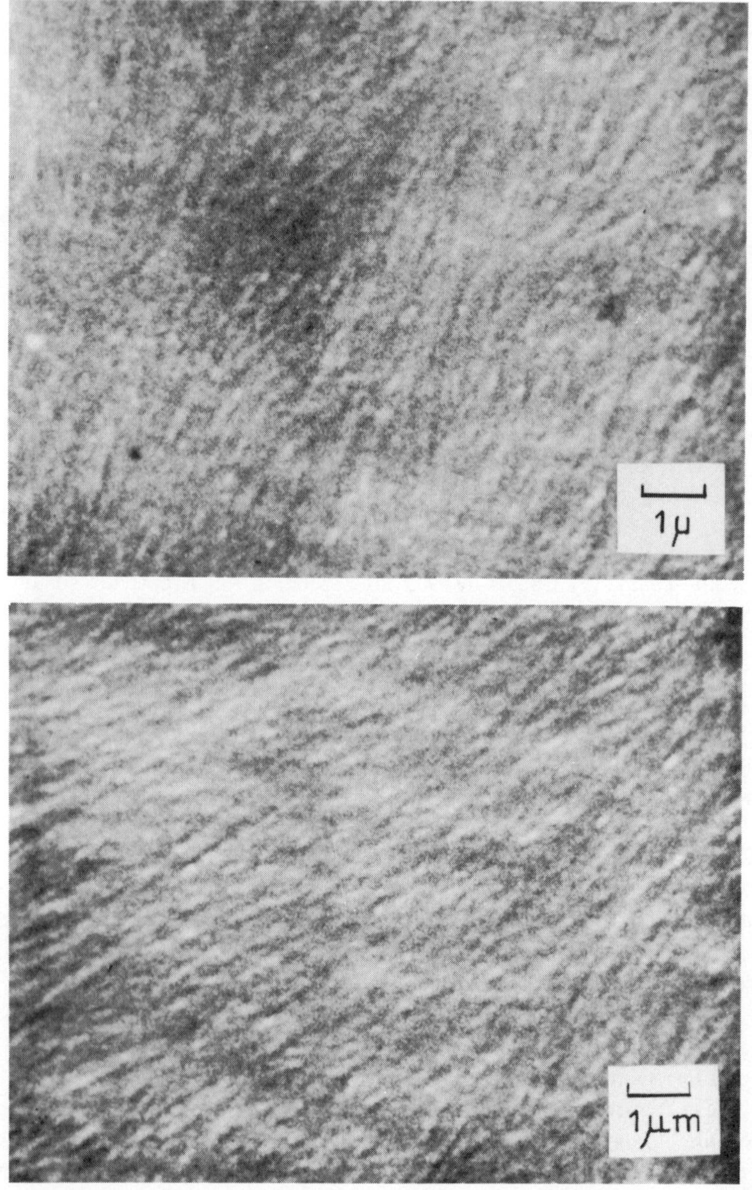

FIG. 17. Nodular alignment along the crack propagation direction observed by SEM of rapid crack growth regions of fully cured CT200–HT901 epoxy resins. ($1\mu = 10^{-6}$ m.)

not simply related to bulk structure or the fracture event.[93] Only the study of the deformation of thin films by TEM by Morgan and O'Neal[85] can be considered a bulk study unaffected by fracture or etched surfaces. However, care is required in accepting these results because the very small (5 nm) structures could result from an out-of-focus image[94] or simply be relevant to thin films only. Also, if these structures are real the density difference required to image them would be large (~5–10%), and if large volume fractions of these structures were claimed for bulk specimens this could contradict the SAXS results mentioned earlier.[60,66] Density differences may be enhanced by heavy-element staining in the bulk followed by ultramicrotomy and TEM of the resulting sections; however, this has rarely been done. Causton[95] has carried out this procedure on a number of amine- and anhydride-cured DGEBA and triglycidyl ether of glycerol (TGEG) resins stained with osmium tetroxide or counter-stained with uranyl acetate or lead citrate. In all cases fine 5–10 nm mottle structure is seen. However, the chemistry of staining in epoxides is unclear and the extent to which the technique is structurally specific is not known.

It is likely that uneven curing due to incomplete mixing of components contributes to some of the structures observed.[43,44] It is interesting that the large 20–90 μm structure reported by Cuthrell in an anhydride system[80] has a similarity to the large 'dimple' structures seen by Misra *et al.* in an amine system.[86] The latter study revealed that the 'dimples' contained 25–200 nm structures which were considered to be microgel particles. Similarly, Lock[96] observed 0·1–70 μm spheres in a DGEBA/anhydride resin put into acetone solution near its gel point. Lock's work also suggested that the fully cured resins were chemically inhomogeneous due to a non-uniform distribution of anhydride.

In spite of these uncertainties and the unfortunate lack of detail in publications on the methods of sample preparation and history, some constructive observations may be made. Generally, amine-cured systems produce smaller structures than anhydride-cured systems. In both the size can be dramatically reduced by increasing the reaction temperature[29,63,84] and in the case of amine systems, changing the reaction stoichiometry.[84,87–89,93] In the high- and low-$\langle n \rangle$ DGEBA/anhydride systems discussed earlier, the nodular distributions also change during cure, as shown in Fig. 18. These distributions are closely log-normal as shown in Fig. 19, and the same type of distribution is found in amine-cured systems.[84] Such a distribution

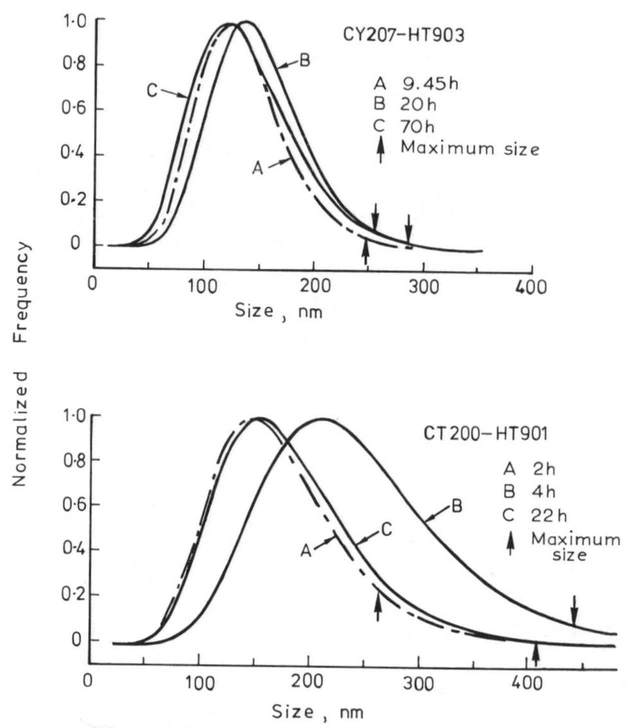

FIG. 18. Nodule size distributions for the CT200–HT901 and CY207–HT903 systems during reaction; the arrows indicate the maximum size observed in these log-normal distributions. (CT200–HT901: A = 2, B = 4 and C = 22 h at 125°C; CY207–HT903: A = 9·5, B = 20, C = 70 h at 120°C following a gelation phase of 20 h at 80°C).

could result from a broad molecular aggregate (or nodule) population or variable grouping of small aggregates (or nodules). Moreover, in the high-$\langle n \rangle$ system particularly, the distribution changes during cure from smaller to larger to smaller sizes, and this parallels the room-temperature density change during cure (Fig. 11).

The possibility that the observed features are related to bulk microstructure is supported by the close similarity between the size of features seen by microscopy and the scatterer sizes required to explain the light and X-ray scattering results, although these may also be open to various interpretations. Also, the features observed on fracture surfaces are not static entities bearing no relationship to the fracture

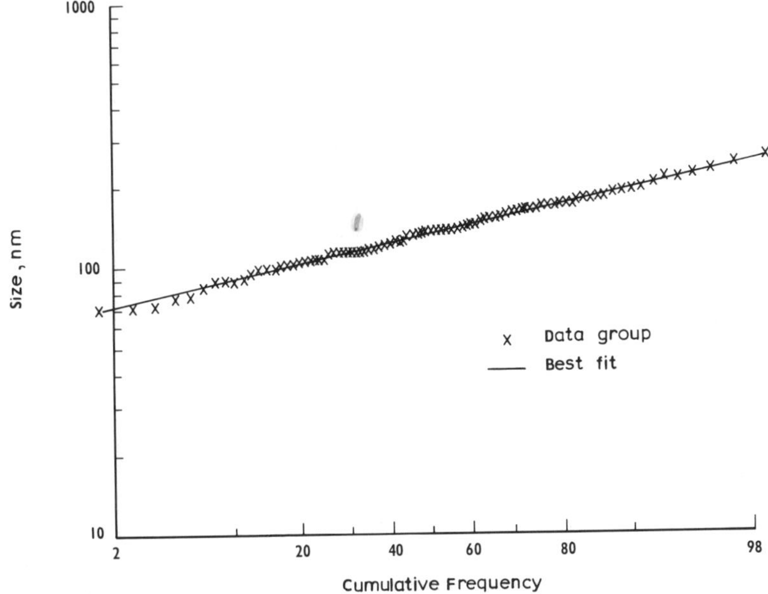

FIG. 19. Cumulative log-normal plot of the nodular size distribution of Fig. 16c.

event. As shown in Fig. 17, nodular features can be oriented and in general they have been observed to align in the direction of crack propagation,[29,87,88,90–93] particularly at higher crack velocities. However, these observations may be connected with the fracture process, and the possibility of microcrazing cannot be ruled out (as discussed later).

Support also comes from the recent work of Kelly and co-workers[97,98] in their attempt to use model inhomogeneous systems to appraise the significance of nodular microstructure. They used a range (60–180 nm) of emulsion-polymerized polystyrene/divinylbenzene 'latex' spheres which were blended and swollen in styrene monomer which was subsequently polymerized. Fracture surfaces produced the expected morphology and samples containing no microgel spheres showed no structure. They also addressed the question raised by Kreibich and Schmid[99] as to whether inhomogeneous epoxy networks should exhibit a two-phase glass-to-rubber transition, particularly when allowed to relax structurally below T_g. No evidence exists for this in anhydride-cured DGEBA systems,[72] where the presence of differential scanning calorimetry (DSC) endothermic peaks and steps in epoxy resins and polystyrene can be explained by the kinetic aspects of structural relaxation.[72,100] However, in the better defined two-phase system of Kelly and Trainor some DSC features were reported which are consistent with relaxation of a single phase whose subsequent transition is superimposed upon a single transition representing the system as a whole.[98]

Similar claims supporting multiphase structure in epoxy resins have been made on the basis of dynamic mechanical measurements. These initially arose from a study on a DGEBA/TETA system by Kenyon and Nielsen,[101] who found that lower room-temperature shear moduli were obtained at higher TETA concentrations. This was explained by a two-phase system where the modulus was determined by the weaker (lower crosslink density) phase within which a stronger (e.g. nodule) phase existed. This was expanded by Mijovic and co-workers[87,89] with their suggestion that the low-temperature β-relaxation mechanical loss peak of DGEBA/DETA resins contained at least two components. In this case the height, width and location of the peak were said to be influenced by phases in which more or less hindered rotation of glycerol groups ($—O—CH_2—CH(OH)—CH_2—$) was favoured. Their results suggested that increasing DETA leads to a greater nodular crosslink density at the expense of the matrix. Takahama and Geil[92] contested this model and suggested that the observations were consistent primarily with chemical structure differences although they accepted that morphological factors could contribute to the width and shape of these peaks. Structured β-relaxation peaks have also been observed in anhydride-cured DGEBA systems and, although chemical and morphological factors may account for some results, structural relaxation also plays a key role.[102] This illustrates that materials characterization, uniform mixing and thermal history are essential variables to tie down before embarking on such structure–property studies.

In summary, whilst some of the features observed in microscopy studies could be artifacts or result from gross non-uniform mixing of components, other observations suggest that many of these features are characteristic of the bulk or in some way related to bulk microstructure. Clearly, many of the supporting studies may be open to alternative interpretations, but in general they do supply evidence that inhomogeneity exists.

4. RELATIONSHIPS BETWEEN PHYSICAL PROPERTIES AND MICROSTRUCTURE

Our discussion so far has shown that non-ideal network formation with the development of more or less well defined microstructure is possible in epoxy resins. The resulting matrix may be incompletely connected and exhibit non-uniform spatial distributions of chemical products, crosslinks and density. If this is the case, bulk and microscopic physical properties could be influenced by this structure if it presents large enough bulk variations, discontinuities or defects or if it has a significant effect on matrix cohesion and deformation processes.

Does microstructure really matter? This question was addressed to inorganic glasses by Uhlmann and his review demonstrated that in phase-separated glasses many properties could be affected, including density, elastic moduli, mechanical strength, thermal expansion and conductivity, electrical conductivity, dielectric loss, diffusion and chemical durability.[103] The same could be expected to some degree in epoxy resins. Few studies to date have been able to explore possible relationships, but some suggest that they do exist.

4.1. Correlations During Reaction

Section 3.2.2 demonstrated that valuable information may be gained by following changes in properties during reaction. This has been done for some properties of the high- and low-$\langle n \rangle$ DGEBA/anhydride systems discussed earlier and the results compared with chemical changes during cure.[8,20,61,64,72] The extent of change of properties, including density, glass transition temperature (T_g) and longitudinal acoustic modulus, were compared with the extent of change of key chemical groups, including anhydride consumption and ether and aromatic ester formation. First-order rate of change kinetics were found for many of these and Table 5 summarizes the rate constants found; Figs 20 and 21 illustrate the correlations graphically. These show that density correlates with aromatic ester formation [complete esterification reactions (iv) and (v)] whilst longitudinal acoustic modulus correlates with ether formation [etherification reaction (vi)]. In the CY207–HT903 system T_g correlates well with ether formation and, although this also appears to be the case for the CT200-HT901 system during the early part of reaction, during the later stage better correlation occurs with aromatic ester formation. It should also be recalled that the scattering inhomogeneity size obtained from dis-

TABLE 5

SUMMARY OF THE CHEMICAL GROUP AND PHYSICAL PROPERTY EXTENT OF CHANGE FIRST-ORDER RATE CONSTANTS FOR THE CT200–HT901 AND CY207–HT903 SYSTEMS DURING REACTION AT 125°C AND 120°C RESPECTIVELY

Chemical group or physical property	$k \times 10^{-5}$ (s^{-1})	
	CT200–HT901 (cure)	*CY207–HT903 (postcure)*
Anhydride	8·6	2·9
Aromatic ester	6·5	4·0
Density	5·8	4·2
Ether[a]	5·0	1·05
Glass transition temperature[a]	4·2	1·1
Longitudinal acoustic modulus (5–10 GHz)[b]	4·4	0·7

[a] First-order behaviour up to initial 50% of change.
[b] Half-order kinetics are more closely followed but k is obtained from the 50% point on the first-order plot.

symmetry measurements closely followed anhydride consumption during the initial stage of reaction and the change in fracture surface nodular size followed the room-temperature density during cure.

Taking these observations and those of the chemical kinetics and pre-existing order studies into account, it is possible to outline a model connecting chemical structure, inhomogeneity and physical properties which is consistent with the observations.

(i) In $n > 0$ DGEBA resins, hydrogen-bonded molecular aggregates exist whose size and concentration is related to the free-hydroxyl group concentration, the temperature and the presence of diluents or polar solvents.

(ii) These aggregates have a higher local concentration of free-hydroxyl groups than their surroundings and so act as primary reaction nuclei for reactions involving anhydride ring-opening. This produces an inhomogeneous reaction of a colloid type.

(iii) Esterification reactions are favoured within and around these nuclei and etherification reactions proceed more slowly de-

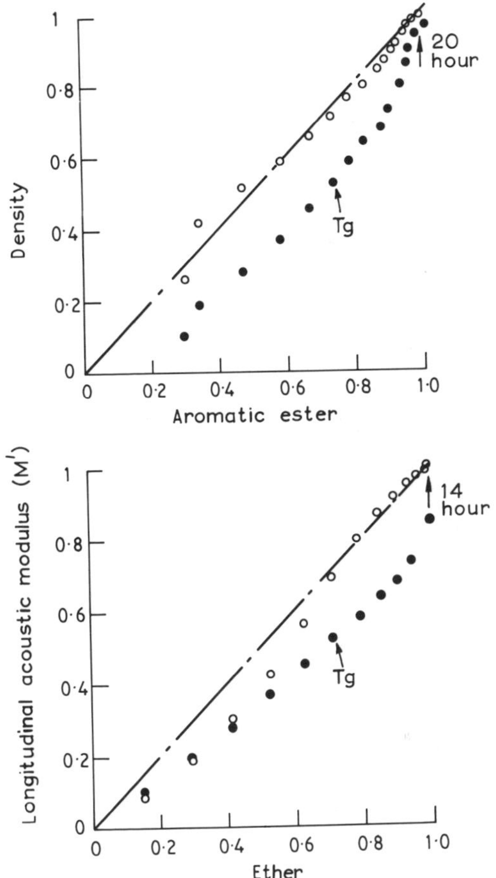

FIG. 20. Correlation of the extent-of-change parameters during cure for density (and glass transition temperature)/aromatic ester and longitudinal acoustic modulus (and glass transition temperature)/ether for the CT200–HT901 system.

pending on local conditions. However, because spatial separation of the $n = 0$ and $n > 0$ fractions has occurred (to some extent), a higher proportion of ester crosslinks is expected within and around nuclei whilst a higher proportion of ether branches and crosslinks is expected in the regions between and at the boundaries of the nuclei. So fully reacted materials

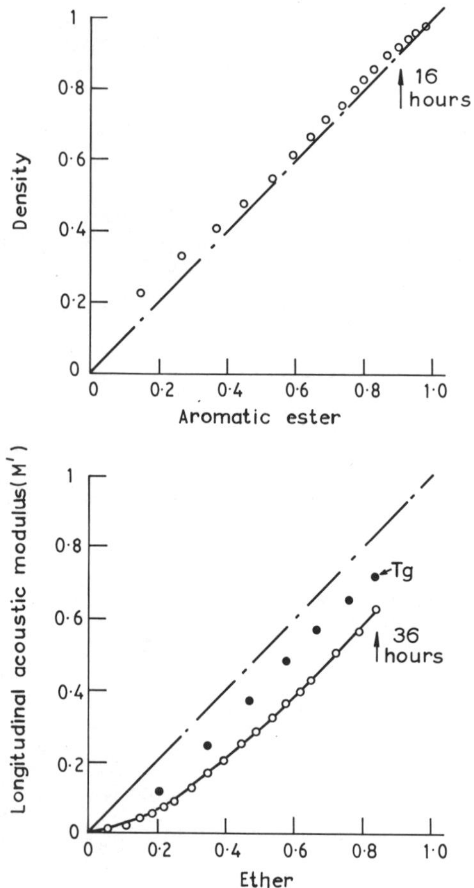

FIG. 21. Correlation of the extent-of-change parameters during cure for density/aromatic ester, longitudinal acoustic modulus/ether and glass transition temperature/ether for the CY207–HT903 system (postcure period).

should exhibit domains containing more ester-crosslink network characteristics with connecting boundaries, or regions containing more ether-crosslink network characteristics.

(iv) Most crosslink reactions involve esterification. These occur more rapidly and will dominate changes in overall density. Etherification reactions are slower and their number will depend on the initial free-hydroxyl group concentration, the extent of aggregate formation and the reaction conditions. These are largely responsible for the connectivity of the matrix and influence properties such as glass transition temperature and modulus.

(v) It is not expected that the network domains will be highly differentiated in the bulk. However, their eventual 'size', density difference from their surroundings and their crosslink density and chain lengths will be directly influenced by the extent, nature and subsequent development of the reaction nuclei. These in turn are primarily controlled by the molecular weight distribution of the DGEBA prepolymer and the reaction temperature.

Similar non-uniformity may arise in other epoxy systems as a consequence of reaction or simply due to incompletely mixed constituents. Alternatively, a homogeneous reaction may be favoured, but the possibilities of achieving an ideal macroscopic network may be limited by network defects or the formation of microgel which is incompletely connected. Whatever the cause, in general the possible range of microstructural types is large and simple structure–property relationships which can account generally for chemical and morphological factors in all epoxy resins should not be expected.

4.2. Influence on Mechanical Properties

The influence of epoxy resin chemistry and microstructure on physical properties has concentrated largely on mechanical properties in view of the mechanical and adhesive use of these materials. Attention was initially drawn to the possible importance of thermoset microstructure by the early work of DeBoer[104] and Houwink,[105] who showed that the tensile strength of some crosslinked resins was two to three orders of magnitude lower than that expected for complete carbon–carbon covalent bonding throughout the bulk and up to an order of magnitude lower than that expected for van der Waals bonding. Houwink considered that this arose from natural bulk defects which were not accidental but resulted from structural features of a statistical nature. This of course received support from Griffiths' study of fracture based on stress concentration at structural defects and the notion of an inherent flaw size.[106] Later, by means of linear elastic fracture mechanics, it was found that fracture energies in glassy linear polymers were at least 100 times greater than those expected for purely brittle fracture[107] and for epoxy resins they have been reported to be a factor of two to three times greater (see references in ref. 108). This is accounted for in modern theories of fracture[109–112] by the ability of the matrix to undergo plastic deformation. Consequently it is important to understand the role of the network and any associated microstructure in providing defects or in influencing deformation processes. These topics have been reviewed in part elsewhere[87,108,113,114] and only structural aspects will be considered here.

4.2.1. Deformation Processes

Brittle crack propagation, even in epoxy resins, involves localized viscoelastic and plastic energy-absorbing processes at the crack tip. The two processes of central importance are crazing and shear-yielding and both involve localized plastic deformation and yielding arising from local stress concentrations and the strain softening behaviour of the material.[114] The essential difference is that shear-yielding occurs at constant volume whereas crazing involves cavitation, resulting in an increase in volume. If sufficiently localized, both may be precursors to brittle fracture.

In fully cured, homogeneous and highly crosslinked epoxy resins it is unlikely that the degree of tensile drawing of material required for crazing[115,116] would be possible without significant bond breaking. Consequently, although reports of more macroscopic crazing have been made[117,118] the evidence is not compelling and slow local deformation and yield at the crack tip without crazing are generally believed to account for observations such as stick–slip crack propagation behaviour and self-toughening.[114,119–122] However, the physical or structural mechanism of deformation is not understood and a number of studies have suggested that network microstructure[87,88,93,108,120,123] and even microcrazing[85,108,124,125] could be involved. Indeed, Narisawa *et al.*,[125] in deep notched epoxies under stress, observed shear-yield bands which became obscure with increasing amounts of plasticizing resin, but they could not determine if crazing occurred. Also, Cherry and Thomson[126] observed extensive plastic deformation in so-called plasticized epoxies. More recently, Yamini and Young[122] compared the yield stress and Young's modulus of a series of DGEBA/TETA systems with linear amorphous polymers using both Argon's and Bowden's theories of plastic deformation. They concluded that the deformation characteristics appear similar to those in glassy thermoplastics which have similar chemical structure.

The formation of short (~1 μm) microcraze structures involving possible nodular features (~5–10 nm) have been observed by Morgan and O'Neal in highly strained 1–100 μm DGEBA/DETA films.[85] They suggested that craze fibrillation was inhibited by the nodules. However, such a process could only occur if the internodular matrix was very weakly crosslinked and highly deformable. In this context the recent work of Donald and Kramer (ref. 127, and references therein) on the effect of molecular entanglements on craze microstructure in glassy linear polymers is helpful. The craze fibril extension ratio, λ_{cr}, increases with the square root of the length of the polymer chain between physical entanglements, l_e. These λ_{cr} values were compared with the maximum extension ratio for the entanglement network (in which polymer chains neither break nor reptate), λ^e_{max}. It was found that the extension ratios generally fall below λ^e_{max} for low values of l_e but increase to well above λ^e_{max} for high l_e. Also, as l_e decreases below about 20 nm a transition from crazing to shear-yielding occurs.

If these observations are applied to epoxy resin networks, the occurrence of crazing could be expected in a homogeneous network with a sufficiently low crosslink density (i.e. $l_e > 20$ nm). However, as we have seen, the network may be non-ideal and inhomogeneous and it is possible that a less crosslinked or weakly connected phase could control deformation and be limited or impeded by a more highly crosslinked phase. Network defects and steric factors would also contribute. So, nominally highly crosslinked but non-ideal networks could also exhibit craze-like deformation behaviour. Similar views have been expressed by Morgan and co-workers[108,123] and these can be summarized as follows.

(i) Lower crosslink-density networks which are susceptible to molecular flow are often formed as a result of incomplete cure or non-ideal reactions (e.g. homopolymerization and intramolecular ring formation).

(ii) Flow may be enhanced in networks containing spatially inhomogeneous crosslink-density distributions, particularly where a continuous phase of low crosslink density is formed.

(iii) Homogeneous plastic deformation is enhanced by the flexibility of the chains between crosslinks and the free volume available to them.

(iv) Network imperfections will lead to inhomogeneous deformation in the form of crazing or shear-yielding and, in such high strain regions, chain scission and molecular pull-out may occur.

These points indicate that it is not possible to generalize about the deformation characteristics of all epoxy resin networks; each must be treated individually, on the basis of its network structural characteristics, which of course need to be determined.

4.2.2. Causes and Consequences of Nodular Structure

If the discussion above on deformation processes is correct, nodular features on epoxy resin fracture surfaces could fall into at least two distinct groups related essentially to deformable and non-deformable networks. The latter case would correspond to a highly crosslinked inhomogeneous network where fracture proceeded around more highly crosslinked regions and where the continuum or internodular matrix was weaker but still highly crosslinked and difficult to deform. Resultant nodular features would then closely represent the bulk inhomogeneity. This case is closest to that suggested by Mijovic and Koutsky[83,87–89] in their studies of highly crosslinked DGEBA/amine systems. The deformable-network case could in principle lead to many possible structural features in lightly crosslinked homogeneous networks or more highly crosslinked inhomogeneous networks. Structure sizes ranging from intrinsic inhomogeneity levels to broken fibril or mica structure features associated with crazing (or psuedo-crazing) could be observed, and in such cases aggregation and orientation of more primary structural elements could occur. This case is probably closer to the more lightly crosslinked DGEBA/polyamine and TGDDM/DDS systems studied by Morgan and co-workers.[15,85,108,123]

Highly crosslinked homogeneous networks with attendant imperfections would be expected to yield fracture surfaces devoid of microstructure, with the exception of that resulting from the fracture event.

Clear relationships connecting microstructure (nodular or aggregate size and concentration) with mechanical properties do not exist but some studies suggest that they could. Mijovic and Koutsky[87,88] consider that mechanical properties are controlled by nodule size. They have shown that the critical strain energy release rate (G_{Ic}) for crack initiation and arrest varies with the nodule size in a DGEBA/DETA system. Their results (shown in Fig. 22) suggest that maximum G_{Ici} and G_{Ica} values are obtained at a nodule size around 30 nm. In contrast, Kelly and Trainor's study[98] of more distinct model in-

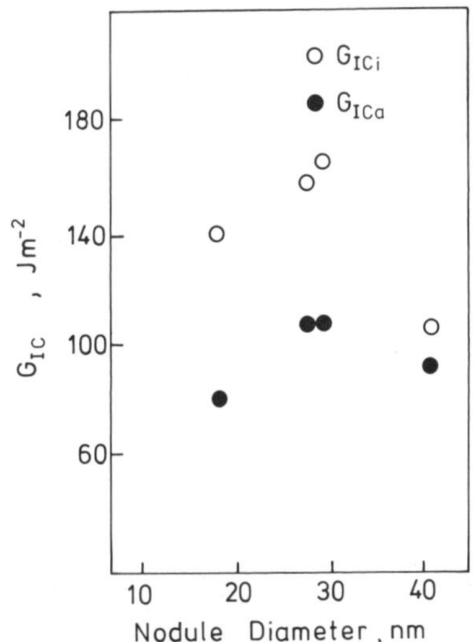

FIG. 22. The variation of critical strain energy release rate for a DGEBA/DETA system, from the work of Mijovic and Koutsky.[87]

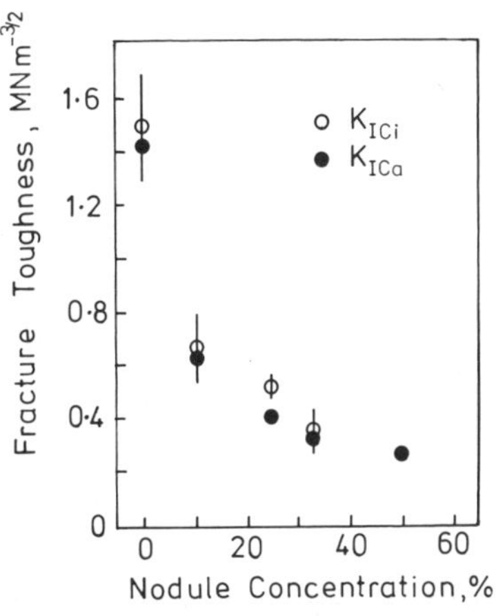

FIG. 23. Variation of the fracture toughness with nodule concentration for a nodule size of 100 nm in a latex sphere/styrene matrix model system reported by Kelly and Trainor.[98]

homogeneous networks based on latex spheres in a polymerized styrene matrix showed that the fracture toughness (K_{Ic}) was practically independent of sphere size in the range 60–180 nm but depended strongly on volume concentration. For a sphere size of 100 nm, Fig. 23 illustrates that the K_{Ici} and K_{Ica} values fall from a styrene-only matrix value of 1·5 MN/m$^{3/2}$ to 0·2 MN/m$^{3/2}$ as the latex sphere concentration is increased to 50%. Interestingly, unstable crack propagation was observed up to a concentration of 33% whilst stable crack propagation occurred at 50% concentration. In DGEBA/amine systems Misra and co-workers[86] claimed that the characteristic Griffiths flaw size varied from ~30 μm to ~140 μm as the amine:epoxy ratio increased from 0·7 to 2. This was close to the structural size of nodule aggregates encountered in etching studies, and flaws related to aggregate interfaces were thought to be responsible. It was also noted that strength increased with the amount of excess amine, perhaps the result of a more deformable network. Finally, Bell's investigation of inadequate mixing in DETA and polyamine-cured DGEBA resins (Section 2.4) indicated that significant improvements in impact and tensile strength could be obtained if fine structure (~10–100 nm) could be avoided.

4.3. Other Properties and Considerations

Little can be said at the moment about the influence of microstructure on properties in general, and this will remain the case unless more attention is directed to assessing reliably the structural characteristics of imperfect networks. However, our discussion has shown that many potential influences exist and that these may be controllable. If this is the case the processing of epoxy resins either to avoid or to enhance network imperfection and microstructure could be invaluable in optimizing physical properties and performance. However, in view of the major adhesion and composite uses of epoxy resins, it is important to appreciate that surfaces and confined volumes of micron dimensions can have significant chemical and physical effects on some resin systems, particularly those prone to forming imperfect networks. This has been demonstrated in DGEBA/anhydride systems, where thin films (~1–10 μm) reacted between KBr discs produced very different reaction kinetics and chemical structure from those in bulk volumes (~10 ml).[8,29] Similarly, filler particles may interfere with reactions and influence final properties (ref. 72, and references contained therein).

Long-term performance and endurance under chemical or physical stress may also be influenced by network structure. Perfect and imperfect networks below their glass transition temperatures are generally metastable and will attempt to relax towards a state of greater equilibrium which usually involves densification and changes in the molecular relaxation and mobility of the network.[100,127–129] However, imperfect networks will have other options for change (e.g. further reactions, phase separation, microstructure) and for translating the effects of such change. Thus diffusion, void formation, mechanical creep and fatigue and similar long-timescale processes could be influenced by the original network structure and its ability to change.

Clearly, the possibilities for network structure–property influences and relationships are myriad but it is necessary now to begin to separate the classical perfect network and chemical structure contributions from the imperfect network and microstructural aspects.

5. CONCLUDING REMARKS

This review has shown that it is possible for epoxy resin networks to contain microstructure but that it cannot be expected universally. A range of near-ideal and imperfect networks may be formed, and the latter can occur in systems which apparently react homogeneously. Inhomogeneous crosslinking reactions are also possible and these may result from pre-existing order in the epoxy prepolymer or occur as a consequence of reaction due to intramolecular reactions or the aggregation of reaction products. In those cases where evidence exists for this, the role of free-hydroxyl groups in promoting certain reactions (e.g. proton donation or involvement in etherification reactions) or in assisting molecular aggregation through intermolecular (and intramolecular) hydrogen-bonding cannot be ignored. Indeed, in non-catalyzed DGEBA/anhydride resins the free-hydroxyl group concentration is the principal variant controlling reaction and microstructure. Of course, uneven mixing of reactants can also result in inhomogeneous reactions and inhomogeneous networks at microscopic and macroscopic levels. In the future more emphasis should be placed on uniform mixing so that the intrinsic behaviour and properties of epoxy resin networks can be achieved and more reliable structure–property relationships discerned.

In such studies the importance of understanding the reaction characteristics and of characterizing the resultant network and microstructure is essential. Reaction behaviour can be tested using various kinetic and statistical treatments focused on a number of key chemical variables which can be determined experimentally. Techniques such as light scattering and small-angle X-ray scattering may also be used constructively, both during reaction and in appraising fully reacted materials. However, as we have seen, interpretational difficulties exist in some areas and it is necessary that these be resolved. Similarly, improved methods of macroscopic and microscopic network characterization are required, particularly to identify inhomogeneous distributions of crosslinks as distinct from density variations. In this regard the structural evidence from electron microscopy, especially from fracture surfaces, has to be interpreted with care. The features observed may be representative of bulk structure but they may also be representative of the deformation and failure process of the network. Confirmation of such structure by bulk scattering techniques should be sought before reaching firm conclusions.

Interestingly, recent small-angle neutron-scattering (SANS) from polystyrene/divinylbenzene networks[130] suggests that this technique could be used to distinguish non-random crosslinking reactions and assess the formation and extent of aggregate structures. Indeed, Bai[133] and Wu and Bauer[131,134] have taken the first step in applying SANS to partially deuterated epoxy resin systems. Bai's study[133] of Epon 825 (monodisperse DGEBA) cured with deuterated and protonated mPDA at 15·8 phr followed the observations of Gupta *et al.* of nodular structure in this system[132] (cf. Table 4). In all cases he obtained a constant scattering intensity over a scattering vector range from 0·016 to 0·220 Å^{-1} (i.e. real-space correlation distances from 400 to 30 Å) and he concluded that the spatial distribution of crosslinks (i.e. mPDA related) was homogeneous. In contrast, Wu and Bauer[131,134] used partially deuterated DGEBA resin cured with a number of difunctional and trifunctional primary amines. Pronounced scattering peaks were observed to arise during cure in the scattering vector range from around 0·07 to 2·00 Å^{-1} (real-space distances from 100 to 3 Å). The scattering from both unswollen and acetone-swollen networks was consistent with both chemical and topological network inhomogeneity in spite of SAXS results which showed that density fluctuations in the materials were negligible.

This review has also shown that hard-sphere/soft-surrounding two-phase models of inhomogeneity are probably not fully representative of all epoxy resin microstructure. Indeed, structures which are diffuse, ill-defined and chemically intermixed are possibly more common. Thus, when interpreting structural information, very small density differences between structural entities and their surroundings should be anticipated. These could be associated with minor or more significant chemical and crosslink differences or very different distributions of network imperfections. Consequently, all of the factors influencing microstructure and the connectivity and cohesion of the matrix need to be considered. In this regard, the recent studies of model inhomogeneous networks are encouraging and informative. However, it would be constructive if similar studies could also be performed on less discrete model structures and SANS studies extended to these and a range of different epoxy resin networks.

Finally, with improvements in revealing the bulk microstructure and network characteristics of epoxy resins, reliable structure–property relationships may be forthcoming. However, it seems likely that such relationships may not be very general because the spectrum of options is potentially vast. Nevertheless, useful guidelines could become available which may allow both structure and properties to be optimized through a suitable choice of resin system and processing conditions.

ACKNOWLEDGEMENTS

The author is grateful for the collaborative involvement of Dr J. V. Champion and Dr M. J. Richardson in various aspects of his work on epoxy resins. Also, part of the work reported here was conducted at the Central Electricity Research Laboratories and is published with the permission of the CEGB.

REFERENCES

1. LEE, H. and NEVILLE, K., *Handbook of Epoxy Resins,* McGraw-Hill, New York, 1967.
2. MAY, C. A. and TANAKA, Y. (Eds), *Epoxy Resins: Chemistry and Technology,* Marcel Dekker, New York, 1973.
3. FISHER, M., LOHSE, F. and SCHMID, R., *Makromol. Chem.,* **181** (1980) 1251 (also available as RAE Library Translation 2067).
4. KREIBICH, U. T., LOSHE, F. and SCHMID, R., in *Nonmetallic Materials and Composites at Low Temperatures,* Eds A. F. Clark, R. P. Reed and G. Hartwig, Plenum Press, New York, 1979, p. 1.
5. LABANA, S. S. and DICKE, R. A. (Eds), *Characterization of Highly Cross-linked Polymers,* Amer. Chem. Soc. Symposium Series, **243** (1984).
6. CHOMPFF, A. J. and NEWMAN, S. (Eds) *Polymer Networks, Structure and Mechanical Properties,* Plenum Press, New York, 1971.
7. JONES, F. R. (Ed.), *Crosslinking and Network Formation in Polymers: Materials, Methods and Applications,* Ann. Nat. Conf. PRI, London, 1982.
8. STEVENS, G. C., *J. Appl. Polym. Sci.,* **26** (1981) 4259.
9. BATZER, H. and ZAHIR, S. A., *J. Appl. Polym. Sci.,* **21** (1977) 1843.
10. BATZER, H. and ZAHIR, S. A., *J. Appl. Polym. Sci.,* **19** (1975) 601.
11. FLIPPEN-ANDERSON, J. L. and GILARDI, R., *Acta Cryst.,* **B37** (1981) 1433.
12. BANTLE, S., HASSLIN, H. W., MEER, H. U., SCHMIDT, M. and BURCHARD, W., *Polymer,* **23** (1982) 1889.
13. LUNAK, S. and DUSEK, K., *J. Polym. Sci., Symp.,* **53** (1975) 45.
14. CHARLESWORTH, J. M., *J. Polym. Sci., Polym. Phys. Ed.,* **17** (1979) 1557.
15. MORGAN, R. J. and O'NEAL, R. J., *J. Macromol. Sci.—Phys.,* **B15** (1978) 139.
16. FISCH, W. and HOFMANN, W., *J. Polym. Sci.,* **XII** (1954) 497; *Makromol. Chem.,* **54–56** (1961) 8; *Plast. Technol.,* **7** (1961) 28.
17. FISCH, W., HOFMANN, W. and KOSKIKALLIO, K., *J. Appl. Chem.,* **6** (1956) 429.
18. FISCH, W., HOFMANN, W. and SCHMID, R., *J. Appl. Polym. Sci.,* **13** (1969) 295.

19. Tanaka, Y. and Kakiuchi, H., *J. Appl. Polym. Sci.* **7** (1963) 1063; **7** (1963) 1951; *J. Polym. Sci. Part A,* **2** (1964) 3405.
20. Stevens, G. C., *J. Appl. Polym. Sci.,* **26** (1981) 4279.
21. Dusek, K., Ilavsky, M. and Lunak, S., in *Crosslinking and Networks,* Ed. K. Dusek, Wiley Interscience, New York, 1975, p. 29.
22. Kwei, T. K., *J. Polym. Sci.,* **1A** (1963) 2985.
23. Tanaka, Y. and Mika, T. F., in ref. 2, p. 144.
24. Bokare, U. M. and Gandhi, K. S., *J. Polym. Sci., Polym. Chem. Ed.,* **18** (1980) 857.
25. Flory, P. J., *Principles of Polymer Chemistry,* Cornell University Press, London, 1953.
26. Dusek, K. and Prins, W., *Adv. Polym. Sci.,* **6** (1969) 1.
27. Dusek, K. and Ilavsky, M., *J. Polym. Sci., Polym. Phys. Ed.,* **21** (1983) 1323.
28. Ilavsky, M., Bogdanova, L. M. and Dusek, K., *J. Polym. Sci., Polym. Phys. Ed.,* **22** (1984) 265.
29. Luttgert, K. E. and Bonart, R., *Prog. Coll. Polym. Sci.,* **64** (1978) 38.
30. Funke, W., *Adv. Polym. Sci.,* **4** (1965) 157.
31. Solomon, D. H., *J. Macromol. Sci. (Rev.) Part C,* **1** (1967) 179.
32. Labana, S. S., Newman, S. and Chompff, A. J., in *Polymer Networks, Structure and Mechanical Properties,* Eds A. J. Chompff and S. Newman, Plenum Press, New York, 1971, p. 453.
33. Funke, W., in ref. 7, p. 6.
34. Dusek, K., *J. Polym. Sci. Part C.,* **16** (1967) 1289.
35. Dusek, K. and Patterson, D., *J. Polym. Sci., Part A,* **6** (1968) 1209.
36. Gordon, M., *Proc. Roy. Soc. Ser. A.* **268** (1962) 240.
37. Macosko, C. W. and Miller, D. R., *Macromolecules,* **9** (1976) 199, 206.
38. Dusek, K., Ilavsky, M. and Lunak, S., *J. Polym. Sci., Symp.,* **53,** (1975) 29.
39. Charlesworth, J. M., *J. Polym. Sci., Polym. Phys. Ed.,* **17** (1979) 1571.
40. Mark, H. and Raff, R., *High Polymeric Reactions, Their Theory and Practice,* Interscience, New York, 1941.
41. Waite, T. R., *Phys. Rev.,* **107** (1957) 463.
42. Waite, T. R., *J. Chem. Phys.,* **28** (1958) 103.
43. Ghaemy, M., Billingham, N. C. and Calvert, P. D., *J. Polym. Sci., Polym. Letts Ed.,* **20** (1982) 439; also ref. 7, p. 50.
44. Bell, J. P., *J. Appl. Polym. Sci.,* **27** (1982) 3503.
45. Rotz, C. A. and Suh, N. P., *Polym. Eng. Sci.,* **16** (1976) 664.
46. Hoburg, J. F. and Melcher, J. R., *J. Fluid Mech.,* **73** (1976) 333.
47. Lang, J. H., Hoburg, J. F. and Melcher, J. R., *Phys. Fluids,* **19** (1976) 917.
48. Wendorff, J. H. and Fisher, E. W., *Kolloid Z. Z. Polym.,* **251** (1973) 876, 884.
49. Fisher, E. W., Wendorff, J. H., Dettenmaier, M., Lieser, G. and Voigt-Martin, I., *J. Macromol. Sci.—Phys.,* **12** (1976) 41.
50. Wiegand, W. and Ruland, W., *Prog. Coll. Polym. Sci.,* **66** (1979) 355.

51. ROBERTSON, R. E., *J. Polym. Sci., Polym. Phys. Ed.*, **19** (1981) 1277.
52. LINDENMEYER, P. H., *Polym. Eng. Sci.*, **21** (1981) 958.
53. YEH, G. S. Y., *Crit. Rev. Macromol. Chem.*, **1** (1972) 173; WANG, C. S. and YEH, G. S. Y., *J. Macromol. Sci.—Phys.*, **B15** (1978) 107, 119; GUPTA, M. R. and YEH, G. S. Y., *J. Macromol. Sci.—Phys.*, **B15** (1978) 119.
54. GEIL, P. H., *J. Macromol. Sci.—Phys.*, **B12** (1976) 173; *Polym. Materials, ASM*, 1977.
55. FLORY, P. J., *Pure Appl. Chem., Macromol. Chem.*, **8** (1972) 1; *J. Macromol. Sci.—Phys.*, **12** (1976) 1.
56. RENNINGER, A. L., WICKS, G. G. and UHLMANN, D. R., *J. Polym. Sci., Polym. Phys. Ed.*, **13** (1975) 1247; UHLMANN, D. R., RENNINGER, A. L., KRITCHEVSKY, G. and VANDER SANDE, J., *J. Macromol. Sci.—Phys.*, **B12** (1976) 153; MEYER, M., VANDER SANDE, J. and UHLMANN, D. R., *J. Polym. Sci., Polym. Phys. Ed.*, **16** (1978) 2005.
57. BALLARD, D. G. H., WIGNALL, G. D. and SCHELTEN, J., *Europ. Polym. J.*, **9** (1973) 965; WIGNALL, G. D. and LONGMAN, G. W., *J. Mater. Sci.*, **8** (1973) 1439.
58. *Organization of Macromolecules in the Condensed Phase* Farad. Disc. Chem. Soc., Roy. Soc. Chem., London, 1979.
59. FISCHER, E. W., STROBL, G. R., DETTENMAIER, M., STAMM, M. and STEIDLES, N., ref. 58, p. 26.
60. UHLMANN, D. R., ref. 58, p. 87.
61. STEVENS, G. C., CHAMPION, J. V., LIDDELL, P. and DANDRIDGE, A., *Chem. Phys. Letts*, **71** (1980) 104.
62. STEVENS, G. C., CHAMPION, J. V. and LIDDELL, P., *J. Polym. Sci., Polym. Phys. Ed.*, **20** (1982) 327.
63. BOGDANOVA, L. M., BEIGOVSKII, I. M., IRZHAK, V. I. and ROSENBERG, B. A., *Polym. Bull.*, **4** (1981) 119.
64. STEVENS, G. C., CHAMPION, J. V. and LIDDELL, P., unpublished manuscript.
65. DUSEK, K., PLESTIL, J., LEDNICKY, F. and LUNAK, S., *Polymer*, **19** (1978) 393.
66. MATYI, R. J., UHLMANN, D. R. and KOUTSKY, J. A., *J. Polym. Sci., Polym. Phys. Ed.*, **18** (1980) 1053.
67. SCHMIDT, R. L., *J. Colloid Interface Sci.*, **27** (1968) 516.
68. STACEY, K. A., *Light Scattering in Physical Chemistry*, Butterworths, London, 1956.
69. PINNOW, D. A., CANDAU, S. J., MACCHIA, J. T. and LITOVITZ, T. A., *J. Acoustic. Soc. Amer.*, **43** (1967) 131.
70. SICOTTE, Y. and RINFRET, M., *Trans. Faraday Soc.*, **58** (1962) 1090.
71. VAN DER HULST, H. C., *Light Scattering by Small Particles*, Wiley, New York, 1957.
72. STEVENS, G. C. and RICHARDSON, M. J., *Polymer* **24** (1983) 851.
73. DEBYE, P. and BUECHE, A., *J. Appl. Phys.*, **20** (1949) 518.
74. ARUTUNYAN, CH. A., TONOYAN, A. O., DAVTYAN, S. P., ROZENBERG, B. A. and ENIKOLOPYAN, N. S., *Dokl. Akad. Nauk*, **214** (1974) 832.
75. BLYACHMAN, E. M., NICKITINA, A. A., ZELENINA, N. L. and SHEBCHENKO, Z. A., *Vysokomol. Soed.*, **A16** (1974) 1031.

76. Solomon, D. H. and Hopwood, J. J., *J. Appl. Polym. Sci.*, **10** (1966) 981, 1893.
77. Solomon, D. H., Loft, B. C. and Swift, J. D., *J. Appl. Polym. Sci.*, **11** (1967) 1593.
78. Solomon, D. H., *J. Macromol. Sci. Rev.*, **C1** (1967) 179; also in *Step-Growth Polymerizations,* Marcel Dekker, New York, 1972, Ch. 1.
79. Erath, E. H. and Robinson, M., *J. Polym. Sci., Part C,* **3** (1963) 65.
80. Cuthrell, R. E., *J. Appl. Polym. Sci.*, **11,** (1967) 949; **12** (1968) 955, 1263.
81. Basin, V. Ye, Korsunskii, L. M., Shokal'skaya, O. Yu. and Aleksandrov, N. V., *Polym. Sci. USSR,* **14** (1972) 2339.
82. Stratsev, V. M., Chalykh, A. Ye., Nenakhov, S. A. and Sanzharovskii, A. T., *Polym. Sci. USSR,* **17** (1975) 960.
83. Racich, J. L. and Koutsky, J. A., *J. Appl. Polym. Sci.,* **20** (1976) 2111.
84. Aspbury, P. J. and Wake, W. C., *Br. Polym. J.*, **11** (1979) 17; see also Aspbury, P. J. and Phillips, M., in *Adhesion,* Vol. 1, Ed. K. W. Allen, Applied Science Publishers, London, 1977.
85. Morgan, R. J. and O'Neal, J. E., *J. Mater. Sci.*, **12** (1977) 1966; *Polym. Plast. Tech. Eng.*, **10** (1978) 49; *Polym. Eng. Sci.*, **18** (1978) 1081.
86. Misra, S. C., Manson, J. A. and Sperling, L. H., *Amer. Chem. Soc. Div. Org. Coat. Plast. Chem. Prep.*, **39** (1978) 152.
87. Mijovic, J. S. and Koutsky, J. A., *Polymer,* **20** (1979) 1095.
88. Mijovic, J. S. and Koutsky, J. A., *J. Appl. Polym. Sci.,* **23** (1979) 1037.
89. Mijovic, J. S. and Tsay, L., *Polymer,* **22** (1981) 902.
90. Schmid, R., *Prog. Coll. Polym. Sci.,* **64** (1978) 197.
91. Takahama, T. and Geil, P. H., *Makromol. Chem. Rapid. Commun.*, **3** (1982) 389.
92. Takahama, T. and Geil, P. H., *J. Polym. Sci., Polym. Phys. Ed.,* **21** (1983) 1247.
93. Yamini, S. and Young, R. J., *J. Mater. Sci.*, **15** (1980) 1823.
94. Thomas, E. L. and Roche, E. J., *Polymer,* **20** (1979) 1413; see also ref. 58, p. 122.
95. Causton, B. E., private communication.
96. Lock, M. W. B., Ph.D. Thesis, University of Surrey, 1972.
97. Kelly, F. N., Swetlin, B. J. and Trainor, D. R., *Macromolecules,* Eds H. Benoit and P. Rempp, Pergamon Press, Oxford, 1982.
98. Kelly, F. N. and Trainor, D. R., *Polym. Bull.*, **7** (1982) 369.
99. Kreibich, U. T. and Schmid, R., *J. Polym. Sci., Symp.* **53** (1975) 177.
100. Stevens, G. C. and Richardson, M. J., *Polym. Commun.*, **26** (1985) 77, and references contained therein.
101. Kenyon, A. S. and Nielsen, L. E., *J. Macromol. Sci. A,* **3** (1969) 275.
102. Stevens, G. C. and Richardson, M. J., unpublished manuscript.
103. Uhlmann, D. R., *J. Non-Cryst. Solids,* **49** (1982) 439.
104. DeBoer, J. H., *Trans. Farad. Soc.*, **32** (1936) 10.
105. Houwink, R., *Trans. Farad. Soc.,* **32** (1936) 131.

106. GRIFFITHS, A. A., *Phil. Trans. Roy. Soc.*, **A221** (1920) 163.
107. BERRY, J. P., in *Fracture Processes in Polymeric Solids*, Ed. R. Rosen, Interscience, New York, 1964.
108. MORGAN, R. J., MONES, E. T. and STEELE, W. J., *Polymer*, **23** (1982) 295.
109. DUGDALE, D. S., *J. Mech. Phys. Solids*, **8** (1960) 100.
110. IRWIN, G. R., *Trans. Amer. Soc. Mech. Eng., J. Appl. Mech.*, **24** (1957) 361.
111. RICE, J. R., in *Fracture—an Advanced Treatise*, Ed. H. Liebowitz, Academic Press, New York, 1968.
112. ANDREWS, E. H., *J. Mater. Sci.*, **9** (1974) 887.
113. YOUNG, R. J. in *Developments in Polymer Fracture—1*, Ed. E. H. Andrews, Applied Science Publishers, London, 1979.
114. KINLOCH, A. J. and YOUNG, R. J., *Fracture Behaviour of Polymers*, Applied Science Publishers, London, 1983.
115. LAUTERWASSER, B. D. and KRAMER, E. J., *Phil. Mag.*, **A39** (1979) 369.
116. KRAMER, E. J., in *Developments in Polymer Fracture —1*, Ed. E. H. Andrews, Applied Science Publishers, London, 1979.
117. LILLEY, J. and HOLLOWAY, D. G., *Phil. Mag.*, **28** (1973) 215.
118. VAN DER BOOGART, A., in *Physical Basis of Yield in Glassy Polymers*, Ed. R. N. Haward, Institute of Physics, London, 1966.
119. GLEDHILL, R. A., KINLOCH, A. J. and SHAW, S. J., *J. Mater. Sci. Letts*, **14** (1979) 1769.
120. GLEDHILL, R. A. and KINLOCH, A. J., *Polym. Eng. Sci.*, **19** (1979) 82.
121. KINLOCH, A. J. and WILLIAMS, J. G., *J. Mater. Sci.*, **15** (1980) 987.
122. YAMINI, S. and YOUNG, R. J., *J. Mater. Sci.* **15** (1980) 1814, 1823.
123. MORGAN, R. J., MING KONG, F. and WALKUP, C. M., *Polymer*, **25** (1984) 375.
124. GLEDHILL, R. A., KINLOCH, A. J., YAMINI, S. and YOUNG, R. J., *Polymer*, **19** (1978) 574.
125. NARISAWA, I., MURAYAMA, T. and OGAWA, H., *Polymer*, **23** (1982) 291.
126. CHERRY, B. W. and THOMSON, K. W., *J. Mater. Sci.*, **16** (1981) 1913, 1925.
127. GOLDSTEIN, M. and SHIMA, R. (Eds), *The Glass Transition and the Nature of the Glassy State*, Vol. 279, Ann. N.Y. Acad. Sci., 1976.
128. O'REILLY, J. M. and GOLDSTEIN, M. (Eds), *Structure and Mobility in Molecular and Atomic Glasses*, Vol. 371, Ann. N.Y. Acad. Sci., 1981.
129. STRUIK, L. C. E., *Physical Ageing in Amorphous Polymers and Other Materials*, Elsevier, Amsterdam, 1978.
130. FERNANDEZ, A. M., WIDMAIER, J. M., SPERLING, L. H. and WIGNALL, G. D., *Polymer*, **25** (1984) 1718.
131. WU, WEN-LI and BAUER, B. J., *Polym. Comm.*, **26** (1985) 39.
132. GUPTA, V. B., DRZAL, L. T., ADAMS, W. W. and OMLOR, R., *J. Mater. Sci.* **20** (1985) 3439.
133. BAI, S. J., *Polymer*, **26** (1985) 1053.
134. WU, WEN-LI and BAUER, B. J., *Polymer*, **27** (1986) 169.

8

Silane Coupling Agents

J. Comyn

School of Chemistry, Leicester Polytechnic, Leicester, UK

1. INTRODUCTION

This chapter is about silane coupling agents and the proliferation of scientific interest which they have aroused since about 1980. There have been a number of papers showing the beneficial effects of silane coupling agents on a wide variety of substrates, but the majority of papers have described the use of modern instrumental techniques such as Fourier transform infrared spectroscopy (FTIR), inelastic electron tunnelling spectroscopy (IETS), and carbon-13 nuclear magnetic resonance (NMR) spectroscopy to examine the interaction of silanes with mainly glass, silica and metal substrates.

Silane coupling agents were developed in the 1940s to pretreat glass fibres for use in composites, with a view to improving their water resistance. Their structure may be represented by $R—Si(OR')_3$, where R is a functional group which can chemically react with the matrix resin and R′ is usually a methyl or ethyl group. R may be an epoxide- or amine-containing group if the matrix resin is an epoxide, or a vinyl-containing group where the matrix is a polyester. A list of the more frequently mentioned silane coupling agents is given in Table 1.

The success of silanes in pretreating glass fibres seems natural, in that by using a silicon compound on a silicate network such as glass, the old and probably unwritten chemical principle of 'like-on-like' is being exploited. This simple principle offers no hope for the pretreatment of metals with silanes, but nevertheless it does seem that silanes

TABLE 1

SOME SILANE COUPLING AGENTS MENTIONED IN THE TEXT

Structure	*Name*	*Abbreviation*
Aminosilanes		
$H_2NCH_2CH_2CH_2Si(OC_2H_5)_3$	3-Aminopropyltriethoxysilane	APES
$CH_3NHCH_2CH_2CH_2Si(OC_2H_5)_3$	*N*-Methylaminopropyltriethoxysilane	MAES
$CH_3NHCH_2CH_2CH_2Si(OCH_3)_3$	*N*-Methylaminopropyltrimethoxysilane	NMMS
$H_2NCH_2CH_2NHCH_2CH_2CH_2Si(OCH_3)_3$	*N*-(2-Aminoethyl)-amino-propyltrimethoxysilane	AAMS
$(CH_3O)_3Si(CH_2)_3NH(CH_2)_2NH(CH_2)_2NH_2$	Trimethoxysilylpropyldiethylenetriamine	DETMS
H_2N–[benzene ring]–$Si(OCH_3)_3$	Aminophenyltrimethoxysilane	APMS
$HCl\cdot CH_2{=}CH$–[benzene ring]–$CH_2NH(CH_2)_2NH(CH_2)_3Si(OCH_3)_3$	Cationic vinylbenzyl silane	CVBS
Epoxy silanes		
[epoxide ring, O]–$CH_2O(CH_2)_3Si(OCH)_3$	3-Glycidoxypropyltrimethoxysilane	GPMS
[epoxycyclohexyl ring, O]–$CH_2CH_2Si(OCH_3)_3$	2-(3,4-Epoxycyclohexyl)ethyl-trimethoxysilane	ECMS

Double-bonded silanes		
$CH_2{=}C(CH_3)COO(CH_2)_3Si(OCH_3)_3$	3-Methacryloyloxypropyltrimethoxysilane	MAMS
$CH_2{=}CHSi(OC_2H_5)_3$	Vinyltriethoxysilane	VES
$CH_2{=}CHSi(OCH_3)_3$	Vinyltrimethoxysilane	VMS
Chlorosilanes		
$ClCH_2CH_2CH_2Si(OCH_3)_3$	3-Chloropropyltrimethoxysilane	CPMS
$Cl{-}C_6H_4{-}CH_2CH_2Si(OCH)_3$	4-Chlorophenylethyltrimethoxysilane	CPEMS
Others		
$HSCH_2CH_2CH_2Si(OCH_3)_3$	3-Mercaptopropyltrimethoxysilane	MPMS
$HSi(OC_2H_5)_3$	Triethoxysilane	HES
$CH_3Si(OCH_3)_3$	Methyltrimethoxysilane	MMS
$CH_3Si(OC_2H_5)_3$	Methyltriethoxysilane	MES
$CH_3CH_2Si(OCH_3)_3$	Ethyltrimethoxysilane	EMS
$C_6H_{11}{-}Si(OCH_3)_3$	Cyclohexyltrimethoxysilane	CMS

are successful as coupling agents on a wide variety of substrates, where they may improve the strength of the initial bond, or more importantly they may improve durability in the presence of water.

Silane coupling agents have recently been reviewed in a book by Plueddemann.[1] As current theories of their mechanism of action were covered both there and by Erikson and Plueddemann,[2] only a brief outline is given here.

The *chemical bonding theory* is the oldest and best known theory, which postulates that chemical bonds are formed between the silane molecule and the interface. This involves hydrolysis of the alkoxy groups, followed by condensation of the trisilanols which are formed; trisilanols may condense with one another and with surface hydroxyl groups (Fig. 1). In this way a polysiloxane network is formed which is covalently bonded to the surface. With glass or silica, interfacial Si—O—Si bonds would be formed, but it would obviously be difficult to distinguish these spectroscopically from transfacial bonds of the same type. Interfacial Si—O— metal bonds which might be formed on applying silanes to metals are regarded as being susceptible to hydrolysis.[3]

The *deformable layer theory* supposes that an interfacial layer is formed in which stresses between resin and substrate may be relieved without bonds being ruptured. Good wetting of the substrate is regarded as an important condition for adhesive bonding and the *surface wetting theory* claims this to be the key factor with silanes. The

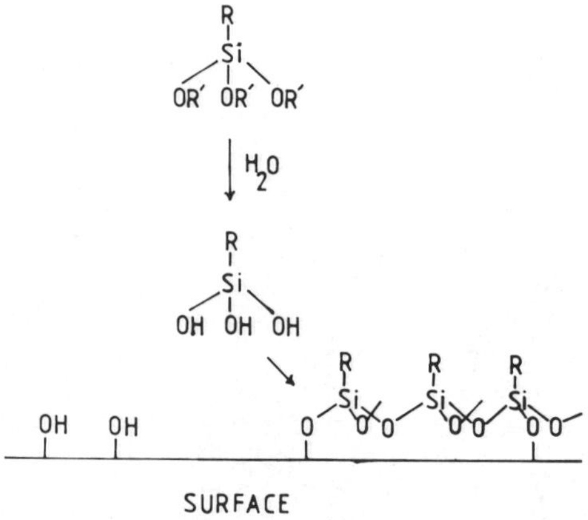

FIG. 1. Reaction of a trialkoxysilane at a surface containing hydroxyl groups.

low viscosity of silanes would assist wetting, and once in intimate contact with the substrate van der Waals' forces would be operative.

The *restrained layer theory* supposes that an interfacial region is formed which has a modulus intermediate between that of the adhesive and the substrate. Finally, the *reversible hydrolysis mechanism* proposes the reversible breaking and reformation of bonds at the interface as a mechanism of stress relaxation.

Another group of coupling agents which can be used for a range of substrates, and which are considered here, are the titanates and zirconates. There are other systems which have been left out either because it is not at all clear that they are coupling agents (e.g. they may act as corrosion inhibitors), or because the amount of scientific information on their mechanism is very limited. Such systems include 1,2-diketones for steel,[4] some nitrogen heterocyclic compounds on copper,[5,6] and some cobalt compounds for the adhesion of brass-plated steel tyre cords to rubber.[7]

2. IMPROVEMENTS IN BONDING WITH COUPLING AGENTS

There has been a series of papers by Walker demonstrating the effectiveness of silane coupling agents in the adhesion of paints to metals. Epoxide and polyurethane coatings were applied to mild steel or aluminium surfaces,[8,9] which had been treated by brushing with solutions of some silanes (MAMS, ECMS, GPMS, MPMS and AAMS) in wet acetone. After being dried in an oven the two-part paints were applied by spraying. Adhesion was assessed by a torque test or a direct pull-off test. The torque test involved bonding a cylinder to the abraded paint surface and removing it 24 h later with a torque spanner. The adhesive used was based on the diglycidyl ether of bisphenol A and a polyamide hardener. In the direct pull-off test, cylinders were bonded to both sides of a painted specimen in an aligning jig, and were subsequently removed in a tensile testing instrument. Some painted panels were tested after immersion in distilled water for 1500 h and some after further drying in air for 48 h. Bonding of cylinders to the wet panels used a cyanoacrylate adhesive.

Some results of these tests are shown in Table 2, where it can be seen that in all cases treatment with a silane significantly increases the initial joint strengths. All joints are weakened on exposure to water

TABLE 2

ADHESION OF EPOXIDE PAINTS TO DEGREASED METALS AS MEASURED IN THE DIRECT PULL-OFF TEST[a]

Metal	*Silane*	*Dry joints*		*1500 h in water*		*Subsequent drying for 48 h*	
		Strength (MPa)	*Debond (%)*	*Strength (MPa)*	*Debond (%)*	*Strength (MPa)*	*Debond (%)*
Mild steel	None	19·9	60	7·2	100	10·9	100
	ECMS	27·3	20	17·3	100	21·8	90
	GPMS	30·0	0	14·3	100	15·6	100
	MPMS	27·1	40	10·9	30	16·4	100
	AAMS	32·0	0	28·1	100	29·2	10
Al	None	21·3	90	5·6	100	11·2	100
	MAMS	30·2	0	12·0	30	19·6	50
	AAMS	31·1	0	11·4	30	20·5	40
Cd	None	25·1	0–30	17·4	30–40		
	MAMS	29·1	0	30·4	0		
	MPMS	29·1	0	26·3	10–30		
Cu	None	27·3	10–60	20·2	20		
	MAMS	27·4	0	21·9	30		
	MPMS	29·7	0	23·8	30		
Zn	None	20·5	60–100	13·4	10–40		
	MAMS	20·5	70	11·6	60–100		
	MPMS	31·0	0	29·1	0		

[a] Data after Refs. 10 and 11.

for 1500 h, but the reduction is much less for silane-treated metals, and all show some recovery on drying. Also included in Table 2 are values of the amount of paint film removed, i.e. the percentage of paint film removed in the bonded area during testing. With steel there is a distinct tendency for this to increase to 100% on exposure to water.

Some results[8] of joint strength against immersion time showed that there is an initial rapid loss in the first 12–24 h, which is followed by a more gradual decrease. The present author considers that the initial loss probably reflects the diffusion of water through the paint to the silane/metal interface.

These percentages of apparent interfacial failure were assessed visually. In an attempt to ascertain the real locus of failure some areas of apparent interfacial failure, using aluminium treated with AAMS, were examined by X-ray photoelectron spectroscopy (XPS). Both the metal and the back of the paint film were examined and no aluminium was detected in either. Also no titanium was detected on the paint film, which contained a titanium dioxide filler. It thus appears that failure occurs in a non-pigmented boundary layer of paint. Silicon was detected on both surfaces but at very low levels, <1·6% on the paint and <0·8% on the metal.

Cadmium, copper and zinc are generally regarded as difficult metals to paint and here Walker demonstrated[10] that silanes can improve initial adhesion and water durability. Some results are included in Table 2 for MAMS and MPMS. Here on exposure to water MPMS is superior to the other silanes, so indicating some degree of specificity.

Recent papers by Walker[11,12] have demonstrated the benefits of using silanes for some different paint systems on aluminium and mild steel.

Walker's work has demonstrated that silanes are effective on a range of metal surfaces, and that in general their effect is non-specific.

3. INTERACTION OF SILANES WITH GLASS AND RELATED MATERIALS

Emadipour and co-workers[13] embedded single strands of E-glass in an epoxide resin formed from DGEBA and methylnadic anhydride (MNA) with benzyldimethylamine (BDA) as accelerator, and measured the force needed to pull out the fibres. Interfacial failure was obtained by properly selecting the embedded length of glass (~2 mm). Some fibres were pretreated with silanes (AAMS, APES, and MAES) applied from aqueous solution (1% or 5%) at pH 5. (Here and throughout this chapter composition percentages are by weight.) In each case treatment with 1% solutions led to significant increases in interfacial shear strengths (these were 14·3% with MAES, 16·9% with APES and 62·0% with AAMS); however with the 5% solution the interfacial shear strengths were actually lowered, by −35·93, −19·5 and −13·8% respectively. The loss of interfacial strengths in the presence of a thicker layer of silane may be due to the fact it consists of three different layers, namely:

(1) A glass/silane interface with covalent chemical bonding.
(2) An interphase of silane oligomers with increasing crosslink

density towards the glass/silane interface. The silane layers most distant from the glass surface are almost entirely un-crosslinked. When the silane layer is thick, mechanical failure occurs in this interphase.

(3) A silane/epoxy interface with covalent chemical bonding.

Thus a maximum bond strength would probably be obtained with a monolayer of silane, covalently bonded to both glass and the epoxide, but these cannot be obtained with current processing technology.

The inhomogeneity of the interphase region has been demonstrated for APES on S-glass by Di Benedetto and Scola[14] using ion-scattering spectroscopy and secondary-ion mass spectrometry (SIMS). Changes in peak sizes at several sputtering depths showed a region of nearly constant composition extending from a depth of 0·5 nm to about 14 nm, but between 16 and 24 nm a dramatic increase in the nitrogen level was evident. In this region the major peaks corresponded to APES, indicating the presence of oligomers. From 24 nm to the glass, properties changed yet again.

Probably the most important property of silanes is not in enhancing dry bond strengths, but in reducing the fall in strength in the presence of water. Figure 2 shows the effect of immersion in water at 95°C on the interfacial strength of E-glass after treatment with 1% solution of AAMS, and then bonding to an epoxide. It can be seen that an initial very rapid loss of strength is followed by a period where the rate of loss is much reduced. However, the silane-treated interface retained its strength better than an untreated one. It seems that water attacks bonds between glass and silanes, and also possibly hydrolyses the siloxane network in the interphase.

In a more recent paper[15] Koenig and Emadipour examined the adhesion between glass monofilaments and an epoxide resin. The filaments had been pretreated with a 1% aqueous solution of the aminosilane AAMS. The initial interfacial shear strength was 27 MPa but on immersion in water at 25°C this fell to *ca* 8·5 MPa in about 30 days and then tended to level out so that after 270 days the value was 4·0 MPa. This behaviour was related to water extraction of oligomers from the silane interphase.

A recent paper by Plueddemann and Pape[15a] examined the use of binary mixtures of silanes on glass. There seems to be no distinct advantages in using these mixtures but neither does there seem to be any disadvantage.

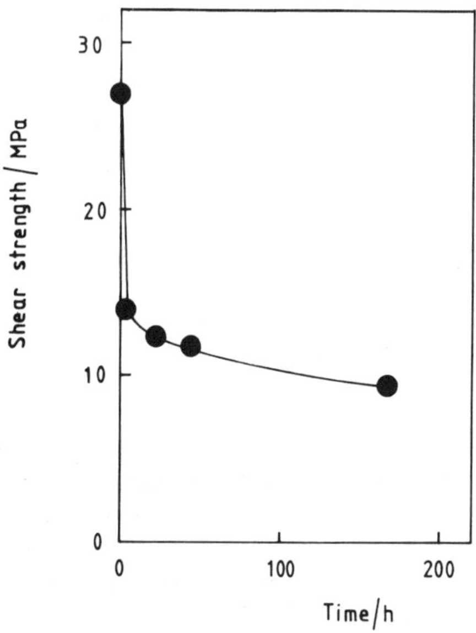

FIG. 2. Effect of immersion in water at 95°C on the interfacial strength of glass fibres treated with *N*-(2-aminoethyl)-3-aminopropyltrimethoxysilane and embedded in an epoxide.[13]

Koenig and his co-workers have used a number of spectroscopic techniques to explore the glass/silane interface. These have included Fourier transform infrared (FTIR) spectroscopy, magic-angle carbon-13 nuclear magnetic resonance (NMR) and Raman spectroscopy. The advantages of FTIR in the study of interfaces largely come from it being computer-linked. The spectra of very thin layers can be improved by multiple scanning, and difference spectra, say between a treated and an untreated surface, can be obtained. In difference spectra, modes which increase in intensity can be presented as peaks with those which are depleted as valleys, so giving assistance in the assignment of bands to products or reactants. The principles and methods of FTIR and its application to polymers have recently been reviewed by Koenig.[16]

One paper[17] used FTIR to look at a heat-cleaned high surface area silica which had been treated with a 1% aqueous solution of APES. Subtraction of the silica spectrum from that of the treated silane allowed that of the surface-absorbed species be recorded. The spectrum (Fig. 3) contained no bands at 2975, 1391 or 958 cm^{-1}, so

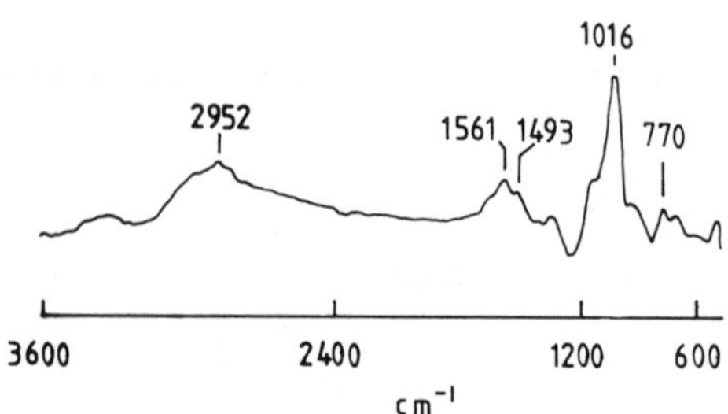

FIG. 3. FTIR spectrum of polymerized APES on a high-surface-area silica.[17]

indicating the complete removal of $SiOC_2H_3$ groups by hydrolysis. Some condensation polymerization is indicated by the appearance of bands at 770 and 1016 cm^{-1} assigned to Si—O—Si linkages. The positions of the amino bands at 1493 and 1561 cm^{-1} indicate strong hydrogen-bonding, and it was suggested that these are due to the asymmetric and symmetric deformation modes of the protonated amine in an $—SiO^- \overset{+}{N}H_3—$ group at the silica surface. Most amino groups appear to be adsorbed, thus both ends of the APES molecule seem to be quite strongly attached to the surface. Peak intensities indicated adsorption as a monolayer.

The FTIR difference spectrum of E-glass fibres treated with a 2% aqueous solution of APES was very similar to the spectrum of highly crosslinked polyaminosiloxane,[18] indicating that APES forms a polysiloxane on the surface of E-glass.

In a more recent paper Naviroj *et al.*[19] examined FTIR spectra of APES deposited from solutions at different pH values on silica wafers or E-glass. The thickness of the adsorbed layer varied with pH, attaining a maximum at the natural pH of APES solution (10·4). However, the really interesting features were the changes in the amine deformation bond at around 1500–1600 cm^{-1}, and to a lesser extent changes in the SiOH mode in the range 900–934 cm^{-1}. These changes are shown in Table 3, where shifts in the silanol peak are associated with varying strengths of hydrogen bonds. The amine group can thus be free or hydrogen bonded, or it may, when protonated, form an ion-pair with either chloride or bicarbonate anions.

Silanes do not form continuous films on glass but they form agglomerates. This was demonstrated by Bascom,[20] who showed that APES, VES, CPMS and CPEMS on glass and metal surfaces form polysiloxane coatings, which can partly be removed by rinsing with

TABLE 3

SPECTRAL CHANGES ASSOCIATED WITH THE DEPOSITION OF APES ON TO GLASS OR SILICA SURFACES AT VARIOUS pH VALUES[19]

pH	*FTIR peaks* (cm^{-1})		*Proposed structure*
	Amine deformation	*Silanol*	
2	1610, 1505	902	$\overset{+}{N}H_3Cl^-$ HO—Si—OH OH
10·8 (air-dried)	1575, 1488	930	H $\overset{+}{N}$—H(HCO_3^-) HO—Si—O- - - - -H OH H
10·8 (N_2-dried)	1601	934	H $\overset{+}{N}$—H HO—Si—O- - - - -H OH H
10·8 (heated)	1591		NH_2 —Si—O—Si—
12	1595	925	NH_2 O—Si—O O^-

organic solvents or water. The term 'sea–island' structure has been used to describe the appearance of agglomerates in electron microscopy.[21] Schrader[22] reported that silane coupling agents form heterogenous layers consisting of physically and chemically adsorbed fractions on glass. The outer fraction (about 95%) can be extracted by water at room temperature but the inner fraction requires boiling water to remove it. The surface fraction is firmly bound and survives for up to 100 min in boiling water.

Chiang and Koenig[23] looked at FTIR spectra of the diamino- and triamino-silanes AAMS and DETMS applied from aqueous solutions on fumed silica. The silica (Cab—O—Sil) had a high specific area (approx 400 m^2/g). Spectra of a bulk sample of the polymerized diaminosilane showed methoxy groups to be absent, and bands at 1113 cm^{-1} and 1034 cm^{-1} indicated polysiloxane formation. On heat treatment, difference spectra of the adsorbed triaminosilane showed an increase in intensity at 1096 cm^{-1} and decreases at 1015 and 940 cm^{-1} indicating a higher degree of polymerization. The shift of the amino band from 1598 cm^{-1} on adsorption indicates that these groups are hydrogen-bonded to the silanols.

To gain more information about the interaction of amine groups with silica, triethylenetetramine (TETA) was selected as a model compound and mixed with silica powder. TETA has a band 1604 cm^{-1} due to the —NH_2 groups which shifts to 1582 cm^{-1} when adsorbed on silica. Similarly a band at 1457 cm^{-1} for —CH_2— groups shifts to 1473 cm^{-1} and the NH stretching mode shifts from 3282 cm^{-1} to 3250 cm^{-1}.

With AAMS on silica, a small shoulder remains at 940 cm^{-1} after heat treatment for 18 h at 50°C or 3 h at 130°C, showing that condensation of the silanols is incomplete. However, if toluene replaces water as the solvent at application, the absence of bands at 940 cm^{-1} and in the 3800–3100 cm^{-1} region, corresponding to Si—OH groups, now indicates complete conversion to polysiloxanes. Adsorption from toluene is clearly different from that from water. Like the monoamine APES, the diamino- and triamino-silanes are strongly hydrogen-bonded to silica. These compounds can be adsorbed by both their silane and amine moieties on the silica surface.

The interaction of MAMS with E-glass fibres has been studied by Graf *et al.*[24] using diffuse-reflectance FTIR. A significant portion of the carbonyl groups are hydrogen-bonded to Si—OH and H_2O. On heating in air at 120°C for between 1 and 2 h a shoulder appears at 1734 cm^{-1} which is due to polymerization or oxidation of the C=C bond. This silane is completely hydrolysed when applied to glass fibres, the asymmetric CH stretching mode of the methoxy group being absent.

The water resistance of some silane coupling agents, applied from aqueous solution to E-glass fibres has been studied by Ishida and

Koenig[25] using FTIR. On exposure to water, the amount of MAMS remaining on glass decreases exponentially at 80°C and after 600 h about 17 molecules per nm^2 remain firmly attached (Fig. 4). The behaviour is quite different with VMS, where only a slight amount of desorption takes place in the first 600 h. Then the concentration decreases approximately linearly to a very low level, where it seems to level out. The differences in behaviour may be due to the formation of cyclic oligomers with MAMS. CMS shows a high resistance to hot water, the amount of adsorbed silane remaining unchanged within experimental uncertainty limits after 5000 h.

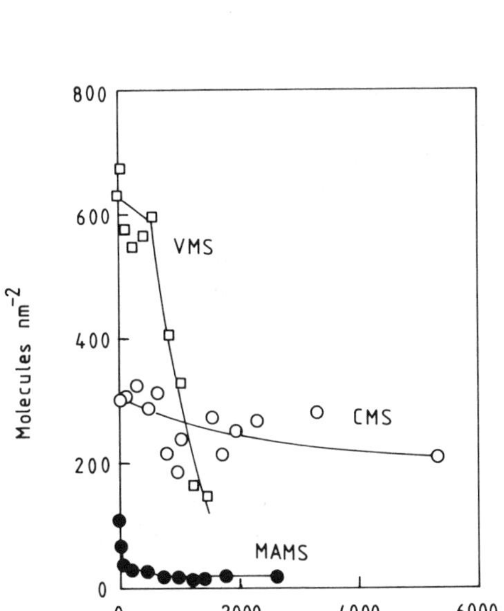

FIG. 4. Desorption of three silanes from E-glass fibres on immersion in water at 80°C. VMS was applied from a 4% aqueous solution, but CMS and MAMS were applied from 1% solution. After Ishida and Koenig.[25]

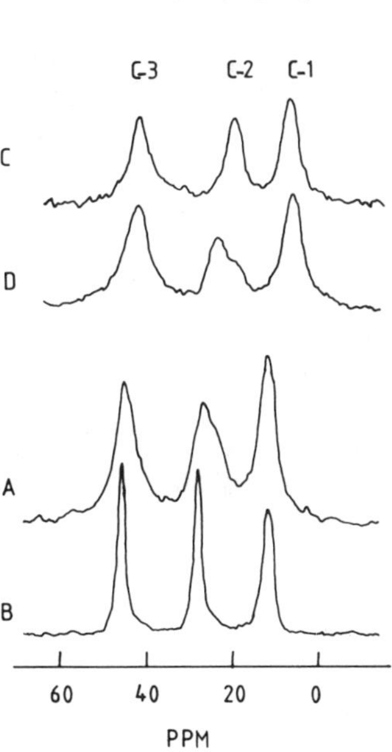

FIG. 5. Carbon-13 NMR spectra of hydrolysed APES. Polymer (A) initially dried at room temperature; (B) dried at 130°C; (C) adsorbed on high-surface-area silica and dried at room temperature; (D) adsorbed on silica and heated at 130°C for 1 day *in vacuo*.[27]

The hydrolytic stability of CMS is probably associated with the adsorption of cyclohexylsilanetriol in crystalline layers.[26] This triol can form large single crystals without polymerization.

A recent paper by Culler *et al.*[26a] has used diffuse reflectance FTIR to examine the adsorption of APES on oxidized silicon powder prepared by heating for 1 h at 200°C in air. The spectrum of oxidized silicon shows, amongst other features, a weak but sharp band at 3744 cm^{-1} which is due to free surface Si—OH groups. This peak is removed when the substrate is treated with silane solutions of 1, 3 and 5% concentration, but persists at lower coverages (0·1 and 0·005%), indicating condensation of surface Si—OH groups at high silane levels. The spectra of submonolayer quantities of the silane have two strong peaks at 1111 cm^{-1} and 1072 cm^{-1}, the latter being assigned to interfacial chemical bonds. This was tested by removing some of the unbonded silane by immersion in water at 70°C for 3 h. It was seen that bands at 1133 and 1045 cm^{-1} were removed and that at 1111 cm^{-1} reduced, but the strong band at 1068 cm^{-1} remained.

Carbon-13 nuclear magnetic resonance (NMR) spectroscopy is another technique which has been applied by Koenig *et al.*[27] to study the interaction of APES and AAMS with Cab—O—Sil fumed silica, an advantage of the technique being that there are no spectral peaks for glass. Magic-angle sample spinning, proton decoupling and cross polarization were all used to improve resolution, and 2000 scans were taken for polymerized silanes whilst with silica-adsorbed samples the figure was 10 000.

Spectra which appear in Fig. 5 all show three peaks which are due to the propyl carbons. The absence of ethoxy-group peaks indicates complete hydrolysis. Peak width is related to molecular motion, in that a sharp peak reflects a high level of molecular motion. Before heat treatment, molecular motion is restricted by hydrogen-bonding of the amine groups, but when this is destroyed the consequent liberation of the propyl groups leads to peak narrowing. A possible chemical reaction is shown in Fig. 6. The initially dried polymer has broader peaks than the heat-treated polymer, probably indicating the presence of cyclic oligomers. Protonation of the amino group would cause a large upfield shift of the C-2 atom and the shoulder on the C-2 peak in the spectrum of the initially dried polymer is probably due to this.

When the polymer is adsorbed on fumed silica all peaks are shifted to higher fields by about 2 ppm when compared with the bulk polymer; these shifts are doubtless due to the presence of glass. The shift in the

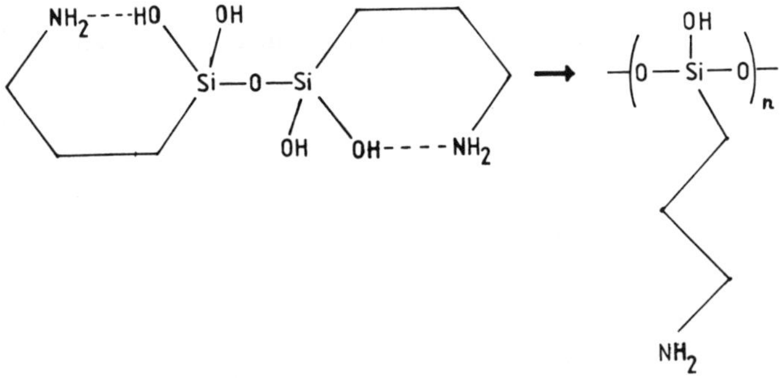

FIG. 6. Possible chemical reaction giving increased propyl group mobility so leading to a narrowing of NMR peaks.[27]

C-2 peak at 20·7 ppm (compare 25·9 ppm in aqueous solution) is due to the electric field induced by a protonated amino group. On heat treatment the height of this peak decreases and a new peak appears at 25·0 ppm. There thus seem to be two types of APES on the silica surface. These may be (i) hydrogen-bonded and free amino groups or (ii) covalently bonded and free silanes; the first explanation gained Koenig's favour.

The carbon-13 spectrum of the neat liquid diaminosilane AAMS has sharp lines and here the methoxy carbons are evident. Peak widths are increased for the bulk polymer and increased still further with adsorption on fumed silica, where all peaks have broad upfield shoulders. AAMS has two amino groups which can both become protonated and attached to the silica surface. Carbon-13 NMR is capable of detecting less than a monolayer of silane coupling agent on glass or fumed silica, and its use has now been extended to GPMS, MPMS and MAMS on glass.[28]

Zaper and Koenig[28a] have recently reported carbon-13 NMR spectra of some further silanes, both after polymerization and after adsorption on fumed silica from 1% aqueous solution. The silanes studied were MPMS, APES, AAMS, GPMS, MAMS and VES. In all cases, well-resolved spectra were obtained which were readily assigned to the different carbon atoms, and the absence of alkoxy-group peaks confirmed complete hydrolysis both in solution and on the silica surface.

Koenig *et al.* had previously obtained Raman spectra of silanes, hydrolysed silanes, and silanes adsorbed on silica and glass.[29,30] More recently they have[31] studied the resonance Raman spectra of a silane coupling agent containing a phthalocyanine ring adsorbed on glass. This substance is a monosilanol and so might be expected to adsorb as a monolayer. Resonance Raman spectroscopy requires that the exciting UV or visible radiation has a frequency which overlaps or is very close to an absorption peak of the compound under investigation. Less than a monolayer of adsorbed silane can be detected by this technique, and a strong Raman line at about $1545\,cm^{-1}$ has a frequency which depends on film thickness and is dependent on heat treatment for the silane on silica powder but not on E-glass. This line is clearly sensitive to intermolecular interactions and warrants further investigation.

Sounik and Kenney[31a] have indicated that silanes with the phthalocyanine group may have advantages as coupling agents for the following reasons.

(1) The silicon atom is octahedral rather than tetrahedral, and hence $3d_{x^2-y^2}$ and $3d_{z^2}$ orbitals are unavailable.
(2) The phthalocyanine ring is aromatic and generally planar or near-planar and it is therefore chemically and thermally stable. It is almost hydrophobic.
(3) The phthalocyanine group can form hydrogen-bonds and van der Waals interactions with glass surfaces.

An example of such a compound is shown in Fig. 7; the Si—OH group is capable of condensing with surface hydroxyl groups. It will be interesting to see how these substances perform in adhesive joints and composites.

The aminosilanes APES, AAMS and DETMS can react with atmospheric carbon dioxide[32] to form bicarbonate salts. Their formation has been studied by FTIR, which shows that the salt was mainly formed after the silanes had made contact with the glass and were being left to dry. Also, only primary amine groups seem to react in this way. Fortunately salt formation is not likely to be a problem in coupling to an epoxide matrix, as heating for 2 min at 100°C causes dissociation.

Eckstein and Dreyfuss[33] added a number of amines including APES to samples of polybutadiene, which were then pressed and heated against soda-glass microscope slides. By later peeling the cured elastomer from the glass, the work of adhesion was measured and values appear in Table 4. At the highest concentration APES is clearly

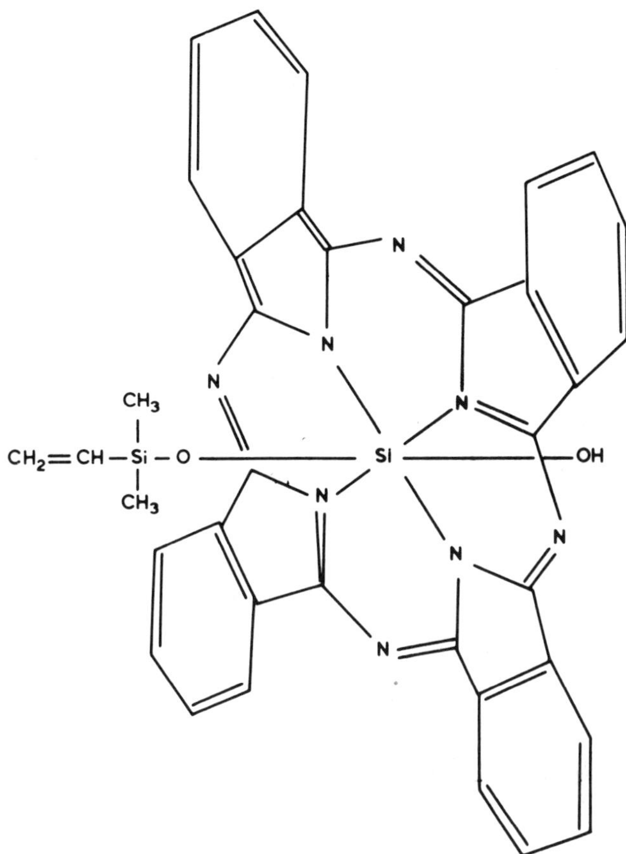

FIG. 7. A coupling agent based on phthalocyanine.

a very efficient coupling agent, giving a work of adhesion almost 40 times larger than any other amine. The addition of $1{\cdot}85 \times 10^{-6}$ mol of APES per 100 g of rubber is sufficient to provide a surface monolayer of silane, but this is clearly insufficient for effective coupling. This may be because some silane reacts with the rubber or because its rate of diffusion to the interface is too slow.

Mica is an aluminosilicate that undergoes high basal plane cleavage forming a planar non-porous surface. It has been used as a substrate for the silanes MAMS and cationic CVBS by Favis *et al.*[34] These were applied from aqueous solutions to mica of specific surface area

TABLE 4

WORK OF ADHESION FOR PEROXIDE-CURED POLYBUTADIENE TO GLASS, IN THE PRESENCE OF AMINES[33]

Amine	*[Amine] (10^{-4} mol per 100 g)*	*Work of adhesion (J/m^2)*
None	0	6·2
Aniline	4·0	6·2
1,3-Diaminobenzene	3·8	13·0
Piperazine	4·0	7·9
1,4-Diaminobutane	4·0	11·0
APES	0·0185	6·2
	0·46	22·3
	2·0	490·0
Et_3SiOH	3·9	6·1

0·612 m^2/g and, after rinsing, the amount of adsorbed silane was measured by CHN analysis. Both silanes were adsorbed in a number of discrete steps, the build-up of CVBS on mica being shown in Fig. 8, where it is seen that build-up starts at zero time. With MAMS, however, there is an induction period before adsorption commences. Each plateau corresponds to monolayer coverage, the area occupied by each CVBS and MAMS molecule being 0·32 nm^2 and 0·24 nm^2

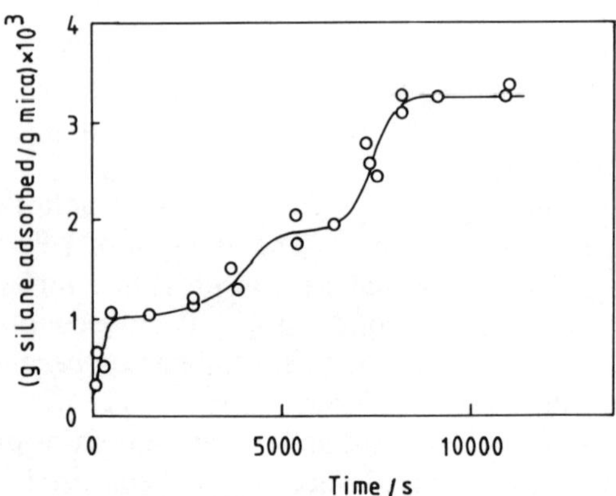

FIG. 8. The building-up of layers of CVBS on mica at pH 3·2. Initial conditions were [CVBS] = 0·25 g/dm^3, 0·5 g CVBS per 100 g mica.[34]

respectively. This implies that both silanes are close-packed and oriented perpendicular to the mica surface.

A technique for building up adsorbed multilayers of silanes has been developed by Netzer *et al.*[35,36] Their method depends on adsorbing a silane containing a long hydrocarbon chain which terminates in a group which can be activated prior to the adsorption of a second layer. The specific compound used was 15-hexadecenyltrichlorosilane (HTS) applied from an *n*-hexadecane/CCl_4/$CHCl_3$ mixture. It was adsorbed on glass, aluminium or silicon ATR (attenuated total reflection) plates. After adsorption the vinyl groups were converted to —CH_2—CH_2—OH groups by treating first with a solution of diborane (B_2H_6) in tetrahydrofuran followed by a solution of alkaline hydrogen peroxide. Their reaction scheme is shown in Fig. 9.

Contact angle data for *n*-hexadecane and water on glass covered with up to four layers of HTS indicated a lowering of molecular organization caused by the progressive accumulation of defects and irregularities as the layers built up. Octadecyltrichlorosilane (OTS) is without an activatable group and here the contact angles are unchanged after exposure to the same sequence of treatments as HTS; it appears that OTS is capable of bridging over surface irregularities.

ATR spectra for HTS on silicon show a band at 3200–3300 cm^{-1} due to C—OH and Si—OH groups, so indicating that the condensation reactions are incomplete. Reflectance spectra on aluminium mirrors

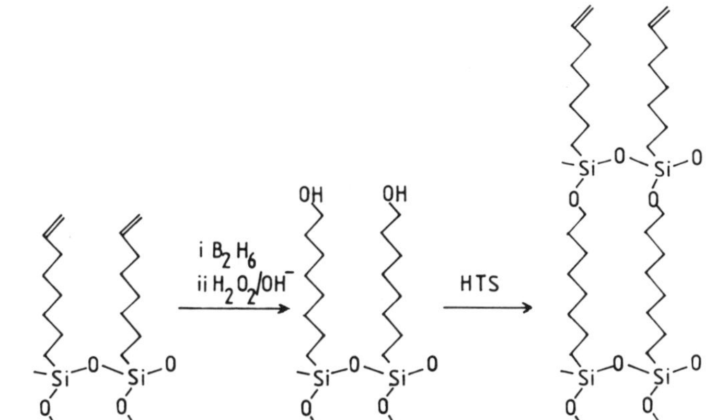

FIG. 9. Building-up layers of HTS by the method of Netzer *et al.*[35] For clarity the methylene chains have been shortened.

show increasing imperfection in the organization of the methylene chains. Only vibrations perpendicular to the mirror plane can be detected, so if the methylene chains are all normal to the surface the C—C but not the C—H bonds should be detected. With a monolayer of HTS there is a weak —CH peak near 2900 cm^{-1}. The C=C group is clearly observed at 1642·5 cm^{-1}, and this group can be converted to —CH_2CH_2OH without affecting orientation. With the second layer of HTS, increasing imperfection of the paraffin chains is indicated by an approximately fourfold increase in the CH_2 bands.

4. INTERACTION OF SILANES WITH METALS

Boerio has used reflection–absorption infrared spectroscopy (RAIR) with Fourier transform techniques to study silanes on mechanically polished metal mirrors. Aqueous solutions of APES (1%) at various pH-values were applied to iron mirrors for 30 min[37] and reflection–absorption IR spectra are shown in Fig. 10. There are quite large changes with pH: at about 8–9·5 strong absorptions near 1135 cm^{-1} are evident, whilst at higher pH the dominant bonds are near 1500 and 1600 cm^{-1}. The films formed at lower pH are polysiloxanes; this is indicated by the intense 1135 cm^{-1} band (Si—O—Si stretch). The weaker bands near 1600 cm^{-1} and 1500 cm^{-1} are due to asymmetric and symmetric modes of —$\overset{+}{N}H_3$ groups, possibly contained in cyclic zwitterions (I).

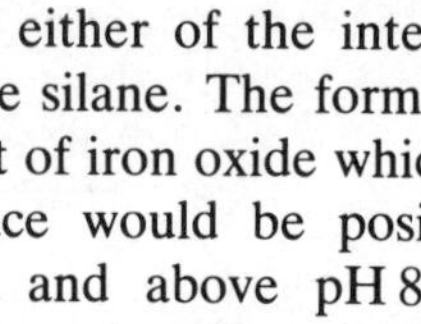

$$\begin{array}{ccccc} & & CH_2 & & \\ & / & & \backslash & \\ CH_2 & & & & CH_2 \\ | & & & & | \\ Si & & & & {}^{+}NH_3 \\ / \; / & \backslash & & & \\ & & O^- & & \end{array} \qquad (I)$$

There is clearly a transition of some kind at about pH = 9·5 which may be a characteristic either of the interaction between silane and substrate or solely of the silane. The former could be associated, say, with the isoelectric point of iron oxide which is near pH 8·5; i.e. below pH 8·5 the oxide surface would be positively charged due to the adsorption of protons, and above pH 8·5 it would be negatively charged due to ionization of surface —OH groups. Figure 10 shows that at high pH there is little absorption at 1135 cm^{-1} indicating that any formation of Si—O—Si bonds is slight. Boerio *et al.*[38] point out that in this case Si—O—Fe bonds may be formed. The results indicate

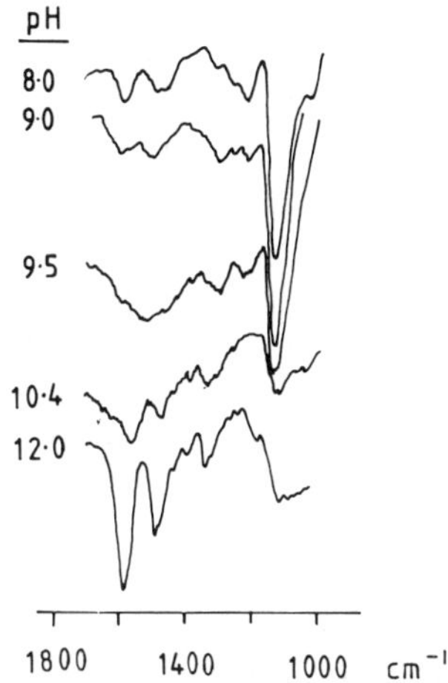

FIG. 10. Reflection–absorption IR spectra of 3-aminopropyltriethoxysilane adsorbed on iron mirrors from aqueous solutions at different pH values. Scale expansion at pH 8·0, 9·0 and 9·5 is twice that at pH 10·4 and 12·0.[37]

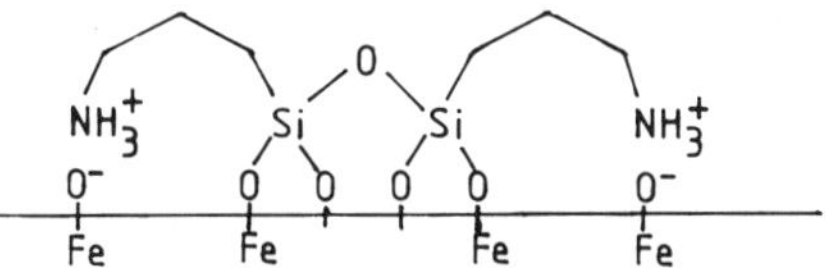

FIG. 11. Proposed structure of APES films formed on iron mirrors at high pH.[38]

that at lower pH APES forms polysiloxane films on iron mirrors, but above pH 9·5 adsorption involves participation of both ends of the molecule as shown in Fig. 11.

The effect of pH on adhesive bonding was investigated by preparing and testing some single lap joints. Strips of iron were immersed in 1% aqueous solutions of APES for 30 min and then dried. Adhesive bonding was by a tertiary-amine cured epoxide adhesive. Some joints were immersed in water at 60°C for up to 60 days. After immersion all silane-treated joints were stronger than untreated ones, and those pretreated at pH 8 were stronger than ones treated at pH 10·4 or pH 12·0, making it seem that formation of $—O^- \overset{+}{N}H_3$ ion-pairs impairs durability.

Films formed at pH 8·0, 10·4 and 12·0 were also examined by XPS. Only a single nitrogen 1*s* peak was observed with a binding energy of 399·6 eV, a value characteristic of $—NH_2$ rather than $—\overset{+}{N}H_3$, so giving rise to some doubt about the assignment of $—\overset{+}{N}H_3$ bands in the infrared.

Boerio and co-workers have also examined the interaction of some silanes with aluminium alloy 2024. The examination of APES deposited on polished aluminium mirrors[39,40] from a 1% aqueous solution gave RAIR spectra very similar to those on iron. Bands near 1000 cm^{-1} and 1500 cm^{-1} may be assigned to the $—\overset{+}{N}H_3$ group, and XPS data gave a N 1*s* binding energy of 401·5 eV which now supports this assignment. A band at 1080 cm^{-1} was assigned to the Si—O—Si stretching mode, indicating that most of the silanols had condensed to form polysiloxanes.

A later paper[41] used XPS to look at APES on aluminium alloy 2024, which contains about 4·5% of copper with lesser amounts of silicon and magnesium. Ellipsometry showed the freshly polished mirrors to have a 5 nm oxide coating and the APES film to be about the same thickness. High-resolution nitrogen 1*s* XPS spectra for APES adsorbed at pH 8·5 gives a broad peak which in fact consists of components at 399·8 and 400·6 eV. The 399·8 band may be assigned to free amino groups or amino groups co-ordinated to copper ions, and support for the latter comes from the copper Auger line at 339·0 eV which is typical for amines co-ordinated for copper. This can be compared with 336·0 eV for copper metal and 337·0 eV for cuprous oxide.

The shape of the aluminium 2*p* peak changed significantly with adsorption, deconvolution indicating components at 77·9, 76·1, 74·0 and 71·4 eV. Although the bands at 77·9 and 76·1 eV may be related to the formation of aluminosilane groups, they are more reasonably assigned to copper 3*p* in the cuprous state. The shapes of the oxygen 1*s* band at 531·1 eV and the silicon 2*p* band at 101·7 eV were also examined for evidence of Al—O—Si but none was forthcoming.

It is a widely held view that oxide hydration is an important weakening mechanism for adhesively bonded aluminium. Boerio *et al.*[42] examined the effect of APES and MAMS on the stability of aluminium oxides using RAIR. The former silane was deposited from 1% aqueous solutions, but methanol was the solvent for MAMS. Hydration stability was assessed by immersion in deionized water at 60°C. The air-formed oxide had a single band at 960 cm^{-1} but in water at 60°C this was quickly replaced by pseudo-böhmite having strong bands near 3400, 3100 and 1080 cm^{-1} and weaker bands near 1640 and 1375 cm^{-1}. The rate of hydration was followed using the band near

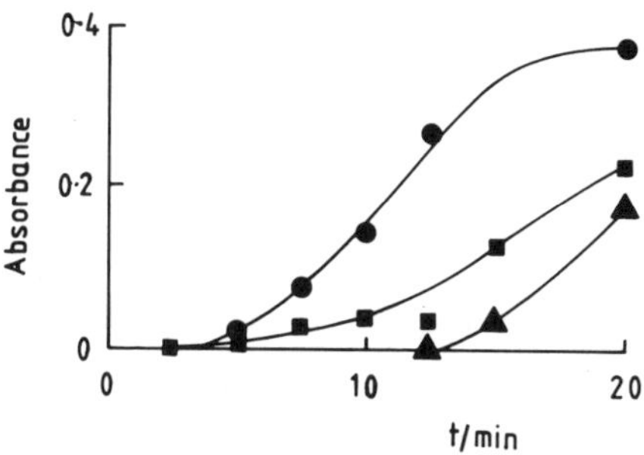

FIG. 12. Intensity of absorbance band near 1080 cm^{-1} as a function of immersion time in water at 60°C: ●, unprimed Al; ■, Al primed with APES at pH 7; ▲, Al primed with APES at pH 10·4.[42]

1080 cm^{-1}. Thin layers of APES inhibit hydration (Fig. 12) and it is claimed that 'aluminosiloxane bonds are also probably formed with the oxide'.

With MAMS, immersion in water at 60°C removes most of the silane, and development of the 1080 cm^{-1} band indicates some hydration of the oxide. However, if the freshly coated mirror is first treated at 100°C for 15 minutes, very little dissolution of the film occurs, presumably due to extensive polymerization. Such a film is an extremely effective inhibitor for the hydration of aluminium.

So far in this discussion little direct evidence has been presented of covalent bonds such as Si—O—Si or Si—O—metal *across the interface.* Some evidence for such bonds comes from Gettings and Kinloch[43] who detected an ion of mass 100, assigned as $FeSiO^+$, in the SIMS spectrum of a steel surface treated with a 1% aqueous solution of GPMS. This ion was absent when APES or a (styrene–functional amine hydrochloride)silane was employed. GPMS gave joints of superior durability when immersed in water at 60°C for up to 1500 h. Gettings and Kinloch claim that 'this is strong evidence for the formation of a chemical bond, probably —Fe—O—Si— between the metal oxide and polysiloxane primer.'

The substance 3-aminopropyldimethylethoxysilane has only one alkoxy group, and it may therefore react with a substrate without the complication of transfacial Si—O—Si groups. Diffuse-reflectance FTIR spectra of this substance on alumina and titania have recently been reported by Naviroj *et al.*[43a] Spectra of treated and untreated

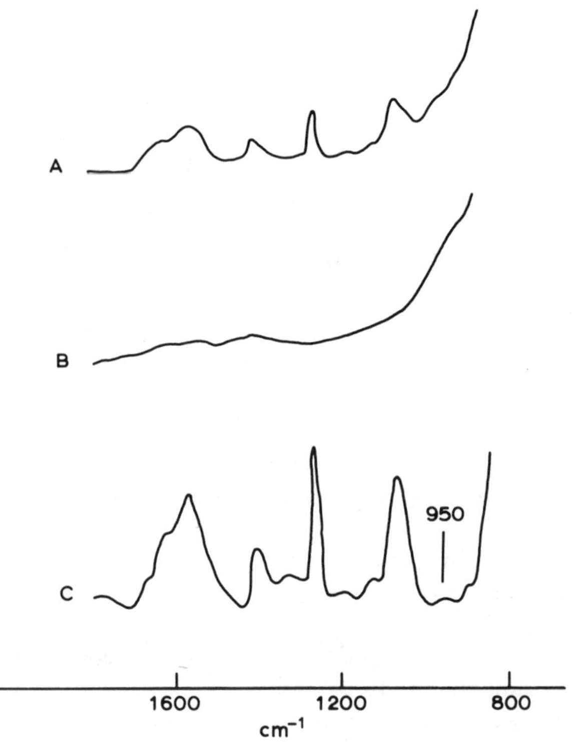

FIG. 13. FTIR spectra of (A) TiO_2 and (B) TiO_2 treated with $NH_2(CH_2)_3Si(CH_3)_2OC_2H_5$; (C) difference spectrum.[43a]

titania are shown in Fig. 13, along with the difference spectrum. The weak band at 950 cm^{-1} was tentatively assigned to interfacial Si—O—Ti groups, and this was confirmed by using silica as the substrate and a titanate [isopropyltri(isostearoyl)titanate; see Section 6] as the coupling agent. The neat titanate has no bands around 950 cm^{-1} but the difference spectra (treated minus untreated silica) fortunately reveals a peak at 950 cm^{-1}. The silane on alumina reveals a weak peak at 963 cm^{-1} attributed to Si—O—Al groups.

Hayes and Evans[44] heated silica and titania substrates in the presence of hexamethyldisilazane, which they consider would form $(CH_3)_3SiO$— groups covalently bound to the surfaces. In examining surfaces by SIMS, a variety of ions were detected, but none indicated covalent bonding at the interface.

Sung and Sung[45,46] have used FTIR to look at APES and VES adsorbed on sapphire (as a model surface for aluminium) and on E-glass fibres. The advantages of using sapphire are its transparency to IR up to 1500 cm^{-1} and its high refractive index, which makes it useful as an internal reflection crystal. Spectra clearly show the formation of a polysiloxane film by a reduction in peaks associated with the ethoxy groups and the emergence of peaks due to Si—O—Si. However, with the vinylsilane the existence of Si—OH peaks indicates that polymerization is incomplete. With the aminosilane, IR peak shifts for amino groups again indicate that they are protonated. The ratio of the intensity of the $—NH_2$ peak near 1600 cm^{-1} to that of the $—CH_2—$ stretching band at 2940 cm^{-1} increases for thinner films of APES, indicating that the films are inhomogeneous.[47]

Allen and Stevens[48] also found that the IR reflectance spectra of GPMS films on aluminium varied with thickness. The OH band near 3400 cm^{-1} moved to a lower frequency with thinner films, indicating a loss of freedom possibly associated with hydrogen-bonding to the substrate.

Aminosilane-coated sapphire was also examined by XPS,[45,46] and the results were in close agreement with those of Boerio, nitrogen 1*s* peaks at 399·0 and 401·3 eV indicating $—NH_2$ and $—\overset{+}{N}H_3$ groups.

RAIR of GPMS deposited on aluminium alloy mirrors from 1% aqueous solutions with a few drops of acetic acid added has been reported by Boerio *et al.*[49] The virtual absence of bands associated with $Si—O—CH_3$ groups near 820 and 1085 cm^{-1} and a reduction in intensity at 1190 cm^{-1} indicate high levels of hydrolysis, and a band at 1105 cm^{-1} illustrates the formation of Si—O—Si groups. Without the addition of acetic acid, spectra are qualitatively the same, but a lower level of hydrolysis is indicated. The similarity of GPMS spectra on iron or aluminium mirrors indicates that any interaction with the metal oxides is not strong.

Allen *et al.*[50] have studied the adhesion of polydimethylsiloxane rubber to aluminium oxide. The compound used (Dow Corning 3140) cures by the hydrolysis of alkoxy groups, so in this respect its adhesion properties reflect those of silane coupling agents. Using surface-energy data for liquids on the bulk materials, it was shown that polymer physically adsorbed on aluminium oxide is unstable in the presence of water; that is, water is thermodynamically able to bring about debonding. If curing the elastomer against alumina caused some changes to the polysiloxane surface, such as forming chemical bonds, then the interface might become stable.

Allen and co-workers[51] have also examined IR spectra of bulk-polymerized samples of GPMS, and the results are of some relevance to the behaviour of coupling agents. Samples were made by hydrolysing GPMS in water at pH 3·85 and then refluxing with added methanol. After removing the methanol/water mixture by distillation, a liquid remained which could be cured in a mould at 40°C for 24 h or spread onto a KBr disc for IR transmission spectroscopy. Unhydrolysed methoxy groups were found in the final polymer, but most of the hydroxyl groups were not accessible to isotopic exchange with D_2O. The accessible OH groups appear to be located at the surfaces. This seems a strange result, not only because OH groups are readily exchangeable, but because polysiloxanes are highly permeable to water. 'Strong resistance to deuteration' has also been noted by Ishida *et al.*[52] for polymerized samples of APES.

Inelastic electron tunnelling spectroscopy (IETS) has been used by a number of experimenters to examine organosilanes adsorbed on aluminium oxide. The technique records vibrational spectra of a monolayer adsorbed on a metal oxide, and aluminium has been widely used as the substrate because it can easily be vacuum-evaporated. Silanes can be applied to the oxidized metal from solution or from the vapour phase, and devices are completed by evaporation of a top electrode which is usually of lead. After cooling the device to the temperature of liquid helium (4·2 K) a small modulation voltage superimposed on a slowly increasing voltage is applied to it. The modulation current produces a second harmonic voltage across the device which is proportional to d^2I/dV^2. This small second harmonic is recovered by a lock-in amplifier and plotted on the y-axis of a chart record. The bias voltage V is applied to the axis.

Most electrons (>99%) tunnel elastically through the device, but a small number interact with vibrational modes. It is these interactions which are detected and which appear in an IR-like spectrum on the chart recorder. Hansma has recently edited a book which reviews the principles and practices of IETS.[53]

Brewis *et al.*[54] looked at the triethoxysilanes with H—, CH_3—, CH_2=CH— and $NH_2CH_2CH_2CH_2$— functional groups and the trimethoxysilanes with C_2H_5—, CH_2=CH—, 3-glycidoxypropyl and

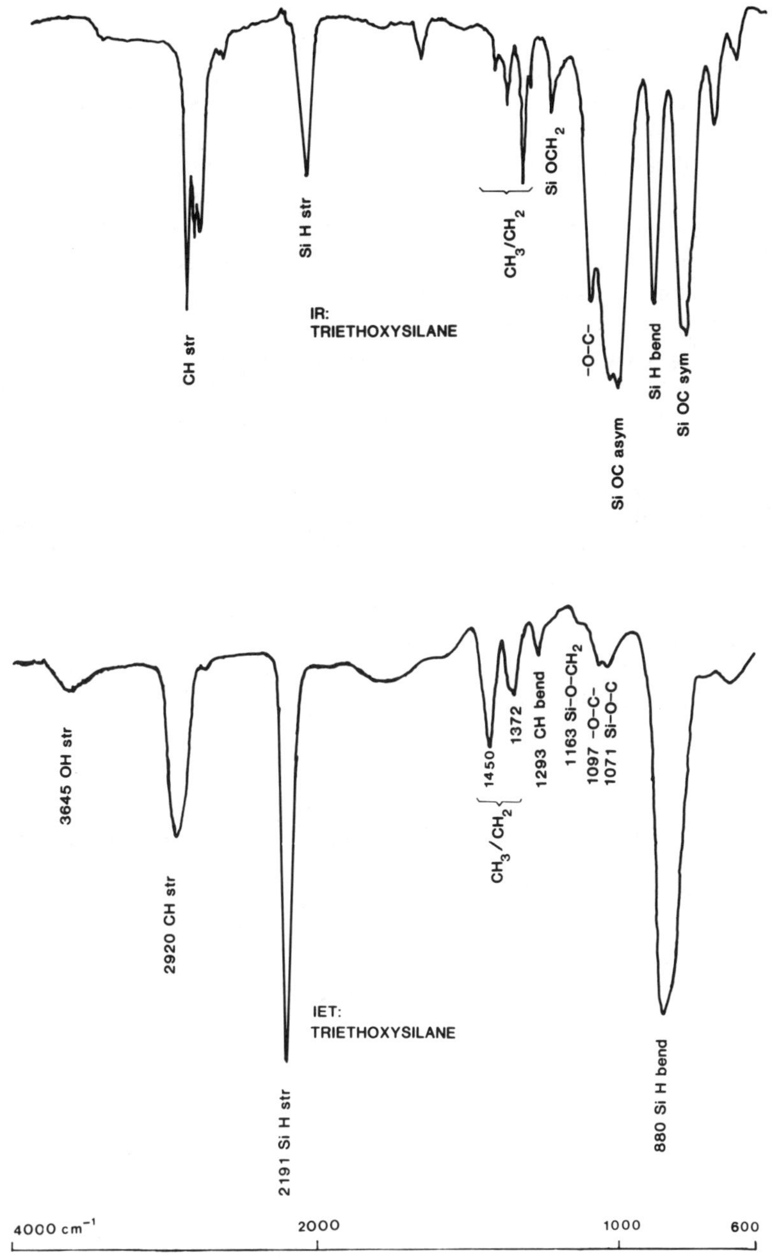

FIG. 14. IR transmission spectrum on a KBr disc and IET spectrum of HES, vapour-phase doped on aluminium oxide.[54]

3-mercaptopropyl functionalities. These were prepared by vapour-phase doping in the presence of atmospheric water, which may play a role in adsorption. The IR and IET spectra of the simple alkoxysilane HES appear in Fig. 14 and show many common features. Assignment of IET spectra basically rests on comparison with IR group frequencies, but both IR and Raman modes can be observed. There is no complete theory on selection in IETS; however, there is some experimental evidence in support of the view that where bonds are aligned perpendicular to the oxide surface, enhanced interaction with tunnelling electrodes gives a relatively intense peak.

HES gives a strong sharp peak at 2191 cm^{-1}, which is interpreted by the Si—H bonding being perpendicular to the oxide surface. The Si—H bonding mode at 880 cm^{-1} is similarly strong. It would appear that this molecule is adsorbed tripod-like by the three ethoxy groups onto alumina. Chemical changes do not seem to occur when triethoxysilanes are adsorbed on aluminium oxide in IET devices. This view is upheld by the work of Diaz *et al.*,[55] who noted using IETS on aluminium that some ethoxy groups remain intact for VES and APES doped from 1% solutions in anhydrous benzene. Spectra of the four triethoxysilanes studied all have peaks due to $Si(OC_2H_5)_3$. APES gives a well defined IET spectrum (Fig. 15), where the NH stretch at 3255 cm^{-1} is at a lower frequency by about 50 cm^{-1} than in the IR. This might indicate hydrogen-bonding, which could be between amine and the surface, or within the APES molecule.

In contrast, the trimethoxysilanes appear to react chemically in IET junctions, as the Si—O—C peaks which are present in the IR are either very weak or absent from IET spectra. This is probably due to hydrolysis of the methoxy groups. However, there is no clear evidence that the hydrolysed silanes then polymerize, possibly because Si—O—C and Si—O—Si groups absorb in similar regions of the spectrum so that any changes would be unclear.

GPMS was the silane which Kinloch and Gettings found very effective on steel and which gave evidence by the $FeOSi^+$ ion of bonding at the interface. It was also the most reactive silane functional group in IET devices on aluminium oxide. The IET spectrum (Fig. 16) shows no evidence of an epoxide group at about 939 cm^{-1}, but there is evidence for >C=C< (1595 cm^{-1}) and >C=O (1646 cm^{-1}) groups which are not present in the original molecule. The following routes were proposed for these changes.

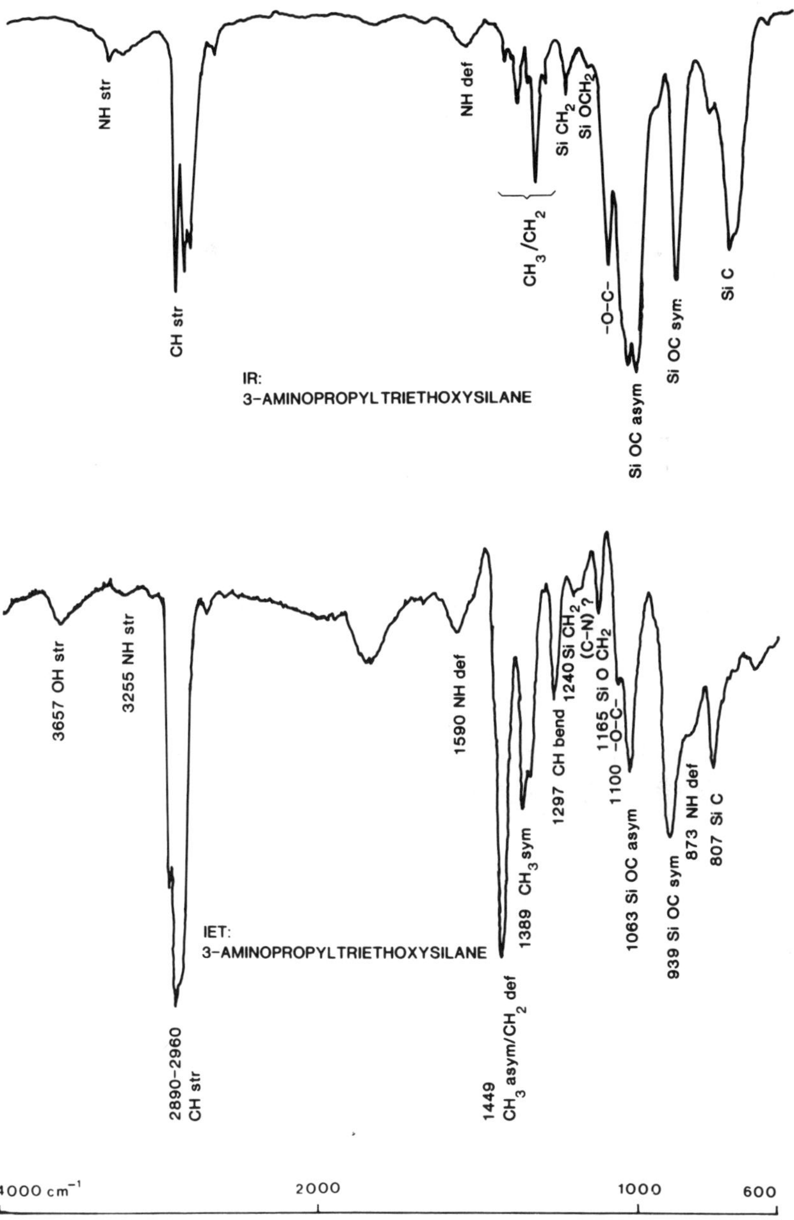

FIG. 15. IR transmission spectrum on a KBr disc and IET spectrum of APES, vapour-phase doped on aluminium oxide.[54]

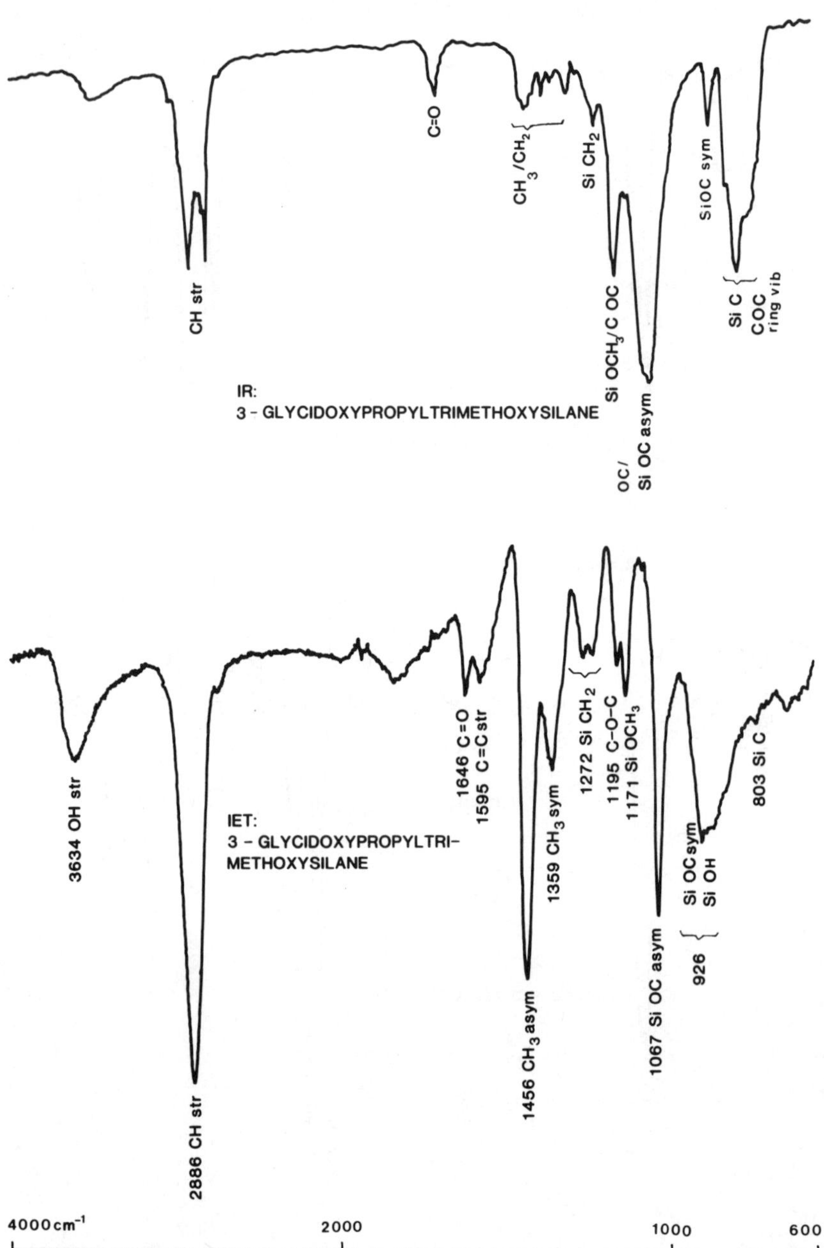

FIG. 16. IR transmission spectrum on a KBr disc and IET spectrum of GPMS, vapour-phase doped on aluminium oxide.[54]

(a) *Conversion of epoxide groups to ketones*

$$\underset{\text{(epoxide)}}{CH—CH_2} \xrightarrow{H_2O} —\overset{OH}{\overset{|}{C}}H—\overset{OH}{\overset{|}{C}}H_2 \xrightarrow{-H_2O} \underset{\text{enol}}{—CH{=}CH—OH} \longrightarrow \underset{\text{keto}}{CH_2—CHO}$$

$$—\overset{OH}{\overset{|}{C}}H—\overset{OH}{\overset{|}{C}}H_2 \xrightarrow{-H_2O} \underset{\text{enol}}{—\underset{OH}{\underset{|}{C}}{=}CH_2} \longrightarrow \underset{\text{keto}}{—CO—CH_3}$$

(b) *Initiation of the anionic polymerization of epoxide groups by the methoxide ion, and a termination process for polymerization which leads to carbon–carbon double bonds.*

$$CH_3O^- + R—\overset{O}{CH—CH_2} \longrightarrow CH_3O—\underset{R}{\underset{|}{C}}H—CH_2—O^-$$

$$\text{∿OCH}_2\text{—(cyclic intermediate with } CH_3, {}^-O, CH_3) \longrightarrow \begin{matrix} \text{∿OCH}_2—CH{=}CH_2 \\ \text{or} \\ \text{∿OCH}{=}CH—CH_3 \end{matrix} + HO\underset{CH_3}{\underset{|}{C}}HCH_2O^-$$

The latter reaction mechanism has been proposed by St Pierre and Price[56] and Dege *et al.*[57] for the anionic polymerization of propylene oxide.

Furukawa *et al.*[58] obtained IET spectra of aminophenyltrimethoxysilane (a mixture of 1,3- and 1,4-isomers with possibly some 1,2- as well) vapour-phase doped in a vacuum system onto aluminium oxide. Frequency shifts in the $—NH_2$ deformation mode [1620 cm^{-1} (IR) to 1575 cm^{-1} (IET)] and of the surface OH stretching mode (3580 cm^{-1} in a blank junction to 3650 cm^{-1} after adsorption of the silane) probably indicate hydrogen-bonding. Some junctions were exposed to saturated water vapour, and the spectra re-recorded after removal of excess water by vacuum pumping. They showed a gradual hydrolysis of the $Si—O—CH_3$ groups by a reduction of intensities of CH_3 stretching modes in the 2800–3000 cm^{-1} region. Another diminishing peak at 1185 cm^{-1} is due to the OCH_3 rocking mode. Again, there were no notable indications of the formation of Si—O—Si groups.

Furukawa *et al.*[59] have also studied APES deposited on aluminium oxide in a vacuum system. They agree with Brewis *et al.* that the adsorbed silane retains ethoxy groups and that there is possibly hydrogen-bonding between amino and surface hydroxyl groups. The effects of exposure to wet air are that most of the initially adsorbed silane is desorbed, and the rest undergoes hydrolysis. The ethanol produced by hydrolysis is probably oxidized to acetaldehyde.

The interaction of silanes with titanium has received some attention. Boerio[60] observed that APES films from 1% aqueous solutions on titanium are unstable in air and has interpreted spectral changes as showing further polymerization in the presence of moisture. He also reported that APES films on titanium and iron are very similar, but their performance as primers is very different. This may be due to the different isoelectric points of the two metals leading to different molecular orientations.

IET spectra of MPMS, vapour-phase doped onto an air-grown oxide on aluminium, has been reported by Van Velzen.[61] Vapour-phase doping here was in a vacuum system rather than in laboratory air as used by Brewis *et al.*, and differences in the amount of water present may be the reason why Van Velzen observed peaks due to residual methoxy groups at 1452, 2896 and 2928 cm^{-1}. He claims that the weakness of the —OH band at 3615 cm^{-1} when compared with an undoped junction shows that these groups are involved in bonding with the silane, most probably by a condensation reaction with the methoxy groups, but he cannot unambiguously assign any band to the Al—O—Si group. A most interesting feature of this work was that two different metals were used as top electrodes. With lead as the top electrode an —SH stretching mode doublet appears; the 2557 cm^{-1} component was assigned to a free —SH and that at 2492 cm^{-1} to a hydrogen-bonded group. On heating junctions for 1 h at 125°C the free —SH band becomes more intense and the hydrogen-bonding band weakens. Simultaneously the —OH stretching band at 3620 cm^{-1} almost totally disappears. The interpretation of these changes is that hydrogen-bonding occurs between —OH and —SH groups (structure II) but this structure is broken up on heating.

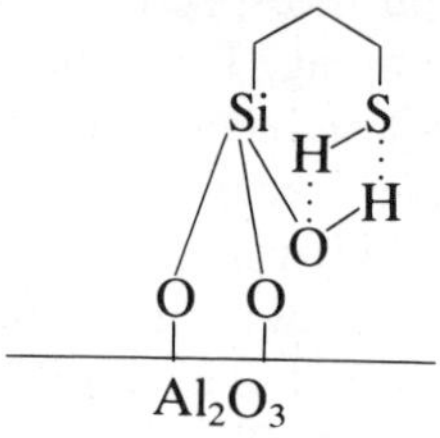

Si S H H O O O Al_2O_3 (II)

With silver as the top electrode the —SH stretching bands are absent, and it was suggested that a silver thiolate (R—S—Ag) is formed. A very weak band at 340 cm^{-1} was tentatively assigned to the —S—Ag vibration. Other spectral changes may also support this case.

Boerio *et al.*[62] have studied the interaction of APES with copper mirrors using RAIR and XPS. The important difference here is that copper dissolves in the APES film. An amino band at 1575 cm occurs which is not seen on iron mirrors, and this indicates the co-ordination of these groups with copper ions. The copper mirrors are covered by a layer of Cu_2O which has a strong IR band at 640 cm^{-1}. This band diminishes when mirrors are immersed in 1% aqueous solution of APES at pH 10·4 showing that the oxide dissolves; a polymeric film is formed which is about 10 nm thick. Examination of silane-treated films by XPS shows a nitrogen 1*s* peak at 399·2 eV which could be due to free amine but is not inconsistent with amine-co-ordinated copper ions. The oxygen 1*s* peak at 531·7 eV is due to siloxane units, but there is a hint of a shoulder at 530·1 eV which could be due to Cu_2O. Copper 2*p* peaks which appear at 932·2 and 932·4 eV are due to copper(I) ions; these are thought to be due to copper in the film rather than on the metal surface (Fig. 17), but on exposure to laboratory air for several days shake-up satellites develop at 943·1 and 962·6 eV showing the oxidation of Cu(I) to Cu(II). The interaction of APES with copper ions is clearly complex and this complexity was also indicated by the copper Auger spectra.

Apart from changes involving copper ions, APES behaves in a similar manner as on iron and aluminium in that a band at 1120 cm^{-1} shows the development of siloxane units and residual Si—OH groups are evident at 920 cm^{-1}. On exposure to a humid atmosphere further polymerization is indicated by a decline in the 920 cm^{-1} band and a shift of the 1120 cm^{-1} band to nearer 1040 cm^{-1}. The assignment of bands at 1570 cm^{-1}, 1470 cm^{-1} and 1330 cm^{-1} to amine-bicarbonate formation was confirmed by preparing some samples in the absence of carbon dioxide. With copper as substrate amine-bicarbonate formation is described as substantial.

Plasma polymerization offers a means of producing thin pinhole-free films upon substrates, and may provide an alternative technique for the application of silane coupling agents to substrates. Plasmas are very energetic, and compounds which are not traditional monomers in the sense of having double bonds, rings or functional groups can be polymerized in this way. Hays and Haaland[63] have plasma-

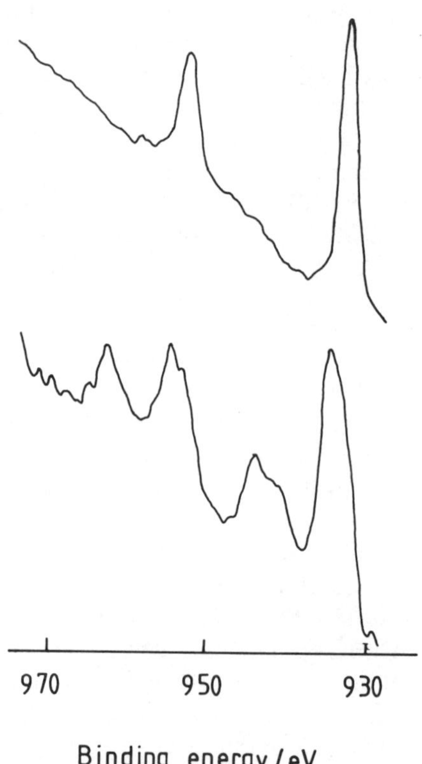

Fig. 17. Copper $2p$ XPS spectra of copper mirrors treated with 1% aqueous solutions of APES. Upper trace: sample stored in a desiccator. Lower trace: exposed to laboratory air.[62]

polymerized the non-functional silane MMS from an argon plasma onto a number of metals. The adhesive strength of plasma films (~1 μm thick) were assessed by bonding onto aluminium studs with an epoxide adhesive, and measuring the force needed for removal. Because failure was at the film–adhesive interface, the lower limit for the strength of the silane–aluminium interface was placed at 33 MPa, and with stainless steel it was 27 MPa. These two metals are both covered with strong oxides; other metals which did not have this characteristic give much lower joint strength (Au 1·4–3·3 MPa, Ag below 3·3 MPa, Cu less than 0·7 MPa).

5. INTERACTIONS OF FUNCTIONAL GROUPS WITH POLYMERS AND RESINS

So far, attention has been directed to the interaction of the trialkoxysilane moieties with solid substrates, but we now examine how functional groups at the other end of the silane molecules interact with resins or polymers.

Chiang and Koenig[18] used FTIR to examine the interaction of the primary and secondary aminosilanes APES and NMMS on E-glass fibres, with the DGEBA/MNA/BDA epoxide resin. Spectra indicated that the secondary amine reacts with the epoxide resin in the absence of the BDA accelerator, but reaction with the primary amine only occurs when the accelerator is present.

When unreacted resin is added to the silane-treated surface, some will penetrate the open siloxane layers which are not highly crosslinked. The degree of crosslinking of the silane increases as the glass is approached. Since over 50% of the silane reacts with the epoxy resin it seems essential that the resin penetrates into the silane layer. Thus the reactions between the resin and the silane build a crosslinked interpenetrating network linking the glass to the matrix.[64]

FTIR spectra showed that glass fibres treated with MAMS cause an increase in the crosslink density of the adjacent epoxide matrix.[64] The amount of unreacted MNA can be measured by following the carbonyl stretching band at 1860 cm^{-1}. The quantities of unreacted anhydride in

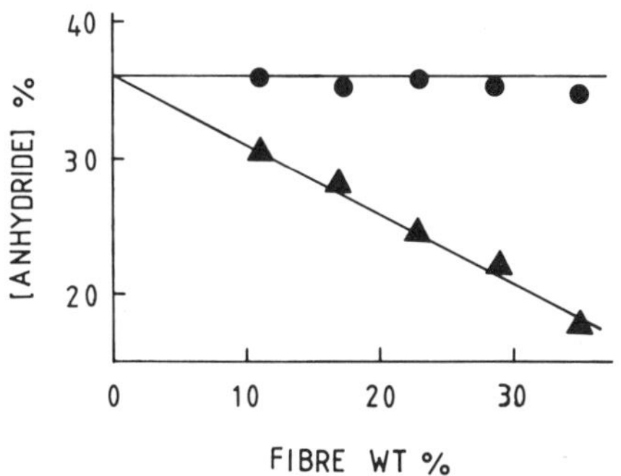

FIG. 18. Dependence of MNA concentration on time for glass fibre reinforced epoxide: ●, untreated fibres; ▲, fibres treated with MAMS.[64]

composites with silane-treated and heat-cleaned fibres is shown in Fig. 18. It was concluded that the total degree of curing of the matrix increases with the amount of silane on the fibre. Thus the matrix–silane interface would seem to have a higher crosslink density than in the bulk, the increase in crosslinking caused by the use of MAMS being about 5–10%.

Ishida and Koenig[65] have also examined the reaction of VMS and MAMS on fumed silica or E-glass with unsaturated polyester resins dissolved in styrene. When adsorbed on fumed silica the C═C bonds of both silanes react with styrene, but on E-glass VMS remains largely unreacted. Untreated E-glass appears to inhibit polymerization of the matrix resin.

Garton[66] deposited APES from 1% aqueous solutions onto germanium IR spectroscopy elements. To the coated element was added a 100 μm film of a mixture of DGEBA, MNA and BDA, which was given a conventional curing cycle. FTIR difference spectra showed that anhydride loss was accelerated by the silane. The valleys at 1780 and 1740 cm^{-1} confirmed a reduction in esterification and residual anhydride whilst the peaks at 1658 and 1563 cm^{-1} showed the production of amide groups.

Reactions with polyamic acids, which are precursors for polyimides, have also been studied. Linde[67] deposited APES from aqueous solution onto a germanium crystal and then covered this with a thick layer of a polyamic acid (structure III) from solution in *n*-methylpyrrolidone. The appearance of carboxylate bands at 1555 and 1335 cm^{-1} in the FTIR spectrum demonstrated that reaction (1) occurred.

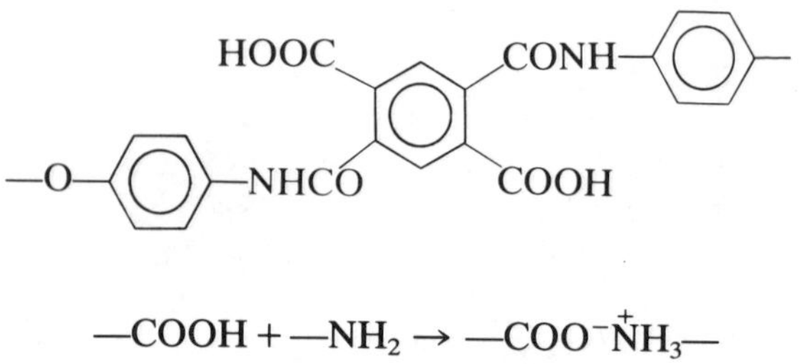

HOOC CONH—⟨◯⟩—

—O—⟨◯⟩—NHCO COOH (III)

$$\text{—COOH} + \text{—NH}_2 \rightarrow \text{—COO}^-\overset{+}{\text{N}}\text{H}_3\text{—} \qquad (1)$$

Heating to 120°C for 5 min induced amide formation (1660 cm^{-1}) whilst prolonged heating yielded the imide (1735 cm^{-1}) by the reaction sequence (2).

Si $\overset{+}{N}H_3\bar{O}OC$ NHCO $\xrightarrow{120°}$

Si NHCO NHCO $\longrightarrow$ (2)

Si N(CO)(CO) + [benzene]

The same polyamic acid was used by Suryanarayana and Mittal[68] with APES deposited from 0·01% aqueous solutions on plasma-oxidized silicon wafers. After thermally curing the wafers at up to 400°C under nitrogen, large increases in peel force were recorded for silane-treated surfaces. The peel force depended on the pH of the silane solution, reaching quite a sharp maximum at about pH 8, which coincides (perhaps fortuitously) with the natural pH of the APES solution. The concept of isoelectric point cannot be of assistance here, as it is at pH 2·2 for the silica surface.

Gähde *et al.*[69] treated glass sheets with the thiol-containing silane MPMS, and then pressed nylon 6 films against this at a temperature of about 250°C. This led to an improvement in peel strength. Later the assemblies were peeled apart and the counterfaces examined by XPS, which indicated that failure seems to be at the silane–nylon interface. Hydrogen-bonding between —SH and —CONH groups may contribute to the strength of this interface.

Sung, Sung and co-workers[70,71] have pretreated sapphire with APES and then bonded molten polyethylene to it. The polyethylene contained minimal amounts of carbonyl groups, making it likely that the interaction between polymer and APES must be physical. The available evidence supported the view that limited interdiffusion occurs; this was based on a slight depression of the melting point of the polymer because of silane miscibility, and measurement of silicone distribution across the interface using energy-dispersive X-ray analysis. On exposure to saturated air at 25°C peel strengths of joints were decreased, and with increasing exposure time a debonded area grew from the edges of the joints. On drying out the debonded areas did not recover, but in the bonded areas the strength was recovered, so implying the reformation of Si—O—Si bonds on drying.

6. TITANATE AND ZIRCONATE COUPLING AGENTS

Titanate coupling agents are effective in increasing the bonding of particulate fillers to polymers.[72–74] They have the general formula $R'OTi(OR)_3$ where R′ is typically an isopropyl group and R is a larger group such as a stearate; R may contain a functional group. It is considered[75] that they react with surface protons in the following manner.

$$R'OTi(OR)_3 + MOH \rightarrow MOTi(OR)_3 + R'OH$$

The following titanates can be effective primers for anodized aluminium alloy bonded with a rubber-modified epoxide adhesive,[76] but strengths depend on drying conditions.

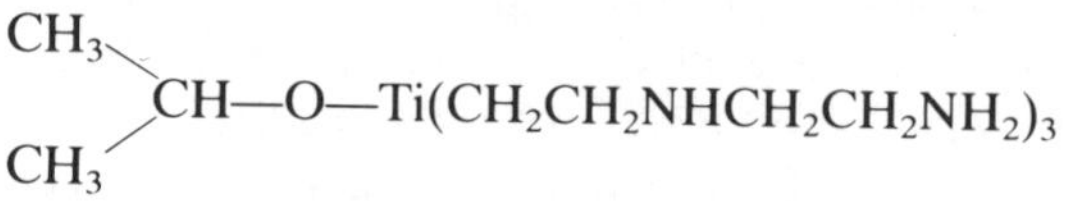

$$(CH_3)_2CH\text{—}O\text{—}Ti(CH_2CH_2NHCH_2CH_2NH_2)_3 \qquad \text{(IV)}$$

Isopropoxytri(ethylaminoethylamino)titanate

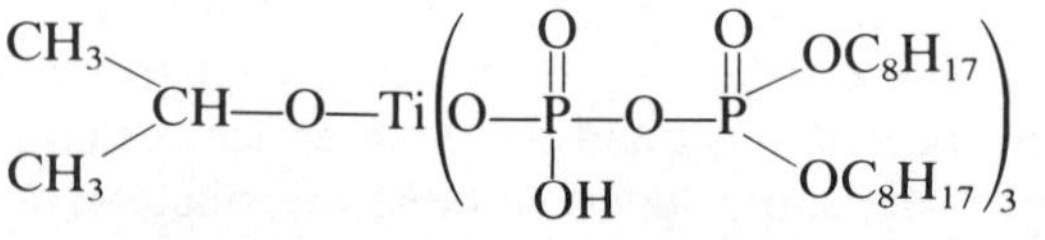

$$(CH_3)_2CH\text{—}O\text{—}Ti\left(O\text{—}\underset{OH}{\overset{O}{\overset{\|}{P}}}\text{—}O\text{—}\overset{O}{\overset{\|}{P}}(OC_8H_{17})_2\right)_3 \qquad \text{(V)}$$

Isopropoxytri(dioctylpyrophosphate)titanate

It can be seen (Table 5) that stronger joints are obtained at lower titanate levels; possibly once a monolayer has been built up further increases lead to weakening.

Sung *et al.*[77] used an isostearyl titanate as a coupling agent for polyethylene on aluminium oxide and found that peel strength increases on heating above 70°C. Sung *et al.*[46] have also examined the FTIR spectrum of compound IV and of isopropoxytri(isostearoyl)titanate

TABLE 5
EFFECT OF TITANATE TREATMENT ON STRENGTHS OF LAP JOINTS ($\frac{1}{2}$ IN OVERLAP) OF ANODIZED ALUMINIUM ALLOY BONDED WITH A RUBBER-MODIFIED EPOXIDE[76]

Titanate	*Drying condition*	*Strength (MPa)*
None	None	9·6
IV; 1% solution in isopropyl alcohol	Room temp.	7·6
	30 min, 66°C	12·1
	30 min, 122°C	8·5
IV; 0·2% solution in isopropyl alcohol	Room temp.	17·6
	30 min, 66°C	20·8
	30 min, 122°C	13·5
V; 1% solution in xylene	Room temp.	13·4
	30 min, 66°C	6·1
	30 min, 122°C	11·3
V; 0·4% solution in xylene	Room temp.	11·6
	30 min, 66°C	12·4
	30 min, 122°C	17·0

adsorbed on sapphire. Spectra of adsorbed and bulk titanates were virtually identical.

According to a later paper[78] some spectral changes were seen on heating the adsorbed titanate to 70 or 140°C; these were the disappearance of peaks in the 1736 cm^{-1} region. It is thought that the high-temperature spectra are authentic for the coupling agent, with the lower-temperature spectra being contaminated by stearate esters which are driven off in heating. Unlike organosilanes, organotitanates do not seem to form polymeric films on sapphire.

Calvert *et al.*[79] coated alumina and silica powders with some solutions of organotitanates and organozirconates in organic solvents, and after rinsing obtained IR spectra. With tricarboxytitanates CH stretching modes at 2850 cm^{-1} and 2950 cm^{-1} were present, indicating adsorption. These were absent in the untreated powders. It was not possible to prepare a tetracarboxytitanate, but on treating the powders with zirconium tetrastearate in benzene, no —CH modes were evident in the spectra, so confirming adsorption via alkoxy and not carboxylate groups.

Some treated powders were soaked in distilled water (containing a small quantity of acetone to aid wetting) at 60°C for 24 h. Values in Table 6 show that over 80% of the coupling agents survived this

TABLE 6
REDUCTION IN COUPLING AGENT ADSORPTION AFTER WATER TREATMENT (24 h AT 60°C)[79]

Coupling agent[a]	*Powder*	*Reduction (%)*
i-PrOTiSt$_3$	SiO_2	17·5
i-PrOZrSt$_3$	SiO_2	11·1
i-PrOTiSt$_3$	Al_2O_3	14·3
i-PrOZrSt$_3$	Al_2O_3	4·2

[a] Abbreviations: i-PrO, isopropoxy; St, stearate.

treatment. These were measured by comparing the intensities of IR bands at 2920 cm^{-1} (CH stretch) and 1110 cm^{-1} (Si—O stretch).

Butt joints in aluminium were constructed using an epoxide adhesive. Without a primer strengths were 10·9 MPa, and with the metal pretreated with a 1% solution of isopropoxytristearoyltitanate in benzene this fell slightly to 9·8 MPa. However, the latter compound is not functional towards the adhesive; using the functional isopropoxytri(9,10-epoxystearoyl)titanate the strength rose to 14·5 MPa.

Reference has already been made to the work of Favis *et al.*, who demonstrated the building up of layers of silanes upon mica. In the same paper[34] they also showed that when compound V was applied to mica from solution in toluene, three layers of molecules could build up. Each titanate molecule occupied 1·43 nm^2; this is more than a silane and reflects the greater bulk of these molecules.

Zirconium propionate is also commercially used as an adhesion promoter for printing inks on polyolefin.[80] It is believed that zirconium carboxylates have polymeric structures, and that they can couple with polymer surfaces.

7. CONCLUDING REMARKS

Over recent years there has been considerable development in analytical techniques which are capable of characterizing adsorbed monolayers. FTIR techniques in particular have provided much new information on adsorbed silanes, but there have been important gains from the use of SIMS, IETS, carbon-13 NMR and XPS.

Using these techniques it has clearly been demonstrated that on glass, silica and metal substrates silanes hydrolyse and then condense to form polysiloxane networks. It is widely believed, and indeed clearly stated in many papers already cited, usually in the introduction, that covalent bonds are formed across the interface. However, the only direct evidence for this is Gettings and Kinloch's[43] detection of $FeOSi^+$ in the SIMS spectrum of steel treated with GPMS, and the detection of Si—O— metal bonds in silane-treated metal oxides by Naviroj, Koenig and Ishida.

Reactions of functional groups with polyester, epoxide and polyimide resins has been clearly demonstrated using FTIR, and the degree of crosslinking of the resin may be different at the interface from that in the bulk. Functional groups may interact with thermoplastics by interdiffusion or hydrogen-bonding.

Aminosilanes have been the most widely studied. Amino groups can react with an epoxide resin, but when deposited on a glass, silica or metal surface they may be linked to the surface or internally cyclized by hydrogen bonds or Coulombic forces involving protonated amine groups. They can also absorb carbon dioxide.

Silanes are effective primers for a range of substrates where they increase initial bond strengths and improve wet durability.

The silane-containing layer at an interface is inhomogeneous, with the degree of crosslinking increasing towards the substrate. Thin layers of silane seem to give the best adhesive bonds, and in the future the use of Langmuir–Blodgett methods, or the procedure of Netzer, Iscovici and Sagiv[35,36] may give improvements in performance through the production of highly ordered monolayers.

REFERENCES

1. PLUEDDEMANN, E. P., *Silane Coupling Agents,* Plenum Press, New York and London, 1982.
2. ERIKSON, P. W. and PLUEDDEMANN, E. P., 'Interfaces in Polymer Matrix Composites' in *Composite Materials,* Vol. 6, Ed. E. P. Plueddemann, Academic Press, New York and London, 1974, Ch. 1.
3. PLUEDDEMANN, E. P., *Int. J. Adhes. Adhes.,* **1** (1981) 305.
4. NICOLA, A. J., JR and BELL, J. P., *Adhesion Aspects of Polymeric coatings,* Ed. K. L. Mittal, Plenum Press, New York and London, 1983, p. 443.
5. PARK, J. M. and BELL, J. P., *Adhesion Aspects of Polymeric coatings,* Ed. K. L. Mittal, Plenum Press, New York and London, 1983, p. 205.

6. Yoshida, S. and Ishida, H., *J. Adhes.*, **16** (1984) 217.
7. Van Ooij, W. J. and Biemond, M. E. F., *Rubber. Chem. Tech.*, **57** (1984) 688.
8. Walker, P., *J. Coatings Tech.*, **52** (1980) 49.
9. Walker, P., *J. Oil Col. Chem. Assoc.*, **65** (1982) 415.
10. Walker, P., *J. Oil Col. Chem. Assoc.*, **66** (1983) 188.
11. Walker, P., *J. Oil Col. Chem. Assoc.*, **67** (1984) 108.
12. Walker, P., *J. Oil Col. Chem. Assoc.*, **67** (1984) 126.
13. Emadipour, H., Chiang, P. and Koenig, J. L., *Res Mechanica,* **5** (1982) 165.
14. Di Benedetto, A. T. and Scola, D. A., *J. Colloid Interface Sci.*, **64** (1978) 565.
15. Koenig, J. L. and Emadipour, H., *Polym. Compos.*, **6** (1985) 142.
15a. Plueddemann, E. P. and Pape, P. G., *Modern Plastics,* (July 1985) 78.
16. Koenig, J. L., *Adv. Polym. Sci.*, **54** (1984) 87.
17. Chiang, C.-H., Ishida, H. and Koenig, J. L., *J. Colloid Interface Sci.*, **74** (1980) 396.
18. Chiang, C.-H. and Koenig, J. L., *Polym. Compos.*, **2** (1981) 192.
19. Naviroj, S., Culler, S. R., Koenig, J. L. and Ishida, H., *J. Colloid Interface Sci.*, **97** (1984) 308.
20. Bascom, W. D., *Macromolecules,* **5** (1972) 792.
21. Sterman, S. and Bradley, H. B., *Proc. 16th Ann. Tech. Conf. Reinf. Plast. Div., SPI,* (1960) 8-D.
22. Schrader, M. E., *Proc. 25th Ann. Tech. Conf. Reinf. Plast. Div., SPI,* (1970) 13-E.
23. Chiang, C.-H. and Koenig, J. L., *J. Colloid Interface Sci.*, **83** (1981) 361.
24. Graf, R. T., Koenig, J. L. and Ishida, H., *Anal. Chem.*, **56** (1984) 773.
25. Ishida, H. and Koenig, J. L., *J. Polym. Sci., Phys. Ed.*, **18** (1980) 1931.
26. Ishida, H. and Koenig, J. L., *J. Polym. Sci., Phys. Ed.* **17** (1979) 1807.
26a. Culler, S. R., Ishida, H. and Koenig, J. L., *J. Colloid Interface Sci.*, **106** (1985) 334.
27. Chiang, C.-H., Liu, N.-I. and Koenig, J. L., *J. Colloid Interface Sci.*, **86** (1982) 26.
28. Cholli, A., Zaper, A. M. and Koenig, J. L., *Amer. Chem. Soc. Div. Polym. Chem. Polym. Prepr.*, **24** (1983) 215.
28a. Zaper, A. M. and Koenig, J. L., *Polym. Compos.*, **6** (1985) 156.
29. Koenig, J., Shih, P. T. K. and Lagally, P., *Mater. Sci. Eng.*, **20** (1975) 127.
30. Shih, P. T. K. and Koenig, J. L., *Mater. Sci. Eng.*, **20** (1975) 137, 145.
31. Ishida, H., Koenig, J. L., Asumoto, B. and Kenney, M. E., *Polym. Compos.*, **2** (1981) 75.
31a. Sounik, D. F. and Kenney, M. E., *Polym. Compos.*, **6** (1985) 151.
32. Culler, S. R., Naviroj, S., Ishida, H. and Koenig, J. L., *J. Colloid Interface Sci.*, **96** (1983) 69.
33. Eckstein, Y. and Dreyfuss, P., *J. Adhes.*, **15** (1983) 193.
34. Favis, B. D., Blanchard, L. P., Leonard, J. and Prud'homme, R. E., *Polym. Compos.*, **5** (1984) 11.

35. NETZER, L., ISCOVICI, R. and SAGIV, J., *Thin Solid Films,* **99** (1983) 235.
36. NETZER, L., ISCOVICI, R. and SAGIV, J., *Thin Solid Films,* **100** (1983) 67.
37. BOERIO, F. J. and WILLIAMS, J. W., *Applns. Surf. Sci.,* **7** (1981) 19.
38. BOERIO, F. J., CHENG, S. Y., ARMOGAN, L., WILLIAMS, J. W. and GOSSELIN, G., *35th Ann. Tech. Conf. Reinf. Plast./Compos. Inst., SPI,* (1980) 23C.
39. BOERIO, F. J., *Org. Coat. Plast. Chem.,* **44** (1981) 625.
40. BOERIO, F. J., *Amer. Chem. Soc. Div. Polym. Chem., Polym. Prepr.,* **21** (1981) 297.
41. BOERIO, F. J., GOSSELIN, C. A., DILLINGHAM, R. G. and BIRKSTRAND, J. M., *15th Natl SAMPE Tech. Conf.,* (1983) 212.
42. BOERIO, F. J., DILLINGHAM, R. G. and BOZIAN, R. C., *39th Ann. Conf. Reinf. Plast./Compos. Inst./SPI,* (1984) Session 4A, 1.
43. GETTINGS, M. and KINLOCH, A. J., *J. Mater. Sci.,* **12** (1977) 2511.
43a. NAVIROJ, S., KOENIG, J. L. and ISHIDA, H., *J. Adhesion,* **18** (1985) 93.
44. HAYES, T. R. and EVANS, J. F., *J. Phys. Chem.,* **88** (1984) 1963.
45. SUNG, N. H. and SUNG, C. S. P., *35th Ann. Tech. Conf. Reinf. Plast./Compos. Inst., SPI,* **23B** (1980) 1.
46. SUNG, N. H., NI, S. and SUNG, C. S. P., *Org. Coat. Plast. Chem.,* **42** (1980) 743.
47. SUNG, C. S. P., LEE, S. H. and SUNG, N. H., *Polym. Sci. Technol.,* **12B** (1980) 757.
48. ALLEN, K. W. and STEVENS, M. G., *J. Adhes.,* **14** (1982) 137.
49. BOERIO, F. J., GOSSELIN, C. A., DILLINGHAM, R. G. and LIU, H. W., *J. Adhes.,* **13** (1981) 159.
50. ALLEN, K. W., GREENWOOD, L. and WAKE, W. C., *J. Adhes.,* **16** (1983) 61.
51. ALLEN, K. W., HANSRANI, A. K. and WAKE, W. C., *J. Adhes.,* **12** (1981) 199.
52. ISHIDA, H., NAVIROJ, S., TRIPATHY, S. K., FITZGERALD, J. J. and KOENIG, J. L., *J. Polym. Sci., Phys. Ed.,* **20** (1982) 701.
53. HANSMA, P. K. (Ed.), *Tunneling Spectroscopy, Capabilities, Applications and New Techniques,* Plenum, New York and London, 1982.
54. BREWIS, D. M., COMYN, J., OXLEY, D. P., PRITCHARD, R. J., REYNOLDS, S., WERRETT, C. R. and KINLOCH, A. J., *Surf. Interface Anal.,* **6** (1984) 40.
55. DIAZ, A. F., HETZLER, V. and KAY, E., *J. Amer. Chem. Soc.,* **99** (1977) 6780.
56. ST PIERRE, L. E. and PRICE, C. C., *J. Amer. Chem. Soc.,* **78** (1956) 3432.
57. DEGE, C. J., HARRIS, R. L. and MACKENZIE, J. S., *J. Amer. Chem. Soc.,* **81** (1959) 3374.
58. FURUKAWA, T., EIB, N. K., MITTAL, K. L. and ANDERSON, H. R., JR, *Surf. Interface Anal.,* **4** (1982) 240.
59. FURUKAWA, T., EIB, N. K., MITTAL, K. L. and ANDERSON, H. R., JR, *J. Colloid Interface Sci.,* **96** (1983) 322.
60. BOERIO, F. J., *Amer. Chem. Soc. Div. Polym. Chem., Polym. Prepr.,* **24** (1983) 204.
61. VAN VELZEN, P. N. T., *Surf. Sci.,* **140** (1984) 437.

62. Boerio, F. J., Williams, J. W. and Burkstrand, J. M., *J. Colloid Interface Sci.,* **91** (1983) 485.
63. Hays, A. K. and Haaland, D. M., *Org. Coat. Appl. Polym. Sci. Proc.,* **47** (1982) 383.
64. Chiang, C.-H. and Koenig, J. L., *J. Polym. Sci., Phys. Ed.,* **20** (1982) 2135.
65. Ishida, H. and Koenig, J. L., *J. Polym. Sci., Phys. Ed.,* **17** (1979) 615.
66. Garton, A., *J. Polym. Sci., Chem. Ed.,* **22** (1984) 1495.
67. Linde, H. G., *J. Polym. Sci., Chem. Ed.,* **20** (1982) 1031.
68. Suryanarayana, D. and Mittal, K. L., *J. Appl. Polym. Sci.,* **29** (1984) 2039.
69. Gähde, J., Loeschcke, I. and Richter, K.-H., *Acta Polymerica,* **34** (1983) 260.
70. Sung, N. H., Kaul, A., Chin, I. and Sung, C. S. P., *Polym. Eng. Sci.,* **22** (1982) 637.
71. Kaul, A., Sung, N. H., Chin, I. and Sung, C. S. P., *Polym. Eng. Sci.,* **24** (1984) 493.
72. Monte, S. J. and Sugerman, G., *Elastomerics,* **115** (1983) 30.
73. Monte, S. J. and Sugerman, G., *Polyurethane: New Paths Prog. Mark., Technol. Proc. SPI 6th Int. Tech./Mark. Conf.,* 1983, p. 131.
74. Monte, S. J. and Sugerman, G., *Polymer Eng. Sci.,* **24** (1985) 1369.
75. Monte, S. J. and Sugerman, G., *Adhesion Aspects of Polymeric Coatings,* Ed. K. L. Mittal, Plenum Press, New York and London, 1983, p. 421.
76. Reinhart, T. J., personal communication cited in ref. 59.
77. Sung, N. H., Kaul, A., Ni, S., Sung, C. S. P. and Chin, I. J., *Proc. 36th Ann. Conf., Reinf. Plast./Compos. Inst., SPI,* Vol. 36, Section 2-B, 1981.
78. Sung, N. H., Kaul, A., Ni, S., Sung, C. S. P. and Chin, I. J., *Adhesion Aspects of Polymeric Coatings,* Ed. K. L. Mittal, Plenum Press, New York and London, 1983, p. 379.
79. Calvert, P. D., Lalanandham, R. P. and Walton, D. R. M., *Adhesion Aspects of Polymeric Coatings,* Ed. K. L. Mittal, Plenum Press, New York and London, 1983, p. 457.
80. Moles, P. J., *Polym. Paint Colour J.,* **173** (1981) 391.

Author Index

Numbers in italic type indicate those pages on which references are given in full.

Subject Index